THE SCIENCE OF ANIMAL HUSBANDRY

Fifth Edition

THE SCIENCE
OF ANIMAL
HUSBANDRY

James Blakely
David H. Bade

A RESTON BOOK
PRENTICE HALL, Englewood Cliffs, New Jersey 07632

Library of Congress Cataloging-in-Publication Data

Blakely, James, *(date)*
    The science of animal husbandry / James Blakely, David H. Bade.—
5th ed.
        p.   cm.
    "A Reston book."
    Includes index.
    ISBN 0-13-794702-X
    1. Livestock. 2. Livestock—United States.   I. Bade, David H., *(date)*.
    II. Title.
SF61.B56 1990
636—dc20

89–33567
CIP

Editorial/production supervision and
  interior design: **Marcia Krefetz and Kathryn Pavelec**
Cover design: **Ben Santora**
Cover photo: *source,* **FPG International;** *photographer,* **A. Upitis**
Manufacturing buyer: **Laura Crossland and David Dickey**

 © 1990, 1985, 1982, 1979, 1976 by **Prentice-Hall, Inc.**
A Division of Simon & Schuster
Englewood Cliffs, New Jersey 07632

Printed in the United States of America

10 9 8 7 6 5 4 3 2

ISBN 0-13-794702-X

Prentice-Hall International (UK) Limited, *London*
Prentice-Hall of Australia Pty. Limited, *Sydney*
Prentice-Hall of Canada Inc., *Toronto*
Prentice-Hall Hispanoamericana, S.A., *Mexico*
Prentice-Hall of India Private Limited, *New Delhi*
Prentice-Hall of Japan, Inc., *Tokyo*
Simon & Schuster Asia Pte. Ltd., *Singapore*
Editora Prentice-Hall do Brasil, Ltda., *Rio de Janeiro*

CONTENTS

PREFACE

This book is written for the student. While initial reservations may have been encountered in the more mature minds of the first teachers who adopted it, there has been eager acceptance of the material, which is designed to make their task a little easier. Worldwide acceptance and adoption by colleges, universities, and schools would indicate more than mere popularity. The fact that most college students keep this book is mute testimony to its value. A steady annual increase in adoptions indicates more instructors with the same evaluation. Now, this new, revised, improved fourth edition incorporates even more features requested by teachers or demanded by the changing times.

Material concerning the production and care of domestic animals has been accumulating since the dawn of animal husbandry. Although the importance of the information is well recognized, it is the feeling of the authors that much of the excitement of learning is conteracted by the volume of material covered and the lack of suitable illustrations in an introductory text. Therefore, the need for a correlated approach to a learning technique is obvious. The three basic elements in teaching a subject are the instructor, the text or materials, and the successful transfer of knowledge from the instructor to the student. It is the intent of this text to provide a condensed, highly illustrated, easy-to-read version of the wonders, as well as the technical aspects, of animal science, while keeping in mind the needs of the instructor concerning visual aids. Each chapter is designed to be covered in the standard classroom time period using a text with numerous illustrations and self-study questions complete with easily revealed answers, followed by thought provoking questions on the same subjects but without available answers. The personal touch of the instructor is the cementing factor in welding together the body of knowledge to be assembled in the classroom. This text and the accompanying instructor's manual will serve as a supply line, filling capable hands with raw materials for further refinement. The authors' approach is simple and direct, aiming

> To provide teaching aids and resource materials *for the classroom teacher.*
>
> To provide educational concepts and resource programs *for the educator/ administrator.*
>
> To provide reference materials and additional self-study guides *for the student.*

All the features that made the first four editions a success have remained. Suggestions from classroom teachers around the world have been incorporated to bring the Fifth Edition up-to-date with modern times and practices. An extensive revision of Chapter 1 reflects some fascinating developments in modern animal husbandry to include biotechnology, marketing, exotic animals, and the use of computers. Minor adjustments were made where needed in each chapter to give the latest information on established and developing breeds. Chapter 2 was extensively reviewed and brought up-to-date with the addition of two breeds and numerous new photographs added to enhance visual appraisal of the changes in types. Hundreds of new illustrations have been added to make this the most highly illustrated and informative textbook on the market today. This textbook has found acceptance in colleges and high school classrooms throughout the United States, Canada, Australia, New Zealand, Great Britain, Singapore, and Africa, and a few copies are even used in the People's Republic of China. This worldwide interest indicates the international applicability to the science of animal husbandry anywhere.

JAMES BLAKELY
DAVID H. BADE

Chapter One

ANIMAL SCIENCE— YESTERDAY AND TODAY

The foundation of animal science is knowledge. The development of the science of animals was begun by pioneers of the past. From the time human beings first attempted to domesticate beasts to today when our exploding population means increasing demands for meat, fiber, and recreation, the knowledge of animal science has been accumulating. Our forefathers lacked modern conveniences such as electricity, rapid transportation, market information, and a host of other tools now taken for granted (Figure 1-1). A hot, dusty, dangerous trail drive often ended with a rock bottom market or death of the animals from hunger, thirst, cold, or disease. The pioneers of yesterday were men of iron will and strong constitution, walking dictionaries of knowledge handed down by their forefathers or learned by trial and error. We are still learning and building onto this foundation.

TRENDS IN ANIMAL SCIENCE

When living, animals are used for milk, wool, draft, transportation, protection, sport, work, and pleasure. When slaughtered, they supply meat and by-products ranging from glue to medicine, confectionery to fertilizer, catgut to chemicals. Few other raw products have such versatility.

1

FIGURE 1-1. Old ways give way to new. (Courtesy of Russ Hill, Copter Cowboy, Odessa, Texas)

Meat has long been the primary food source for people in most countries of the world. A rising standard of living in the United States has carried beef as a status symbol on the coattails of success. With today's exploding population and its ever-increasing consumption of meat per person, it is no wonder that livestock production leads agricultural income in the United States. Although some second thoughts are being entertained concerning the feeding of grain to cattle and sheep, there are still large areas of land suitable only for a harvest of rough plants by animals that can convert a useless product into edible goods in the form of meat.

THE BEGINNING OF ANIMAL SCIENCE

Fossil remains of animals resembling the cow have been found in Asia dating back 3 to 4 million years. This "great ox" was about 6 ft tall, had horns similar to a goat with a spread of 6 to 7 ft, and grazed on twigs and shoots of trees (Figure 1-2). Similar enormous, prehistoric species have been located that are the predecessors of swine, while the horse's ancestor was a small desert animal measuring less than 1 ft high.

In the Old Stone Age (10,000–8000 B.C.) human beings made no attempt to domesticate animals, regarded most with superstition, and ate those that they were lucky enough to hunt down. In the New Stone Age (8000–6000 B.C.), human beings changed from hunters to husbanders of animals by domestication. This was the beginning of animal science.

FIGURE 1-2. The prehistoric ox was 6 ft tall at the withers.

EARLY ANIMAL HUSBANDRY

It is thought that the early mountaineers domesticated the trailable species (cattle, sheep, horses, goats, and dogs) and followed them as they grazed or hunted. The plains farmer domesticated the other species (cats, pigs, chickens, geese, ducks), driving them into their village during the night.

Cattle were admired, worshipped, used, and abused in early times. Egyptian Pharaohs associated themselves with having the "might of a bull" and, like the Canaanites, Romans, and Africans, had gods fashioned from cattle. Early Egyptians made great developments in the art of animal husbandry. Drawings done as early as 2625 B.C. show calves with hobbled front legs and herders delivering calves and branding cattle. The fattening of livestock is referred to in the Bible when the prodigal son ate the "fatted calf." Using animals for sport also developed during this time, with bullfights, including female "bull grapplers"; the riding of bulls; and the dangerous battles of armed youths pitted against a wild herd of cattle in an arena.

For time immemorial, swine have been a part of our world culture. Although not native to America, swine are mentioned as early as 4900 B.C. in writings of Chinese historians and about 1500 B.C. in Biblical writings, and were well known in historical accounts of Great Britain about 800 B.C.

Sheep were probably domesticated during the early Neolithic Age, making them among the first animals to feed the guiding hand of man. Probably first domesticated in central and western Asia, sheep are pictured on Egyptian monuments dated between 5000 and 4000 B.C. Biblical passages abound with mention of sacrificial lambs, so it is well documented that sheep spread with every civilization.

It is known that a small and inferior type of horse, about 3 ft tall, at one time existed on the North American continent but became extinct many years before the voyage of Columbus. A superior type of horse, probably originating in Central Asia, had been domesticated by the ancient Egyptians, Greeks, and western Europeans and used as a beast of burden, for transportation, and for warfare centuries before the discovery of America.

THE INTRODUCTION OF ANIMALS TO NORTH AMERICA

When Columbus reached America, no domestic animals were found. Horses, cattle, sheep, and hogs were brought on his second voyage in 1493. Small importations continued periodically: Vera Cruz brought cattle from Spain to Mexico in 1521; swine were introduced from Cuba to Florida by Hernando De Soto in 1539. The first American importation of sheep was at Jamestown in 1609. The horse, brought by the Spanish Conquistadors who explored much of the Americas in the early sixteenth century, soon became established. The animals imported from Europe were used mainly for milk, butter, hides, and draft. With wild game plentiful, meat was not the citizens' main concern. These animals, under the guiding hands of notables like George Washington and Thomas Jefferson, multiplied and purebred herds developed in the east. Following the American Revolution, livestock moved westward, and by the early nineteenth century was distributed over most of the east, the south, and the far west.

Similar movement of stock created large herds of cattle, sheep, goats, and horses at missions throughout the Spanish territory. Herds of 424,000 cattle at two missions were recorded. Thus large Mexican ranches developed and many still exist.

THE GROWTH OF ANIMALS IN THE UNITED STATES

Cattle in America

Prior to the Civil War, large herds were marketed to the east for hides and tallow through the rendering plants in New Orleans.

During the Civil War, the eastern and northern markets were cut off and cattle multiplied in Texas. In Texas, prices at the end of the war were $3 to $6 per head, while in the north and east, they brought 10 times that amount. The possibility of profit led to the romantic period of the trail drives from Texas to meet the transcontinental railroad lines in Kansas. An estimated 10 million head of hardy, rough, Longhorn cattle traveled the 600- to 1700-mile trail to market (Figure 1-3).

FIGURE 1-3. About 10 million hardy Longhorn cattle traveled the long trail drives from Texas to Kansas. (Courtesy of the Library of the State Historical Society of Colorado)

The contribution the Longhorns made in cattle production was best described by Evetts Haley, Texas cattleman and historian, in:

A Bit of Bragging about a Cow

The cow to which I propose a monument established no blood-lines, set no butter-fat records, and produced no prizewinning beeves for the International Show. But she did take care of herself and her calf out on the open ranges of Texas.

When it did not rain on her range, she, without benefit of weather and crop reports, just got up of her own accord and walked off to where it had. When the water-holes dried in drouth, she prowled the gullies to the head-springs, and when they failed, she did not lie down in despair and die because nobody hauled a supply, but pointed her nose into the breeze and walked until she found some. When snows and blizzards came, she headed for the breaks and thickets, and browsed on brush long before cotton seed was ever pressed into cake.

She did not depend on government trappers for protection, but with sharp horns and sharper senses she fought it out on the ground with packs of coyotes and powerful "loafer" wolves. She scorned legal quarantine lines and lived happily without them, sloughed off her own ticks before vats were conceived, rubbed the warbles out of her back on the rough bark of the mesquites before rotenone was discovered and self-oilers were built for profit, and raised her calves without the benefit of vaccine. She knew no shipping fever for she took herself up the trail a thousand miles to market.

She adjusted her own increase to what the range would honestly carry, and followed an "efficient program of range improvement" just by drinking water and eating grass and walking all over creation while nature had her way.

Already I can hear the carping critics [ask], "What positive 'program' of improvement did she bring?"

She produced the millions of Longhorn steers that marched up the Texas Trail to distant markets and revived this state economically after the Civil War. She quickly converted the Great Plains from a land of unused grass to a productive industry in twenty years' time.

She took . . . imaginative and daring boys out of the cotton patches without public expense, and converted them into genuine adventurers in life and business . . . she inducted more boys into the fascinating mysteries and processes of nature than the public schools, converted more hands to the fine art of working cattle than the rodeos ever will, and prematurely took more reckless . . . boys away from home to healthy life and adventure on the trail than have been reformed by the state's corrective school since.

In spite of the fluctuations of the business barometer, she kept more cowmen from going broke . . . while furnishing . . . people with buyable beef . . . than any breed on record. And at last she stimulated more good stories and stirred the minds of old men with more exciting memories, than all the radios combined . . . and so there should be a scheme—not a "project" or a "plan"—but an underground intrigue if necessary, on the part of the few remaining believers in self-reliance and sturdy independence, to erect a monument to this courageous individualist, the Longhorn Texas cow.*

The introduction of barbed wire in the 1870s, along with both the end of the Indian Wars and the killing of the buffalo, opened the midwest for cattle production. The cattle industry grew almost overnight with the end of the open range announced by the coming of barbed wire. This closing of individual herds meant great strides in animal improvement. Closer supervision, a critical eye, and sound business practices dictated heavier concentrations of more efficient animals on dwindling lands. The quickest means of improvement lay in the practice of mating superior sires with existing herds. The purebred sire now made his contribution; demand outstripped supply; replacements changed in type and temperament. The Longhorn was literally bred out of existence, ushering in the "Golden Purebred Era" and marking the demise of one way of life and the birth of another.

Out of this romantic history arose the dynamic business industry of animal science today. Research, technology, and improved management techniques are being developed and scrutinized in order to keep up with the demands of an affluent and ever-increasing world population for livestock goods and services. Cattle will most certainly play an important role in the unfolding drama of our continuing strategy to glean from nature a bountiful life.

Sheep in America

Columbus also brought sheep in 1493. Cortez introduced Spanish Merinos to Mexico in 1519 and undoubtedly brought other types of sheep as the Spaniards moved into New Mexico opening missions and establishing basic farming and husbandry practices. Spanish accounts of multicolored sheep imported to Mexico are thought to be the forerunners of the multicolored Navajo Indian

*Charles Wayland Towne and Edward Norris Wentworth, *Cattle and Men* (Norman, Oklahoma: University of Oklahoma Press, 1955), pp. 114–115.

sheep that exist today for the purpose of producing a coarse blanket and rug wool. Although accounts are not clear as to the exact type, the first direct importation, consisting of British breeds, was to Jamestown in 1609.

The first importations were made primarily for the purpose of producing wool, but because of predators' attacks, poor management, and lack of shelter, these early breeds quickly deteriorated in quality and yield of wool. The situation became so critical that many towns passed legislation requiring each citizen to contribute one day's labor per year in clearing new pasturage, prohibited exports, and elevated town shepherds to a high status.

In 1783, to improve wool quality and yield, direct importations of the fine wool Merinos from Spain to America were made and continued heavily until 1816. As the westward movement of the colonists progressed over the Allegheny Mountains about 1830, sheep also progressed in quality and numbers.

The Civil War caused a drastic increase in the price of wool because of limited supplies of cotton to both sides, creating demands which in turn stimulated production. By 1884, America could boast of 50 million sheep, a figure not equaled until 1942 during World War II when 56 million head were estimated. Since then, numbers have steadily declined because of competition from synthetic fibers and pressure from other meat products.

Production shifted from the east to the Rocky Mountain region by the turn of this century and finally, to date, to the 11 western states and Texas (Figure 1-4). The sheep, though the colorful days of sheep ranching are now history, still continue to contribute to the American agricultural economy as evidenced by the many breeds and types that produce mutton, lamb, and every grade of wool.

FIGURE 1-4. Sheep herds are still prevalent in the western states and Texas. (Courtesy of the U.S. Department of Agriculture)

Swine in America

The word "swine" is used here synonymously with "hogs" or "pigs" in the English language, although the animal scientist makes distinctions among the various terms for the purpose of communication clarification. The *Sus scrofa* (European wild boar), the *Sus indicus* (Chinese and Siamese swine), the *Sus vittatus* (East Indian swine), and the *Sus wadituaneus* (Neapolitan swine) were used as basic building blocks for the bloodlines and breeds in existence today. The white breeds supposedly owe their characteristics to the *Sus indicus* (Chinese), while black swine resulted from *Sus wadituaneus* (Italian) background. Other colors can apparently be attributed to *Sus scrofa* (European) and *sus vittatus* (East Indian), which have had a dominating influence in many breeds in America.

Hogs were first brought to the New World, from the Canary Islands to what is now the island of Haiti, by Columbus in 1493. Swine multiplied rapidly on the island and served as a continuing supply of stock to accompany expeditions. In 1539, Hernando De Soto launched an expedition from Cuba and landed in Florida with 13 hogs, which numbered 700 only 3 years later. De Soto explored from Florida to Missouri, and it is thought that many of our "razorback" swine (Figure 1-5) are descendants of these nondescript early hogs that wandered off. These same hogs served as foundation stock for later development into the many breeds in existence today. Mongrel sows were often trapped from the wild state and bred to imported boars, which began arriving as early as 1825. The early corn-growing areas of Tennessee, Kentucky,

FIGURE 1-5. "Razorbacks" are thought to be descendants of the early imported hogs which wandered off and reverted back to the wild. (Courtesy of the U.S. Department of Agriculture)

Ohio, and the Connecticut Valley became chief hog-producing centers and marketed their corn through swine.

As agriculture developed west of the Allegheny Mountains, so did the production of swine, a natural complement for corn and other grain farming. Pork was popular because it could easily be smoked, pickled, or otherwise preserved, and it had a high level of energy, because of the fat concentration—a fact not overlooked by the hard-working pioneers.

With the infusion of Chinese, Neapolitan, Russian, and European boars, swine-breeding programs began to evolve. They eventually produced thoroughly genuine American breeds, many of which are held in higher esteem than the ancestors from which they sprang. Most of the breeds raised in America originated from these early breeding programs and continue to draw upon the existing genetic material for improvement within the breed. Swine have continued to make a steady, though fluctuating climb in total numbers. Although styles and types have changed with human demands, both purebred and crossbreeding systems continue to have a substantial impact on the agricultural well-being of America.

Horses in America

Columbus was reported to have brought horses to the West Indies in 1493. However, it was not until Cortez landed on the shores of Mexico in 1519 that the initial herd of 16 head was reintroduced to the Americas. By 1521, more than 1000 head had been imported to support the conquest by the ambitious Spaniards. Hernando De Soto, credited also with the introduction of swine to the United States, had 237 horses in his expedition to Florida in 1539. Following the death of De Soto during the expedition, many of his followers abandoned their horses and returned by boat from the upper Mississippi River. Coronado followed in 1540 bringing more horses. The band led by Coronado no doubt followed a plan similar to that of De Soto, leaving stock that would thrive on the grassy plains, eventually to be prized by American Indians, master horsemen. During the early seventeenth century, the Spanish missionary movement into New Mexico also enhanced the rising horse population, contributing further to wild bands because of lost, strayed, or abandoned animals.

While the wild horse population was busy developing under the guiding hand of the Indian tribes, 17 saddle horses for riding were imported to Jamestown between 1609 and 1611; at least two of these horses were stallions.

Because oxen were primarily used for plowing, the horse was neglected as a useful working animal until 1840 when the buggy appeared on the scene. Larger draft horses were imported from France to pull the larger wagons and freight-moving equipment that followed the buggy. The greater size of some horses led to their use as plow horses, a prosperous move for plantation owners who then developed easy-riding saddle horses and flashy, high-stepping buggy horses, which they used to oversee their large holdings with comfort and speed.

The development of the western range, the trail drives after the Civil War, and the increasing affluence in America led to cow horses, pleasure horses, and racing stock. By 1867, it was estimated that there were 7 million horses in the United States. This number increased to 21 million by 1913, but the coming of the automobile and the tractor paralleled the decline in horse numbers, which by 1960 had fallen to 3 million head. However, the love affair with the horse in America is far from cooling off. The sport of racing and the renewed interest in recreation horses has spurred this proud, magnificent species into popularity once again. Most authorities agree that the 1968 census estimate of 7 million horses will continue its upward trend well into the 1980s.

YESTERDAY—TODAY

Recent advancements in space-age technology have not bypassed animal husbandry. Two startling examples that have been applied to modern raising of domestic animals are the use of the helicopter and the ultralight. The helicopter has become an indispensable tool in many of the western states because of the large acreage involved and difficulty of the terrain. Very accurate wildlife surveys can be made from the helicopter. Income from hunting leases without depletion of existing wildlife is made possible through proper game management. Without the helicopter, this increasingly important revenue-producing form of land management would only be a difficult guessing game.

Predator control with the commercially hired helicopter and, more recently, with the individually owned ultralight (Figure 1-6) has taken on new

FIGURE 1-6. The ultralight is used on some farms and ranches for transportation, crop and animal observation, checking fences, even predator control.

Chap. 1   ANIMAL SCIENCE—TODAY AND YESTERDAY

FIGURE 1-7. The helicopter is a multifaceted tool for the livestock director. (Courtesy of Russ Hill, Copter Cowboy, Odessa, Texas)

significance. Many poisoned baits, cyanide traps, and other devices used to kill wolves, coyotes, and other predators have been banned by the Environment Protection Agency (EPA). Ranchers in many of the western states have taken to the air in ultralights carrying a shotgun to meet the challenge with surprisingly good results. The ultralight is also used to survey pastures and growing crops, to check fences, and for quick, economical transportation on large acreages.

By far, the most established practical use of commercial helicopters is in gathering livestock, primarily cattle, goat, and sheep (Figure 1-7). A "helicopter cowboy" can gather a maximum of 15,000 to 30,000 acres of ranchland in a 4- to 8-hour period, 20 to 40 square miles—in one day with 90% or better retrieval. In comparison, depending on the terrain, a dozen cowboys on horses would need 3 to 5 days to gather in the same-size area. However, helicopters and cowboys on horseback can work together successfully, allowing for even more flexibility.

According to some experts, the cost of operating a helicopter for gathering livestock versus cowboys is actually less. The expense of labor, machinery, trailers, feed, and upkeep of horses on a year-round basis for the customary semiannual roundup cannot easily be justified in comparison to hiring a commercial helicopter expert for only a day or two each year. The pasture saved with the reduction in numbers of horses needed can mean several more head of producing livestock, and as labor gets more scarce and expensive, the helicopter, ultralight, and other devices may be reasonable alternatives. Although old ways will never be eliminated completely, they certainly are giving way to the new.

Some authorities estimate that technological changes, especially biotechnology, will take place 500 times faster in the next two decades than in the past. Use of artificial insemination began in the 1930s, frozen semen in 1951, embryo transfer in 1953, frozen embryos in 1971, and microsurgical embryo splitting in 1982. The future holds exciting promise through genetic engineering by developing vaccines to prevent disease or designing animals to fit specific needs or consumer demand.

Evidence of the importance of rapidly developing technology can be found in the new Institute of Biosciences and Technology Center, an 11-story building under construction adjacent to the Texas Medical Center in Houston, Texas. Research at such institutions and in universities throughout the world are estimated to accelerate the biotechnology industry to a predicted $200-billion business worldwide by the year 2000.

New uses for animals and animal products are arriving on the scene as fast as creative minds can find a way to market them. Some of the more recent innovations include the development of a new synthetic breed of cattle; raising alligators, crawfish, alpacas and llamas, and deer; and numerous new twists to marketing old products in entirely new ways.

Synthetic Breeding

A new synthetic breed of cattle is currently being developed from five-eighths French Salers and three-eighths Texas Longhorn blood. This combination of genetics utilizes the most historically adapted breed of cattle in America—the Texas Longhorn—with a highly proven carcass-quality breed—Salers. The Salorn (Figure 1-8) is an attempt to utilize modern technology with two of the oldest and most adapted breeds in the world. The Texas Longhorn has been in North America nearly 500 years, under a survival-of-the-fittest production system. The Salers has an equally historic record, and both breeds are considered to have a lack of genetic defects, making the gene pool very stable and clean. Today's cattle industry requires a genetic plan totally opposite to genetics designed 100 years ago. This breed is being created to meet the consumer

FIGURE 1-8. A new synthetic breed of cattle currently being developed combines the traits of two ancient breeds. Crosses between French Salers bulls on Longhorn cows (left) will give rise to Salorn (right), an innovation still in the foundation stages. (Courtesy of Horlock Land and Cattle, Riviera, Texas)

Chap. 1 ANIMAL SCIENCE—TODAY AND YESTERDAY

demand for natural lean beef with lower fat and lower cholesterol content, and in addition, to meet concerns regarding longevity, early puberty, calving ease, maternal traits, disposition, and carcass quality. Even the pigmentation of sensitive membranes will be stressed in the color factors. Cattle will not be spotted like the Longhorn, but red in color and pigmented around the eyes, ears, nose, udder, and so on, to produce maximum resistance to pinkeye, cancer eye, and udder burn. This is just one example of biotechnology combining the best qualities of natural selection to meet consumer demands, highest-quality products, and lowest costs of production.

Alligators

Research is being conducted at several universities in the southern United States on the alligator, once considered an endangered species, but now so numerous that they have attracted the eye of agriculturalists on the lookout for a new market. Alligators are being grown commerically, similar to poultry, inside buildings with a controlled environment. They grow fast, producing high-quality meat for a specialty market and hides, which are always in great demand (Figure 1-9).

Alligator ranchers have to feed special meats (which are kept frozen) and that is currently the most expensive drawback to this specialty animal. Researchers hope to be able to develop a cheaper "mill type" feed to simplify this type of husbandry and lower production costs. Alligators can be kept in environmentally controlled areas, similar to hog or poultry houses. Half-water/half-dry areas and a controlled temperature between 85 and 90 °F keeps the reptiles from going into hibernation.

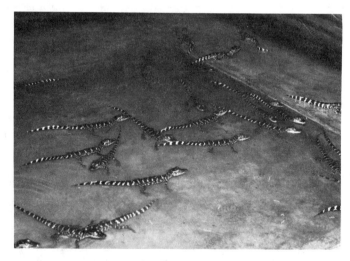

FIGURE 1-9.   Approximately 2-month-old alligators being grown commercially inside a controlled environment building for the specialty market of meat and hides. (Courtesy of Perry L. Little, Sam Houston State University, Huntsville, Texas)

Because of the demand for alligator steaks at $4 to $6 per pound in certain restaurants and hides for fine-quality leather products at $25 to $30 per foot, commercial production holds considerable promise.

Crawfish

In recent years, the popularity of Cajun-style cooking has created a market for crawfish as a specialty item in restaurants, not only in the south, but around the United States. Normally, crawfish are trapped in the wild, but increased activity in the new market has led to commercial production.

Ponds, not to exceed 20 acres in size, that retain water at least 16 in. deep with a flat and level bottom are needed. Levees around the pond should be 36 in. high after settling and wide enough to allow vehicular traffic on at least two sides. The pond should be completely open with no trees, brush, or other vegetation on the levees. Provisions must be made to drain the pond annually with drain pipes large enough to remove at least 3 in. from the pond in 48 hours. Screens on the drain pipes are used to prevent crawfish from leaving the pond.

In the southern part of the United States, stocking usually takes place between April 15 and 30 with 50 to 100 lb per acre of pond-raised crawfish. Twelve to eighteen inches of water is maintained on the pond until the last of May, when the water is drained off. This causes crawfish to burrow into the soil following the water as it percolates downward. When the land has dried enough for preparation, the planting of a forage crop, such as rice, takes place about mid-July. Generally, ponds are reflooded about the first or second week of September and the forage crop acts as feed for the developing crawfish.

Successful crawfish production depends on an adequate supply of high-quality water. For this reason, commercial production is ideally suited for rice-producing areas. However, rice is not harvested as a crop in this situation because crawfish completely devour the forage. Specific varieties of rice have even been developed for this particular production need.

Harvesting of the crawfish is done by hand by setting out numerous cone-shaped stand-up traps made of ¾-in. poultry wire. Baits of beef melt, shad, carp, fish heads, chicken necks, and even artificial baits have been used to attract crawfish inside the trap. Crawfish are simply dumped from the trap into a container and purged in clean water for 24 hours to allow the crawfish to clean themselves of mud and food. Harvesting in this manner continues through the fall months and early spring, when the process is repeated.

First-year producers can expect about 600 lb of crawfish per acre, and established producers are known to produce 1500 lb per acre. The market price fluctuates around $1 per pound, with the major outlets being specialty restaurants. The high quality of pond-raised crawfish results in better prices compared to the wild-caught variety, and some farmers and ranchers see this creative enterprise as a valuable alternative use to intensive production on relatively small acreage. Although not for everyone, it is an interesting development in the field of animal science.

FIGURE 1-10. A male alpaca recently sold for $25,000. The alpaca's high-quality fine wool is a most valuable resource. (Courtesy of Bruce Coleman Inc., New York, New York)

Alpacas and Llamas

Another interesting development in American animal husbandry is the use of alpacas and llamas, although there are only a few breeders in the United States. The alpaca's coat is prized for its fineness. Used to make high-quality woolen materials, this luxury fiber holds some promise for specialty breeders.

About 300 alpacas and llamas were imported into the United States prior to 1920, when an outbreak of hoof-and-mouth disease caused import laws to prohibit the practice. By 1980, the U.S. population had grown to only 8000, while demand is currently about 100,000 per year, mostly as pets, for zoos, and for recreational purposes such as light mountain trail-pack animals. This demand and the price of the wool at $4 per ounce, coupled with the limited supply, make the price of breeding stock an expensive but potentially lucrative enterprise. Although nondescript males, as pets, may sell at $1000 to $1500, young females are selling at around $10,000, with show animals going for as much as $100,000 (Figure 1-10).

New Twist

A new twist to an old market is always an exciting avenue for those who are adaptable to change. Just one such example is found in the use of a whole roast pig: cured, cooked, and complete with an apple in its mouth. In this process, the pig is cleaned, treated with a special curing solution, cured for several days, and then slow cooked in a smokehouse for 18 to 24 hours. The roast pig can be delivered to parties such as are held at conventions featuring a Hawaiian luau and has been shipped by air successfully hundreds of miles from the point of preparation (Figure 1-11).

FIGURE 1-11. Randy Alewel, Warrensburg, Missouri, prepares almost 1000 whole roast pigs per year for the special convention and party market. Such processed hogs are often shipped by air to distant destinations. (Courtesy of Dean Roger Mitchell, Extension Publications, University of Missouri–Columbia)

The fully cooked pig is served on a wooden tray, covered with plastic. After serving, the bones and waste are easily wrapped in the same plastic for disposal. The cost for the completed project is $65, plus the market price of the hog and transportation expenses.

Deer

Venison or deer meat has been a gourmet meal served in European restaurants for many years. Commercial ventures of Red Deer farms are big business in New Zealand, where special fencing, handling equipment, and animal husbandry practices produce a year-round, dependable supply of vension (Figure 1-12).

In other parts of the world, especially the United States, game management programs have converted some large acreages into prime hunting preserves. Deer-proof fencing and strict and ruthless harvesting of inferior animals have allowed game managers to produce trophy-sized bucks. Most preserves provide a guide as a part of the hunting fee package to ensure harvesting of only those animals that measure up to certain standards to preserve superior breeding stock. Some hunting preserves even have portable, refrigerated processing facilities to cut, wrap, and freeze the meat on the spot while preserving the hide and horns for the taxidermist. It is not unusual for a game farm managed in this way to realize $2000 to $5000 per deer harvested. Simi-

FIGURE 1-12. Red Deer are grown commercially for the venison market in New Zealand (Highlands Station, Ford family). (Courtesy of Raewyn Saville, Rotorua, New Zealand)

lar programs have broadened the hunting rights to include quail, pheasants, geese, ducks, and even feral hogs (domestic hogs gone wild).

Agridome

One of the most innovative and unusual uses for domestic farm animals was the establishment of the Agridome in Rotorua, New Zealand. It features 21 breeds of trained sheep participating in two daily, one-hour stage shows to tourists from all over the world. When New Zealand's participation at Expo '70, Osaka, Japan, took place, the display was centered around the New Zealand sheep industry. Godfrey Bowen, a world record sheep shearer and well-known New Zealand breeder, trained eight pedigreed rams to walk on stage to a commentary describing each breed. The exhibition also included moving and the holding of animals by voice and whistle command to working sheep dogs. Sheep shearing by both hand and mechanical shearers was demonstrated. This exhibition was so successful that on his return to New Zealand, Godfrey Bowen, with his friend George Harford, established a special facility called the Agridome in Rotorua, New Zealand, and opened it to the public.

The facility (Figure 1-13) seats 700 people, is set on a 270-acre working farm, and features commentaries through headsets on a multitrack sound system for English, Japanese, Mandarin, French, Spanish, and German. A shearing demonstration by a champion shearer, and the chance to see sheep dogs working both on stage and outside, are highlights of the show. Since the facility opened in 1980, it has drawn over 1 million visitors from around the world. A well-stocked shop featuring a wide range of sheepskin products, woolen goods, natural fiber clothing, and souvenirs, as well as a fine restaurant that serves farm breakfasts, roast lamb lunches, barbeques, and a selection of cabinet foods, make for a creative approach to marketing.

FIGURE 1-13. The Agridome in Rotorua, New Zealand, offers a unique blend of animal husbandry and show business. Over 1 million people have paid admission to see two daily shows featuring trained sheep representing 21 breeds. (Courtesy of Raewyn Saville, Rotorua, New Zealand)

THE COMPUTER AGE

The most complex computer mechanism in the world is the human mind, which analyzes thousands of bits of information daily and often makes split-second decisions that could mean the difference between life and death, solvency, and bankruptcy. It has been calculated that to duplicate the complex computer mechanism of the human mind, it would take an electronically packed three-story building about the size of a city block.

The computer (Figure 1-14), once brought into perspective, should be thought of as nothing more than a simple machine. The difference is that the computer will handle mathematical equations that will take into account as many variables simultaneously as one cares to analyze. Like any machine, it does it without question, without emotion, without tiring, and most often without error. The key to utilizing the computer is to understand that it is just another tool and, just as a tractor eased the physical work of our forefathers, the computer is able to ease the work of sophisticated analysis. Today, farm machinery has taken another giant leap forward. Combines are now being designed, for example, capable of deciding how fast to go according to crop density, self-steering, printing out production performance, and monitoring maintenance requirements—all through on-board computers. Similar exciting developments are becoming commonplace among the most progressive animal producers.

For those unwilling to invest the time in understanding the basics of *database* and *programming*, computers will seem complex. With sufficient patience and motivation to use this electronic extension of the human mind, a

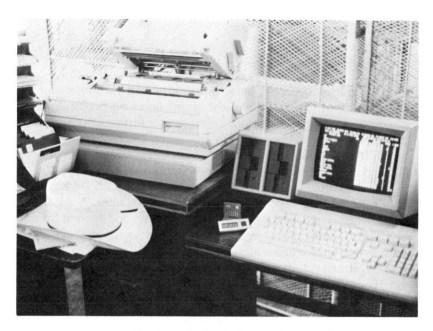

FIGURE 1-14. Computer terminal, disk drive, and printer.

new era in animal science will unfold and, just as tractors replaced the mule in history, so the computer is destined to continue to replace slower, antiquated methods of analysis.

Storage (Memory)

To take some of the mystery out of the computer for the beginner, consider it as nothing more than a certain number of filing cabinets for storage of information (memory). The more filing cabinets, the more memory we have. Computers store information, however, in "electronic filing cabinets," and the information can be searched, retrieved, analyzed, compared, and the end result read on a screen (such as a TV set uses). Memory capacity is expressed in "kilobytes" (K). For example, 64K means storage for some 64,000 characters. A character is roughly equivalent to an average sized word (five letters). Memory is stored in the computer and/or on "floppy disks," hard disks, magnetic tape, and so on, much like music is recorded electronically to be listened to, or in this case, to be viewed later. For example, the most common floppy disk storage system works this way: Disks are placed in a disk drive mechanism (there may be one or several to draw information from), and the machine commanded to search and retrieve the information filed (stored) under some predetermined heading or category. So, basically, the computer is nothing more than an electronic filing cabinet, a TV screen, and, in cases where it is needed, a calculator and typewriter.

Database

As mentioned previously, computers are able to analyze many mathematical equations simultaneously, but the user must realize that the machine does nothing unless it has the basic information with which it can work. One does not push a button and have a machine solve all problems. The proper data must be entered into storage before any information can be analyzed to get answers from questions asked of it. Thus the base of information (data) on which it can draw becomes the foundation (database) on which formulas are entered for analysis. The database organizes information in one place and allows that information to be retrieved very rapidly—when it is needed. The result is that information need only be entered once. A database system does what a librarian does for a library. Without a good system for organizing facts in a library, information could not be found easily, if at all. Putting the information in a known order allows any authorized person to find the desired information. Retrieval, in the case of the computer, is like sending the librarian to get the information for you.

Typical information that could serve as a database for a cattle operation might include for each head of stock, for example, the following stored information:

Frame size	Growth stimulants
Birth weight	Diet
Weaning weight	Weight
Sex	Gain
Breed	Yield grade
Diet	Price
Feed additives	Sire and dam

The database may include simple information such as shown above or an almost unlimited amount of input. Careful analysis of goals and filling the computer with the right database are critical to satisfactory problem solving.

Modeling

The computer using the database is then programmed to predict what would happen to performance if some of the variables from the database were changed. For example, if the calf crop average production values for this year were subjected to a different level of feed additive, diet, genetic influence of sire or dam, or any number of variables stored in the database, what would happen? Rather than wait until next year and try out a different combination, the computer can try every possibility and give predicted responses so that the most reasonable chance for impeedment is assured and the best change in management practices is offered for final approval by the human decision maker. The answer does not guarantee success but merely figures the odds of a winning combination much like a computer figures the odds of each horse in a horse race, based on breeding, experience, and past peformance.

Computers vary in size and price from compact models about the size of the average typewriter, costing only a few hundred dollars, to very large terminal computers housed in large buildings, costing many thousands of dollars. Many farmers and ranchers own small personal computers, which can solve most of their problems, and through the use of a telephone hookup, still be allowed "access" to larger terminal computers located at university and private centers. Thus an inexpensive computer through the use of the common telephone line can "tap" into a more complex and expensive computer, usually for a fee based on amount of time the user is "on line" (connected). Most land grant colleges throughout the United States today have some form of computer service such as this available to farmers and ranchers through the extension service.

Hardware and Software

The computer and necessary components are referred to as hardware. The type of program that is written to perform a specific function is known as software. Think of it as instructions to the machine. Since these instructions are usually condensed to a softer-than-metal substance, like a floppy disk, the term "software" seems appropriate. Just as we might instruct a hired hand to plow a field, feed cattle, or chop wood, we need software to instruct the computer (hardware) on what job function it should concentrate.

One of the major uses of the computer is for ration formulation, using a variety of feed ingredients available. Software programs are written to balance a ration, not only for its nutritive content, but also for the best nutrition at the lowest price. This "least cost" ration formulation is one of the primary uses of computers in animal science.

Another exciting use of the computer is in animal breeding. For instance, a radio transmitter placed in a cow's vagina is being used now by some breeders to monitor temperature data accurately. Information processed by the computer determines when the cow ovulates so that artificial insemination takes place when there is the highest probability of fertilization.

In addition, an instantly accessible pedigree history of sire and dam makes it possible to compute the inbreeding coefficient for an embryo before it is ever conceived. For example, if a certain mating seems desirable but the sire and dam are known to be related, the computer can determine instantly if this mating would generate more than the 6% inbreeding danger level in the offspring. Linebreeding establishments would find this information invaluable, and computers are now being used in an animal version of a computer dating service.

Software is also available to "instruct" computers on analyzing for weight gain predictions, cost of gain, adjusting weaning weights, calculating yield grades, marketing decisions, budgets, records, accounting, inventories, financial-investment analysis, taxes, crops, and many, many others too numerous to mention.

For more detailed and additional information on the use of what is available in the computer software field, contact your extension economist or write for (1) "Survey of Availability of Micro and Mini Computer Software," Department of Extension Economists, University of Florida, Gainesville, Florida 32611, and (2) "Nationwide Survey of Agricultural Software," Forrest Stegelin, Texas A&M University, College Station, Texas 77840.

STUDY QUESTIONS

1. _____ brought cattle, sheep, goats, and hogs to America.
2. Prior to the Civil War, cattle were slaughtered for their _____ and _____ .
3. The type of cattle credited with opening up the animal industry was called _____ .
4. The end of trail drives and the beginning of domestic cattle operations were marked by the development of _____ .
5. _____ , _____ , and _____ are the three main uses of animals today.
6. Longhorn cattle were upgraded using purebred sizes, and the _____ Era arose.
7. White swine breeds are descendants of the *Sus indicus* species originating in _____ .
8. Many swine breeds in the United States today are a result of the early breeding programs using purebred boars from Russia, Europe, and China and _____ sows.
9. Sheep production in the United States has shifted from the east to the _____ and Texas.
10. The Spaniards are credited with the large importations of the _____ .
11. The buggy as transportation led to the importing of _____ horses, which eventually even replaced the oxen for working stock.

DISCUSSION QUESTIONS

1. How many domestic animals can you name that are native to America?
2. What species were introduced to America, by whom, and for what reason?

3. In light of world human population growth and demand for direct human consumption of the large quantity of grain now consumed by livestock, is there a future for animal husbandry? Pretend you are debating a vegetarian on the merits of abolishing versus continuing the general practice of rearing domestic meat animals. Justify your position with more than opinions.

Chapter Two

THE BEEF INDUSTRY

The beef cattle industry has shown tremendous growth in the past three decades. Economic contribution on a worldwide basis has also been tremendous. The beef cattle industry will continue to develop as long as mankind has feed and forages that are useless for direct consumption. As long as we have areas that are unable to sustain growth of crops yet will maintain adequate supplies of rough forages, the beef cattle industry will continue to develop.

CATTLE-RAISING REGIONS

Beef production has steadily increased in the United States in all of the major cattle-raising regions (Figure 2-1). This trend will continue to increase to meet the growing demands for beef.

Northeast Region

The northeast region is of minor importance to commercial beef production. Beef herds are small and are often a supplementary enterprise by part-time

FIGURE 2-1. Major cattle-raising regions in the United States. (Courtesy of the U.S. Department of Agriculture; Agricultural Economic Report 235)

farmers. This area is predominantly dairy; thus beef production will remain stable and is not expected to increase in the future.

Corn Belt and Lake States

An increase in beef production in these areas stems from more animals being home finished with locally grown grains. Beef herds are small, and increases in calf production stem from efficient pasture management on old cropland and conversion of some dairy herds to beef herds. Future expansion is limited by the inability of beef to compete for land with crops that yield a higher profit. Any increase will come from better management of existing herds.

Southeast

The most dramatic increase in beef production has occurred in the southeast. The transition of cropland (especially cotton) to improved pastures has led to a rapidly expanding cow–calf industry. A trend has been to specialization of farms, leading to increasing numbers and quality of existing beef herds. Beef production will continue to increase but at a slower rate, with increases in calves grass fattened and cow–calf operations.

Northern Plains

Beef cow numbers have doubled in the Northern Plains in the last 20 years. Beef herds are becoming a major enterprise rather than a supplementary, part-time venture. Future growth will depend on how well cattle can compete for land with grain crops. The trend is toward forage production.

Southwest

The southwest has always been the largest cattle-producing area. Much of the land is rangeland, timberland, or rough terrain that can be used most efficiently for grazing. A dramatic increase is noted in the finishing feedlot operations, particularly in the Texas Panhandle. A ready source of sorghum grains and a supply of feeder calves have developed the feedlot industry in this region. Future increases will be small because of the maximum use of land suited only for grazing. Some increase may result from a switch from sheep and goats to beef, more efficient management of existing herds, and further increases in the feedlot industry.

Mountain and Pacific Regions

The Mountain region has doubled beef production in the last 20 years, mostly because of expanding feedlot operations, especially in Colorado and California. Beef herds are kept in the mountainous regions and are quite large. Future expansion will come from further increases in feedlot operations and better management of existing herds. Competition for land from crops and urban centers will limit expansion of the Pacific region.

BREED SELECTION

Certain breeds are adaptable to different climatic and feed conditions and therefore have specific advantages to being produced over another breed. These breeds can be divided into four major categories: European breeds, Indian breeds, U.S.-developed breeds, and more recently, the exotics.

Actually, there is no perfect breed because some animals will possess traits desired under one condition that are not suitable for another environment. Selection of a breed is based on personal preference, environmental conditions, adaptability, reproductive efficiency, mothering and milking ability, longevity, size, ability to gain, and other traits that fit personal preferences.

ZOOLOGICAL CLASSIFICATION

Cattle belong to the phylum *Chordata* (animals having a backbone), class *Mammalia* (milk giving), order *Artiodactyla* (even-toed, hooved), suborder *Ruminatia* (cud chewing), family *Bovidae* (hollow horn), genus *Bos* (ruminant quadrupeds). Species are divided into *Bos taurus* (most of our domestic breeds) and *Bos indicus* (humped cattle).

Although other species were frequently mentioned by early writers, today they are mostly disregarded, at least in the United States. For the sake of historical interest, a listing of the early European species from which most of our cattle come seems appropriate.

1. *Bos akeratos.* A hornless breed of northern Europe.
2. *Bos brachycephalus.* A short-headed type thought by some authorities to be the forerunner of such distinguished breeds as the Hereford, Devon, and Sussex.
3. *Bos frontosus.* Linked by fossil remains found in Sweden with the Simmental and other spotted breeds of current Swiss and German fame.
4. *Bos longifrons.* Often referred to as the "Celtic ox"; comparatively small in size, and probably an ancestor of the Brown Swiss.
5. *Bos nomadicus.* Shared the Paleolithic Age with human beings in India.
6. *Bos primigenius.* The Auroch (giant ox), from which the Shorthorn, among others, supposedly descended.

The *Bos taurus*, our most common species today, appears to have evolved from a mixture of *Bos longifrons* and *Bos primigenius*. The *Bos taurus* and the humped cattle now referred to as *Bos indicus* constitute our two existing species.

BREEDS OF BEEF CATTLE

Any discussion of breeds should follow a systematic order. Table 2-1 lists the four divisions and the order of discussion for breeds.

EUROPEAN BREEDS

The European breeds are so called because they came from the traditional European countries that exist today. Countries such as England, Scotland, France, Switzerland, and Germany are big contributors of modern beef breeds.

Although many breeds originated in England, the term *English breeds* generally refers to three predominant bloodlines: Angus, Hereford, and Shorthorn. Legend has it that in the early days of cattle breeding, a prominent animal scientist was invited to report his views on qualities that cattlemen should be selecting for. The audience was about equally divided among Angus, Shorthorn, and Hereford breeders. The inevitable question was asked: "Which, in your opinion, is the better breed?" His answer embodied more truth than di-

TABLE 2-1. THE FOUR DIVISIONS OF BREEDS OF BEEF CATTLE

European Breeds	Indian Breeds	U.S.-Developed Breeds	Exotic Breeds
Angus	Brahman	Amerifax	Ankole Watusi
Hereford		Ankina	Beef Friesian
Shorthorn		Santa Gertrudis	Blonde D'Aquitaine
Milking Shorthorn		Beefmaster	Brown Swiss
Red Angus		Brangus	Charolais
Red Poll		Braford	Chianina
Devon		Charbray	Corriente
South Devon		Red Brangus	Galloway
		Polled Hereford	Gasconne
		Polled Shorthorn	Gelbvieh
		Barzona	Hays Converter
		Braler	Limousin
		Simbrah	Lincoln Red
			Longhorn
			Luing
			Maine Anjou
			Marchigiana
			Meuse-Rhine-Issel (MRI)
			Murray Grey
			Normande
			Norwegian Red
			Piedmont
			Pinzgauer
			Romagnola
			Salers
			Salorn
			Scotch Highland
			Simmental
			Sussex
			Tarentaise
			Welsh Black
			White Park
			Beefalo
			Musk Ox

plomacy when he said, "I like the Hereford on the range, the Shorthorn in the feedlot, and the Angus on the table."

All breeds have one or more characteristics they are noted for. If any one breed had all the qualities to meet every condition demanded of it, all other breeds would soon disappear. This is not the case; thus the need for the following discussion is obvious.

Angus

This breed originated in northeastern Scotland and was imported to the United States in 1873 to cross with Longhorn cattle to improve the beef indus-

FIGURE 2-2. The Angus is solid black and polled and yields an excellent carcass. (Courtesy of Don Shugart Photography, Grapevine, Texas)

try. It is solid black with the distinguishing polled (hornless) characteristic that is considered one of its superior traits. The Angus has a smooth hair coat (Figure 2-2) and is comparatively small in size, with bulls weighing to 1900 lb and cows weighing 1500 lb, generally. The Angus is small at birth and small at weaning. The desirable traits that are attributed to the Angus are cold tolerance, mothering and milking ability, early maturity, little calving difficulty, good rustling ability, high fertility, and perhaps the most outstanding quality—a reputation for producing an excellent-quality carcass with small bones, high muscling, and a low percentage of fat covering. Common criticisms of the Angus are lack of size, overprominent shoulders in some bulls, and bulls not trailing cows in sparse open range as well as bulls from other breeds.

Hereford

Herefordshire, England, is the origin of one of the best known breeds of cattle in the world. The Hereford was imported in 1817 and has been a favorite of western cattlemen ever since. It is medium to large in size, with bulls weighing to 2100 lb and cows to 1700 lb, generally. It has a medium birth weight and a medium to large weaning weight. The color is one of the most unusual of any breed, consisting of a white face and a red body with the white extending over the throat, brisket, flanks, and switch and below the knee and hock (Figure 2-3). Herefords are horned, with horns preferably growing downward and inward. The desirable traits of the Hereford include hardiness, grazing ability, rugged adaptability, reproductive efficiency, good temperament and disposition, heavy bones, and thick flesh.

Herefords have been criticized for low milk production, susceptibility to cancer eye and pinkeye because of eye pigment, and prolapse of the uterus. However, through proper breeding programs, low milk production is a thing of the past in many outstanding herds.

FIGURE 2-3. The "Whiteface" breed or Hereford has been a favorite of western cattle raisers. (Courtesy of Don Shugart Photography, Grapevine, Texas)

Shorthorn

Figure 2-4 shows a typical Shorthorn representative. This breed was also developed in the northeastern part of England and was brought to the United States in 1783—the first purebred major beef breed ever imported. Short, incurving horns are normal. The color of the Shorthorn is unusual in that the association allows three colors for registration: red, white, and roan. Any combination of red, white, or roan is acceptable with the red and roan being the

FIGURE 2-4. The Shorthorn may be red, white, or roan in color. (Courtesy of Don Shugart Photography, Grapevine, Texas)

most popular among breeders. The size is large in comparison to most breeds, with bulls ranging from 2200 lb and cows to 2000 lb. Calves are medium size at birth, and weaning weights are medium to heavy. The distinguishing characteristics of this breed are high milk production, efficient utilization of roughage, good temperament, and rapid rate of gain in feedlot. The Shorthorn is adapted to a farming enterprise where an abundance of feed exists. The Shorthorn has been criticized for a tendency to have a coarse, lower-quality carcass, especially if fed past optimum slaughter weight.

Milking Shorthorn

Figure 2-5 illustrates a typical Milking Shorthorn. This is a dual-purpose breed that had its origin in England from the existing Shorthorn cattle. It is similar in size and color to the Shorthorn but has been selected for its milking ability. It may be either polled or horned, with bulls weighing over 2000 lb and cows to 1700 lb. This breed is more angular and longer than the beef Shorthorn type, carrying less flesh and indicating more evidence of dairy type. However, the offspring can be well fleshed and compete favorably because of high milk production.

FIGURE 2-5. The Milking Shorthorn is a dual-purpose breed developed from the Beef Shorthorn breed. Shown is Crest View Mary, 3rd Champion in 1969 in Schmidt, Nebraska. (Courtesy of the American Milking Shorthorn Society)

FIGURE 2-6. The Red Angus breed was developed from existing Angus herds by selecting from the natural occurrence of about 1 red mutation out of every 500 Black Angus births. The Red Angus has similar characteristics to Black Angus.

Red Angus

The British Isles again are responsible for the stock that produced this breed of cattle (Figure 2-6). The red color is a natural mutation that comes from Black Angus cattle. Even now, 1 of every 500 Angus in black herds is red. This characteristic is recessive so that it breeds true, and thus the Red Angus is easy to maintain as far as color patterns are concerned. It is similar in size and other characteristics to the herd from which it sprang.

Red Poll

The Red Poll (Figure 2-7) also came from England from the counties of Norfolk and Suffolk. It was imported in 1873 by New York breeders for use as a dual-purpose breed, a characteristic for which it is still noted. The color is solid red with some white permissible in the switch and below the under line. The Red Poll, as the name implies, is naturally polled. It is small upon maturity and produces medium-sized calves with medium weaning weights. Bulls range to 2000 lb with cows often weighing to 1500 lb. Its outstanding characteristics are early maturity, good grazing, high milk production, and a reputation of producing carcasses of high cutability. The Red Poll tends to have a large barrel and is lightly fleshed in the loin and hindquarter. This criticism stems from the Red Poll's dual-purpose use with breeding emphasis placed on milk production as well as beef.

FIGURE 2-7.   The Red Poll breed is noted for being polled, red in color, and dual purpose.

Devon

Another dual-purpose breed with its origin in southwestern England from the counties of Devon and Somerset is shown in Figure 2-8. The Devon was introduced in the United States about 1623; it was one of the first breeds introduced into the New World. The color is deep ruby red, ranging to chestnut in color. Because of their color pattern, Devons have been nicknamed "rubies." The Devon has horns that are creamy white with dark, generally black tips, a characteristic that makes it easy to identify from other breeds. It is medium size at maturity and produces medium calves with medium weaning weights.

FIGURE 2-8.   Devons, because of the deep ruby red color, were called "rubies."

FIGURE 2-9. The South Devon produces the famous Devonshire cream. (Courtesy North American South Devon Association)

Its desirable traits are beefiness, dual usefulness, and good milking ability. Its adaptability to this country would appear to be in breeding up and cross on existing breeds. The Devon is also noted for its docility and ability to adapt to temperature extremes. However, the Devon is criticized for slow growth and slow finishing in feedlots.

South Devon

Originating from the west of England, these cattle are not to be confused with the North Devon, which is somewhat similar to the beef Shorthorn. The South Devon (Figure 2-9) is light red or brown and the largest of the British breeds. The origin of the breed is not definitely known, but herds have been established since the sixteenth century. Reports indicate that the South Devon has had a fine performance record for more than 100 years with outstanding qualities of beef, milk, and butterfat. The famous Devonshire cream is also from these cattle. The cattle are in many parts of the United States, with the heaviest concentration in the midwest.

INDIAN BREEDS

Some 30 breeds from India, such as the Nellore, Guzerat, Gir, Zebu, Red Sendhi, and many others go to make up the Zebu cattle.

Brahman

The Brahman breed was developed in this country from a blending of three breeds of Indian cattle: the Gir, the Guzerat, and the Nellore. The first Zebu cattle were brought to the United States about 1849. The cows range in size generally up to 1300 lb with bulls weighing to 2000 lb or even more. The Brah-

FIGURE 2-10. The Brahman is a favorite in hot, humid regions where disease, parasites, and insects are troublesome. (Courtesy of Don Shugart Photography, Grapevine, Texas)

man is a medium-sized breed, producing medium-sized calves with generally light weaning weights. It is horned and varies in color from light gray, mottled, to almost black. It is humped at the back of the neck, which is a continuation of muscles from the shoulder, and has long pendulous ears and loose pendulous skin along the dewlap, sheath, and throat. The Brahman (Figure 2-10) has desirable traits that are possessed by few other breeds. It thrives well under minimal managment, is tolerant of the heat, has excellent mothering ability, is resistant to adverse conditions such as diseases and parasites, and is sought after for crossbreeding purposes because it belongs to another species. The crossing of species gives the greatest hybrid vigor. The breed is criticized for lack of cold tolerance, late maturity, and low fertility in some bloodlines. The Brahman is interesting because it possesses small muscles under the skin allowing it to "twitch" its hide, a favorable factor for insect resistance. A yellow secretion is also noted on the neck, which is reported by some to be a natural insect repellant.

U.S.-DEVELOPED BREEDS

Amerifax

After mating Beef Friesian seedstock with Angus herds for a number of years, a group of cattlemen from several states joined together and founded the Amerifax Cattle Association. Their existing percentage cattle are being used as foundation animals for the new breed. Desired characteristics for the purebred Amerifax (Figure 2-11) are polled and solid red or black in color.

FIGURE 2-11. One of the newest American breeds of cattle, Amerifax are genetically ⅝ Angus and ⅜ Beef Friesian. (Courtesy of Amerifax Cattle Association)

Ankina

The Ankina Breeders, Inc., founded in 1975, establishes minimum performance standards of yearling weight ratios to determine eligibility for registration. Ankina (Figure 2-12) are black or dark brown and polled (scurs are acceptable). Any mating plan that results in ⅝ registered Angus and ⅜ full-blooded Chianina with proof of purity from the registration certificates are acceptable providing the foundation animals are free from genetic defects.

Santa Gertrudis

The first breed developed by a U.S. breeder was the Santa Gertrudis, a name given to it because of a creek that ran through its place of origin, the King Ranch in Texas. This breed derives its bloodlines from the Shorthorn and

FIGURE 2-12. The result of selectively mating the maternal traits of a registered Angus with the terminal sire merits of a full-blooded Chianina is the genetic combination for Ankina. (Courtesy of Ankina Breeders, Inc.)

Chap. 2 THE BEEF INDUSTRY

FIGURE 2-13. The red Santa Gertrudis was the first breed developed in the United States. (Courtesy of Don Shugart Photography, Grapevine, Texas)

Brahman breeds—being ⅝ Shorthorn and ⅜ Brahman. It was recognized as a breed in 1940. The color is a deep, dark cherry red with a smooth, slick, shiny coat, horns, and lose hide similar to the Brahman breed (Figure 2-13). The Santa Gertrudis is large in size with cows up to 1600 lb and bulls up to 2000 lb, generally. It produces medium-sized calves with medium to heavy weaning weights. Its outstanding characteristics include excellent beef conformation, heavy hindquarter development, excellent foraging ability, efficient beef production on grass, resistance to diseases and insects, heat tolerance, and efficient feed conversion under feedlot conditions. The Santa Gertrudis is nervous, matures late, and tends to be coarse if fed beyond optimum slaughter weights. Breeding stock are criticized for low reproductive efficiency in some herds and pendulous sheaths in some bulls.

Beefmaster

The Lasater ranch in Falfurrias, Texas, developed the Beefmaster (Figure 2-14) in about 1931. It achieved breed status in 1954 and consists of ¼ Hereford,

FIGURE 2-14. As its name implies, the Beefmaster was developed for the production of beef without regard to traits such as color. (Courtesy of Don Shugart Photography, Grapevine, Texas)

¼ Shorthorn, and ½ Brahman blood. The interesting point about the Beefmaster is that it has no required color pattern whatsoever, the main emphasis being on beef, as the name would imply. Its color could include brown, reddish-brown, red, red-and-white spots, and all imaginable combinations. The Beefmaster is also not discriminated against if it is horned or polled. The breed is generally medium in size, producing medium-sized calves with heavy weaning weights. The breed was developed by selecting for six traits: disposition, fertility, weight, conformation, hardiness, and milk production. The breed is reputed to be gentle, intelligent, and early maturing so that heifers may be bred at 12 to 14 months. Other qualities praised by breeders are heavy weaned calves under range conditions, production of choice carcasses, necessity of a minimum of management, disease resistance, heat tolerance, few calving problems, and heavy milk production. The goal of the association is to produce the highest-quality red meat per pound of liveweight. The Beefmaster is a relatively new breed and thus is not as uniform as are longer-established breeds.

Brangus

Oklahoma and Texas shared in the development of the Brangus (Figure 2-15). It respresents ⅜ Brahman and ⅝ Angus blood. The color is black with the characteristic hump from the Brahman side of the breeding and the characteristic polled nature from the Angus side. The Brangus is heavy in size and produces medium-sized calves with heavy weaning weights. The desirable characteristics include thick, beefy conformation; good growth rate; heat toler-

FIGURE 2-15. The Black Brangus arose from the crossing of Brahman and Angus cattle.

Chap. 2   THE BEEF INDUSTRY

ance; insect and disease resistance; and good mothering ability. The Brangus is criticized for having a bad disposition and lack of thickness in the hindquarters.

Braford

This breed was also developed in the United States by crossing Herefords and Brahmans with a characteristic ⅝ Hereford and ⅜ Brahman standard. This gives the characteristic medium size with unusual color patterns such as brindle, mottled, and various combinations which one might expect using the Hereford and the Brahman stock available (Figure 2-16). The Braford produces medium-sized calves with heavy weaning weights. The desirable traits are ability to thrive under little management; heat, insect, and disease tolerance; good mothering ability; and efficient conversion of feed, especially forages, to beef. The Braford lacks the cold tolerance necessary for production in extreme northern states.

Charbray

The Rio Grande Valley was the site of the original cross between the Charolais (discussed below) and the Brahman bloodlines that led to the development of the Charbray (Figure 2-17). No definite bloodline standards are set. However, the preferred is ¹³/₁₆ Charolais and ³/₁₆ Brahman. The Charbray is large, with cows weighing up to 2200 lb and bulls often reaching 3200 lb. The color is light tan to cream white, and the Charbray is horned. It produces large calves with heavy weaning weights. Outstanding characteristics include mothering ability, rapid rate of gain, resistance to external parasites and insects, size,

FIGURE 2-16. Crossing the Brahman and Hereford breeds gave rise to the Braford breed. (Courtesy of the International Braford Association)

FIGURE 2-17. Crossing the Brahman and Charolais breeds gave rise to the Charbray breed. (Courtesy of the Charbray Division, American–International Charolais Association)

and scale. Common criticisms include long shallow bodies and too much length in the legs.

Red Brangus

This breed sprang from the Brangus breed, and because of the characteristic black color of the Brangus and the before-mentioned characteristic of red mutations from the Angus breed, it is not unusual to expect the same mutation from Black Brangus. This is how the Red Brangus developed as a mutation from the existing Black Brangus in Texas in 1946. It is similar in size and characteristics to the Black Brangus and is illustrated by Figure 2-18.

Polled Hereford

The Polled Hereford (Figure 2-19) represents the development of an idea spawned in the minds of a small number of midwestern Hereford breeders in the late 1890s, who realized that it was both possible and practical to develop "modern Herefords, minus horns." Warren Gammon from Des Moines, Iowa, sent inquiries to 2500 members of the American Hereford Association in 1901, attempting to locate some naturally hornless purebred Herefords. From the 1500 replies he received, Gammon located and bought 4 bulls and 10 females. Thus the development of the Polled Hereford had its origin in a survey to fulfill the realization of an idea. The Polled Hereford is similar in size and all other characteristics to the Hereford except that it is hornless. They are most often criticized for a lack of muscling in the hindquarter.

FIGURE 2-18. The Red Brangus developed from Black Brangus herds much as the Red Angus developed from Black Angus herds. (Courtesy of the American Red Brangus Association)

Polled Shorthorn

Originally referred to as Polled Durham, this breed is a distinctly American innovation. Ohio is recognized as the major location of the breed's development about 1870. Shorthorn bulls were used on native polled cows, using inbreeding techniques to develop a breed that had the color and conformation of the Shorthorn but that would breed true for the hornless characteristic. Some natural mutations from the ordinary, horned Shorthorn were also used. In 1889, the Polled Shorthorn Association was formed using these animals as the foundation for the breed.

FIGURE 2-19. The idea of "Herefords minus horns" gave us the Polled Hereford breed. (Courtesy of Don Shugart Photography, Grapevine, Texas)

FIGURE 2-20. The Polled Shorthorn developed from the polled offspring in horned Shorthorn herds. (Courtesy of the American Polled Shorthorn Society)

The breed in recent years has had much horned blood infused by breeders who continue to strive for improvement, while retaining, through proper selection, the hornless genotype. These cattle are typical Shorthorns in color, size, and other outstanding characteristics with the exclusion of horns. Figure 2-20 shows a typical specimen.

Barzona

The Bard Kirkland ranch in Arizona during the 1940–1950s gave rise to this newly recognized breed (Figure 2-21). The Hereford, Santa Gertrudis, Angus, and a breed from Africa known as Africander were used to produce a light red breed, bearing the name of Barzona (a combination of Bard and Arizona). It is medium size, ranging up to 1000 lb for a cow and 1500 lb for a bull. The calves are medium size with medium to heavy weaning weights. The Barzona's desirable traits include ability to thrive under harsh conditions, heavy wean-

FIGURE 2-21. The Barzona breed was developed for the Arizona range, but it performs well in most other areas.

Chap. 2   THE BEEF INDUSTRY

ing weights, superior feed conversion, fast growth, heat tolerance, and mothering ability. It is widely adaptable to conditions that include mountainous ranges, dry regions, and other locations where sparse vegetation exists. The Barzona has been criticized for lack of uniformity compared to the more established breeds.

Braler

The Braler (Figure 2-22) is a recently developed association utilizing the "grading up" process developed by numerous other new breeds. A first cross between Brahman and Salers breeds establishes the first generation. The advantages of this cross are obvious. The Salers is reportedly a trouble-free breed, high in fertility, with calving ease. Those characteristics combined with the hybrid vigor from the Brahman cross and the disease and insect resistance of the Brahman make the Bralers a potentially important contributor to semitropical, rough terrain areas.

Simbrah

The American Simmental Association created a registry for the Simbrah Association in 1977. Crosses between Simmental and Brahman cattle in an organized program to create a ⅝ Simmental–⅜ Brahman animal was initiated through the process of "grading up." Cattle that are recognized as ⅝ Simmental–⅜ Brahman blood are certified as purebreds. These cattle are found in 38 states, although the majority of them are registered in Texas (4200), Mississippi (1700), Oklahoma (325), and Florida (170).

The advantages of the Simbrah are said to be longevity, heat tolerance, disease and insect resistance, durability, grazing ability, calving ease, early sexual maturity and milking ability, high carcass yield, favorable lean to fat

FIGURE 2-22. Grand Champion Braler bull. (Courtesy of Don Shugart Photography, Grapevine, Texas)

ratio, and good temperament. The obvious disadvantage to the Simbrah is that many will perceive it to be just another crossbreed, and it has such a wide variety of color patterns that it is difficult to distinguish it as a separate breed.

EXOTIC BREEDS

The term *exotic breeds* as used in this book refers to relatively recent breed arrivals or revivals in the United States. The older term *lesser known* has been replaced by the more mysterious "exotic," although the chances are as great, or possibly more so, that it refers to an old, established breed in improvement.

Ankole Watusi

This breed, often referred to simply as "Watusis," is more of a novelty in the United States and orginated in Africa. The Ankole Watusi has drawn huge crowds at stock shows because of their horn size and length (Figure 2-23). Animals often do not exceed 600 to 800 lb and are very slow to mature. However, at 6 months of age, horn size approximates that of traditional "Mexican steers," making them a potential new source for entertaining, steer roping events. Crossbreeding experiments have not shown any advantage to beef producers, so the horns remain the object of chief interest in this unusual breed.

FIGURE 2-23. Ankole Watusi, the longest and largest horned cattle in the world, originate in Africa.

FIGURE 2-24. The Beef Friesian originated in the Netherlands, where for centuries it has been one of the old, established dual-purpose breeds.

Beef Friesian

A high percentage of all beef consumed in the British Isles comes from Friesians or Friesian crosses. The Beef Friesian (Figure 2-24) is a big, long-muscled animal black and white in color. The breed offers early sexual maturity, high fertility, and ease of calving. The cows are excellent mothers and milk producers, and the slaughter cattle grade well and produce high-yielding carcasses.

Blonde D'Aquitaine

This is a French breed of cattle that is very large in size. It has been used for draft as well as beef, producing large calves weighing up to 98 lb at birth. It is horned, with a fawn color varying from wheat color to slighly red. Desirable characteristics are large size and heavy weaning weight. The Blonde D'Aquitaine is heavily muscled, with a deep chest, wide hips, and well-developed hindquarters. It is used mostly in this country for crossbreeding purposes (Figure 2-25).

FIGURE 2-25. Originally, the Blondes existed in the southwest of France as three breeds: Garonnais of the plain, Le Quercy of the hills, and the Blonde-des Pyrenees.

FIGURE 2-26. The dairy breed Brown Swiss is sometimes called the "American Exotic" breed. (Courtesy of the Brown Swiss Cattle Breeders Association)

Brown Swiss

The Brown Swiss was imported to the United States from Switzerland in 1869 for use as a dairy breed. It is one of the largest and most heavily muscled dairy breeds but is also used for beef breeding. Considered as the "American Exotic" (Figure 2-26), it is known for thriftiness, milking ability, and growth rate. Common criticism is the dairy type, including large barrel and lack of finish and flesh in the hindquarters.

Charolais

Figure 2-27 illustrates the type of cattle known as Charolais, which had its origin in the province of Charolles in France. It was a dual-purpose animal used for draft and meat. The first importation came to Mexico in 1930 through the efforts of Mexican industrialist Jean Pugibet. That importation consisted of 2 bulls and 10 cows. In 1966, further importations were made to Canada, and that same year, the first semen was imported from France to the Untied States. France maintains an embargo on exportation of these cattle, and they have been highly sought after from parts of Mexico where purebreds exist, and semen is also in great demand.

The size is large for mature cattle as are birth weight and weaning weight. Cows range commonly to 1700 lb and bulls to 2600 lb, with calves

FIGURE 2-27. The Charolais was one of the first breeds brought to the United States for use in cross-breeding programs. (Courtesy of Don Shugart Photography, Grapevine, Texas)

weighing up to 100 lb at birth and 600 or 700 lb at weaning. The color is white to cream or light white with some reddish pigmentation to the skin, especially around the nose and eyes and under the belly. The Charolais is horned but some polled animals do exist; it is generally shown de-horned. The outstanding characteristics include excellent muscling in the loin, round, and hindquarters; heavy bones; good mothering ability; rapid growth; high dressing percent; and cold and heat tolerance. The Charolais is sometimes criticized for wide variations in the breed because of a limited supply of breeding stock and up-breeding to breed status. This, in time, will diminish as the breed grows in number, as culling of the breed becomes more active, and as stockmen continue to develop good Charolais herds from outstanding breeding stock. Some difficulty in calving of heifers has been noted, especially if heifers are not allowed to grow out properly.

Chianina

Figure 2-28 shows this breed developed in the Chianina Valley in Italy. It has been a triple-purpose breed for many years in Italy, used for draft, milking, and meat consumption. It was the sacrificial cow of the ancient Romans. The outstanding characteristics of this breed are its tremendous size and fast growth rate. It is extremely large at maturity, some reaching 6 ft at the withers and up to 4000 lb, with heavy calves at birth and at weaning. Its obvious desirable traits are to be used in this country for crossbreeding purposes to increase weaning weights and growth patterns. The Chianina is also noted for its excellent muscling in the hindquarter and loin back area, producing carcasses that are muscular and lean.

Corriente

Corriente cattle are hardy, small-horned, rodeo cattle common to the Sierra Madre Mountains in Mexico (Figure 2-29). Used mostly for roping and bull-

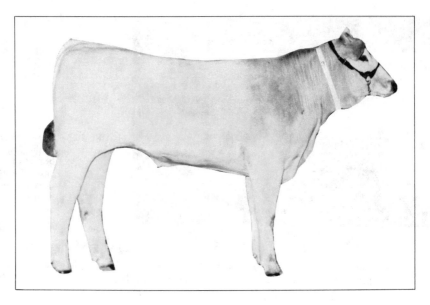

FIGURE 2-28. Italy's Chianina is a triple-purpose breed used for meat, milk, and draft.

dogging, there is evidence that Corriente steers may compete favorably in today's fed, lean beef market when retired as rodeo stock. Some reports have indicated mature weights of 1100 lb, and competitive slaughter grades as compared to other breeds on as much as 40% less feed.

Because so many Corrientes were being imported each year into the United States from Mexico, some enterprising cattlemen decided to meet the demand for rodeo stock by raising and selling their own. Breeders scattered throughout the United States are represented by the North American Corriente Association. The small, hardy cows are said to withstand heat or cold equally well and require only 75% as much winter feed as average-sized breeds.

FIGURE 2-29. Cates Smoker, 1988 Grand Champion Yearling Bull, North American Corriente Association. (Courtesy of N.A.C.A., Casper, Wyoming)

Chap. 2 THE BEEF INDUSTRY

FIGURE 2-30. Galloways have been around a long, long time as a distinct beef breed. English writings as early as 1530 praised "the black cattle of Galloway whose flesh is tender and sweet." (Courtesy of American Galloway Breeders Association)

Bulls are used by some commercial breeders on first-calf heifers because of smaller calves produced and less calving difficulties.

Galloway

The Galloway breed (Figure 2-30) originated in the province of Galloway in Scotland. It was first imported to the United States in the nineteenth century and is similar in appearance to the Angus, being black and polled but having a very long curly-haired coat that distinguishes it from the Angus. It is similar in size and type to the Angus but generally shows more length in the body. It is a very hardy breed with excellent tolerance of cold temperatures. The Galloway is used quite extensively in North Dakota and other areas that have severe cold temperatures. However, lack of heat tolerance limits its use to cold areas. A striking deviation of color pattern in some Galloways gave rise to selection for the trait and development of a separate breed, the belted Galloway.

Gasconne

The Gasconne (Figure 2-31) was developed in mountainous, southern France primarily as a draft breed. It has a reputation for being one of the most fertile French breeds with minimum calving difficulty. It was imported into the United States in the early 1970s at the same time when a large number of breeds were being introduced from Europe. Due to its nature and time of importation, it has not been widely tested or used in this country to date.

Gelbvieh

This breed (Figure 2-32) is native to Austria and the Bavarian region of West Germany and is considered a dual-purpose type of cattle. The size ranges up

FIGURE 2-31. This gray French breed, the Gasconne, is moderate in size and muscling.

FIGURE 2-32. The Gelbvieh comes from Austria and Germany, where they are known as "yellow cattle." (Courtesy of American Gelbvieh Association)

FIGURE 2-33. The Hays Converter is a breed developed in Canada. (Courtesy of Canadian Hays Converter Association)

to 1600 lb for the cow and 2800 lb for the bull. Birth weight is heavy, as is weaning weight. Its color is solid golden red to rust red. Gelbvieh means "yellow cattle." The outstanding characteristics include high calving percentage, good gaining ability, good milking and mothering ability, ease of calving, and large size with good muscling.

Hays Converter

This breed (Figure 2-33), developed in Canada by Senator Harry Hays, the former Minister of Agriculture, evolved from a combination of two crossbred animals, Hereford–Brown Swiss and Hereford–Holstein. It is usually black with a white face but may be red. The Hays Converter has good rustling ability and is noted for its hardiness, growth, and milk production.

Limousin

The Limousin (Figure 2-34) originated in a French province containing high rocky hills. The Limousin ranges from golden wheat color to deep red gold. The horns are light in color. The Limousin is small to medium at birth, developing into a large size at maturity. Cows often reach 1300 lb and bulls 2400 lb. The characteristics that are associated with this breed are high fertility, ease of calving, milking and mothering ability, good rustling ability, longevity, and good rate of gain.

FIGURE 2-34. The Limousin is another exotic breed from France. This steer was Grand Champion, Houston, 1984.

FIGURE 2-35. England's Lincoln Red breed developed from early Beef Shorthorn cattle.

Lincoln Red

The Lincoln Red developed from the Shorthorn breed in Lincolnshire, England. It is deep, cherry red in color and may be horned or polled (Figure 2-35). It is similar to the Shorthorn, but is more beefy. It is medium in size, with medium to large calves at birth and weaning.

Longhorn

Although not generally considered an exotic breed by the modern cattleman, the Longhorn has been around a long time, and recent interest in developing the traits the early Longhorn cattle possessed—such as mothering ability, disease resistance, ability to thrive under rugged conditions, hardiness, fertility, ease of calving, and longevity—has led to the possibility of infusing this blood with modern breeds. Thus it does not seem far-fetched to include the Longhorn (Figure 2-36) as a possible exotic. An association was recently formed, and a

FIGURE 2-36. A renewed interest in crossbreeding with modern Longhorn cattle has led some to consider them as exotics. (Courtesy of the U.S. Department of Agriculture)

Chap. 2   THE BEEF INDUSTRY

lot of interest is being shown. The origin of this breed was through natural selection from Spanish cattle brought into the United States about 1521. The Longhorn is famed throughout the world for qualities that have yet to be forgotten and may be heard from again through modern breeding techniques.

Luing

The Luing (pronounced "ling") is a recently synthesized beef breed that originated on the island of Luing off the west coast of Scotland, where the climate is cold and wet. The breed (Figure 2-37) was started in 1847 by crossing the Shorthorn and the Scotch Highland breeds. According to John Rouse (*World Cattle,* Vol. 1, University of Oklahoma Press, Norman, Oklahoma, 1970), "selection has been for beef conformation and rate of gain on the herd, which is raised in the open the year round. Color has been disregarded in the process and is even more varied than in the Shorthorn; it is red to white, yellow, roan or brindle." It is considered a maternal breed, with a great degree of hardiness.

Maine Anjou

France gave rise to yet another breed (Figure 2-38) that has created interest recently in the United States. The Maine Anjou came into being in the United States in 1970 because of an importation of semen only. This is being continued today to produce large cattle with bulls weighing over 2700 lb and cows approaching 2000 lb. Birth weights are very high, often above 100 lb, producing calves with heavy weaning weights and great size and scale. The Maine Anjou is dark red and white in color with the head predominantly red and the eyes generally surrounded by red color. Desirable traits are size and scale, lean carcass production, ability to adapt to harsh conditions, heavy weaning weight, rapid growth rate, and good milking and mothering ability.

FIGURE 2-37. The Luing was developed in Scotland.

FIGURE 2-38. The Maine Anjou is a French breed that came to the United States by way of imported semen. (Courtesy of Don Shugart Photography, Grapevine, Texas)

Marchigiana

Another breed from Italy is the Marchigiana. It is gray and horned and resembles the Chianina, but is not as large in size. It is noted for efficient beef production and average size of bulls, up to 2650 lb, with cows weighing from 1350 to 1500 lb (Figure 2-39).

Meuse-Rhine-Issel

The Meuse-Rhine-Issel (MRI) probably originated from the same foundation stock as the Dutch Friesian but underwent selection for red instead of black color and for a shorter-legged, heavier-boned, more muscular body type that tends to be more typically dual-purpose when compared to the conformation of the Friesian. About 24% of the cattle in the Netherlands are MRI's, 74% Friesians, and 2% other breeds. Because of its dual-purpose characteristics, the MRI is considered a two-way breed in crossing programs (Figure 2-40).

Murray Grey

The Murray Grey originated in Australia by crossing a light roan Shorthorn cow with an Angus bull. The breed developed by use of the bull calves. It is silver gray in color and like the Angus in type and production of carcass (Figure 2-41). The Murray Grey is small at birth (60 to 70 lb) and small to medium at maturity.

Normande

The stock, which is found in the Normandy region of northwestern France, comes probably from the cattle imported by Viking conquerors in the ninth and tenth centuries. Because of its conformation and milk yields, the Nor-

FIGURE 2-39. Marchigiana is a white breed, a middle-of-the-road type of Italian cattle with regard to frame and muscling. Like the Chianina and Romagnola, it is strictly a beef breed and is not used for milk production. Their white hair and black skin give these cattle much resistance to cancer eye, pinkeye, sunburn, and other skin problems. They are also heat-tolerant, which make them well adapted to many regions of the United States.

FIGURE 2-40. The MRI is a red and white dual-purpose breed found in southeastern Holland.

FIGURE 2-41. The Murray Grey is similar to the Angus in type and carcass. (Photo by author, Australia)

FIGURE 2-42. The Normande are known for their "bespectacled" appearance. (Courtesy of American Normande Association)

mande occupies a quarter of the cattle population of France. In addition, its mothering qualities make it a very suitable breed for crossing for beef production. The Normande bears an interesting and variable color pattern. Most Normandes are mahogany and white or black and white with rings around the eyes giving them a "bespectacled" appearance (Figure 2-42). They have adapted to various management methods and different climates in North America.

Norwegian Red

This dual-purpose breed (Figure 2-43) accounts for about 60% of the cattle population in Norway. Because the breed has developed since World War II from the amalgamation of three older breeds, it is difficult to characterize the Norwegian Red. The FAO publication *European Breeds of Cattle* states that the breed was formed through the fusion of three other strains, the Norwegian Red and White, the Red Tronder, and the Polled Eastland. Calving ease is

FIGURE 2-43. The Norwegian Red is a dual-purpose breed. (Courtesy of North American Norwegian Red Association)

reported to be similar to the British breeds. In crossbreeding programs, the Norwegian Red is considered a maternal breed that could increase milk production and perhaps cause a slight improvement in growth with no increase in calving difficulty.

Piedmont

The Piedmont is a grayish-white breed with black about the head, neck, nose, eyes, and switch. The Piedmont breed originated in northwestern Italy. It is large in size and has been selected for red meat, with double muscling very prevalent in the breed.

Pinzgauer

The Pinzgauer breed was developed in Austria and adjacent areas of Germany and Italy. It is a chestnut brown breed, used for meat and milk. The Pinzgauer is medium to large in size and produces a lean but beefy-type carcass with thick, wide loins (Figure 2-44).

Romagnola

These big, gray cattle (Figure 2-45) are the principal beef breed of northern Italy. They were developed for both meat and draft, are relatively low on the leg, and are very muscular. They have horns and the bulls usually have dark markings on the shoulders and a black patch around the eyes. This breed is

FIGURE 2-44.   The Pinzgauer is a descendent of the European mountain breeds.

FIGURE 2-45. The Romagnola is another of the Italian
breeds with white hair
and black skin.
(Courtesy of Canadian
Romark Association)

considered a terminal sire breed in crossing programs. The cattle have ranked favorably on tests in Canada and in interbreed competition on weight per day of age compared at one year.

Salers

The breed (pronounced "sahlairs") takes its name from the mountainous region of south central France, where the soil and climatic conditions are relatively severe. The Salers (Figure 2-46) is reported to be a trouble-free breed that is above the average of French breeds in fertility and calving ease. The dark-red, dual-purpose cattle are tall cattle, characterized by upright lyre-shaped horns.

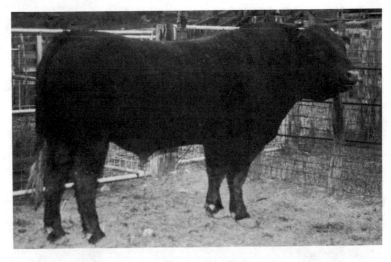

FIGURE 2-46.   The dark-red Salers is noted for its high fertility. (Photo by author, Australia)

Salorn

Salorn is a genetically lean beef breed for today's consumer from two unique foundations, the French Salers and the Texas Longhorn. The combining of these two breeds is an effort to duplicate a similar project by Jan Bonsma, who combined the perfectly adapted Africaner native cattle with Hereford and Shorthorn lines to produce the synthetic Bonsmara breed, which has totally dominated the cattle industry in South Africa.

With over 300 years of testing by nature to prove the absence of genetic functional weaknesses, Longhorn females bred to Saler bulls, because of their impressive growth traits and dominance of major beef carcass contests, will produce a $5/8$ Saler–$3/8$ Longhorn foundation breed. Smooth coated, red in color, of maximum longevity, of high efficiency of production in a wide range of cattle-producing areas of the world, and with low-choice carcasses with a minimum of external fat, the Salorn breed will be promoted as a global approach to the genetic production of lean beef (Figure 2-47).

Scotch Highland

A most unique breed originating in Scotland was imported to this country in 1922 in the form of the small Scotch Highland (Figure 2-48). The Scotch Highland is generally brown in color, has a very heavy coat that is long and shaggy, and has great width in spread of horns. Other colors exist and have no specific requirements in the association. Its outstanding characteristics are ability to survive on poor grazing, production of small calves that decreases instances of dystocia (calving problems), mothering ability, and cold tolerance. It is very popular in the northern latitudes, and its most outstanding characteristic is probably the fact that it produces very small calves and therefore the bulls are bred quite often to first-calf heifers. The Scotch Highland lacks heat-toler-

FIGURE 2-47. This Salorn heifer is an example of genetically engineered production utilizing Saler bulls and Longhorn cows. (Courtesy of Horlock Land and Cattle, Riviera, Texas)

FIGURE 2-48. The brown, heavy-coated Scotch Highland breed is very well adapted to cold areas. (Courtesy of the American Scotch Highland Breeders Association)

ance and is criticized for low feed efficiency, lack of muscling, and slow growth in some animals.

Simmental

The Simme valley of Switzerland gave rise to today's most predominant breed in Europe. In France, it is known as "Pie Rouge" and in Germany, "Fleckvieh." The Simmental is red in color (varying from dark to almost yellow) with white spots and white face (Figure 2-49). It is known as a good milker and rapid gainer, and has a long, heavily muscled body. It is large in size at birth, weaning, and maturity.

FIGURE 2-49. The Simmental breed is perhaps the most popular breed in Europe. (Courtesy of Donnell Ag Genetics, Chickasha, Oklahoma)

FIGURE 2-50. England's Sussex is deep cherry red in color. [Reprinted by permission from John E. Rouse, *World Cattle: Cattle of Europe, South America, Australia, and New Zealand.* (Copyright 1970 by the University of Oklahoma Press, Norman, Okla.), p. 318]

Sussex

The Sussex is another English breed. The color is deep cherry red, with a white switch (Figure 2-50). It is horned and medium in size. The Sussex is used in crossbreeding and yields carcasses well muscled in the loin and hindquarter.

Tarentaise

These hardy, medium-sized mountain cattle (Figure 2-51) are derived from an ancient Alpine strain in southeastern France. It was first described under its present name in 1959. Regarded as a dairy breed in the country of origin, em-

FIGURE 2-51. The Tarentaise are a mountain breed. (Courtesy of American Tarentaise Association)

FIGURE 2-52.   The Welsh Black is among the oldest of the British breeds. (Courtesy of Welsh Black Cattle Association)

phasis in North America has been directed to their beef characteristics. Tarentaise have a solid, wheat-colored hair coat ranging from a light cherry to dark blond. Their horns are lyre-shaped. It is considered a maternal breed.

Welsh Black

One of the oldest British breeds, the Welsh Black dates back in Wales to the fourteenth century. Its development was triple-purpose: draft, milk, and meat. As the name implies, it is black in color. The Welsh Black is horned and has a shaggy hair coat (Figure 2-52). It is adapted to withstand cold weather, and is hardy and medium in size. However, it lacks heat tolerance and is criticized for slow growth past weaning.

White Park

Descendents of the British Isles' Chillingham "wild, white cattle" herd, the 2000-year heritage of White Park cattle (Figure 2-53) is that of natural selection dictated by nature. In England, the cattle were domesticated and served as status symbols for feudal lords, who enclosed them in areas similar to game preserves called "parks." They became known as the British White Park cattle. They were first brought to the United States during World War II to preserve seedstock in case of German invasion and most ended up in Illinois. The cattle are comparable in size to Hereford, Angus, and Shorthorn. They are naturally polled and have black on the points.

FIGURE 2-53. The White Park cattle were once kept in "parks." (Courtesy of White Park Cattle Association of America)

Beefalo

The most recent beef development is the Beefalo, bred by D. C. Basolo of Tracy, California. The hybrid consists of ⅜ buffalo, ⅜ Charolais, and ¼ Hereford (Figure 2-54). The Beefalo is claimed to be hardy, fast growing, able to calve without difficulty, resistant to many common cow diseases and parasites, and able to produce a tasty grass-finished product. As with many of the exotic breeds, only time will tell whether the Beefalo will obtain a place in the industry.

FIGURE 2-54. Joe's Pride, the 20-month-old hybrid Beefalo bull, was recently sold for a record $2.5 million to a Canadian breeding farm. (Courtesy of the World Beefalo Association)

Musk Ox

One of the most unusual forms of animal husbandry exists in the field of musk-ox husbandry and qiviut knitting, taught in workshops throughout Alaska. The musk ox, which less than 20 years ago totaled less than 14,000 animals and was confined to a few sites in the Canadian north and Greenland, has now been established in Alaska, Quebec, and Norway to become a key component in the Arctic's economy. Over 200 Eskimo families in 22 villages on the Alaskan tundra coastline are kept busy designing and knitting garments from the fibers produced from the musk ox.

The musk ox, whose name is misleading since it does not produce a musk odor and is not an ox, produces two coats of hair so thick that no rain or cold can penetrate it. Body heat is retained with two thicknesses of hair: an outer black hair coat, and an inner coat, called qiviut—a soft, gray-brown hair.

The silky underfleece works its way out through the longer coat in the springtime giving the musk ox the appearance of an unmade bed. This qiviut, softer than cashmere, is harvested by simply pulling it from the musk ox. Qiviut weighs almost nothing, yet is one of the warmest fibers in nature. Some bull oxens can produce as much as 7 lb of qiviut, so fine and fleecy, it will make a stack 6 ft high yet will produce 100 scarves.

The garments made from the musk ox are among the finest in the world. For information on products available and more information on musk-ox husbandry, write Oomingmak, Musk Ox Producers' Cooperative, P.O. Box 80291 HO, Fairbanks, Alaska 99708.

STUDY QUESTIONS

1. The _____ region of the United States has had the most dramatic increase in cattle production.
2. The largest cattle-producing region has always been the _____ .
3. The humped cattle are the _____ species.
4. "English breeds" generally refers to _____ .
5. The most famous breed for carcass quality is the _____ .
6. A breed that allows three colors in the registry association is the _____ .
7. The greatest hybrid vigor is obtained by crossing _____ .
8. The first breed developed in the United States was the _____ .
9. Many purebred herds of _____ exist in Mexico.
10. A giant breed of cattle from Italy is the _____ .
11. The Charolais and _____ originated in France.
12. The "American Exotic," known as a dual-purpose breed, is the _____ .

Chap. 2   THE BEEF INDUSTRY

13. The two breeds best noted for breeding to first-calf heifers are _____ and _____ .

14. The dominant breed in Europe that is gaining popularity in the United States is the _____ .

15. A triple-purpose breed is used for _____ , _____ , and _____ .

DISCUSSION QUESTIONS

1. Why are there so many breeds of cattle?
2. Develop a plan for providing the maximum hybrid vigor in a crossbreeding program for cattle. Name and justify your reasons for selecting the breeds to be used.
3. How could you maintain hybrid vigor indefinitely?
4. Choose an exotic breed and defend your choice as if you were trying to convince your banker for a loan on the merits of the sterling qualities of this breed.
5. Choose an established purebred breed of cattle and justify your reasons for the selection of this registered breed in the same way as described in Question 4.

DIRECTORY OF BEEF CATTLE ASSOCIATIONS

AMERIFAX
Amerifax Cattle Association
P.O. Box 149
Hastings, Nebraska 68901

ANGUS
American Angus Association
3201 Frederick Boulevard
St. Joseph, Missouri 64506

ANGUS, RED
Red Angus Association of America
Box 776
Denton, Texas 76201

ANKINA
Ankina Breeders, Inc.
5803 Oakes Road
Clayton, Ohio 45315

ANKOLE WATUSI
Ankole Watusi International
 Registry
Star Route Box 45
Hebron, North Dakota 58638

BARZONA
Barzona Breeders Association of
 America
P.O. Box 631
Prescott, Arizona 86302

BEEFALO
American Beefalo Association
200 Semonin Building
4812 U.S. Highway 42
Louisville, Kentucky 40222

BEEFMASTER
Beefmaster Breeders Universal
Suite 350, GPM South Tower
800 Northwest Loop 410
San Antonio, Texas 78216

Foundation Beefmaster
  Association
200 Livestock Exchange Building
Denver, Colorado 80216

BELTED GALLOWAY
The Belted Galloway Society, Inc.
P.O. Box 5
Summitville, Ohio 43962

BLONDE D'AQUITAINE
American Blonde D'Aquitaine
  Association
Rt. 2, Box 21A
Porum, Oklahoma 74455

BRAFORD
International Braford Association,
  Inc.
P.O. Box 1030
Fort Pierce, Florida 33450

BRAHMAN
American Brahman Breeders
  Association
1313 La Concha Lane
Houston, Texas 77054

BRAHMENTAL
American Brahmental Association
P.O. Box 956
Floresville, Texas 78114

BRALERS
American Bralers Association
P.O. Box 40472
Houston, Texas 77240

BRANGUS
International Brangus Breeders
  Association
9500 Tioga Drive
San Antonio, Texas 78230

BRANGUS, RED
American Red Brangus Association
Dept. A, Box 1326
Austin, Texas 78767

BROWN SWISS
Brown Swiss Cattle Breeders
  Association of the U.S.A.
800 Pleasant Street
P.O. Box 1038
Beliot, Wisconsin 53511

BUFFALO
National Buffalo Association
Judi Hebbring, Executive Director
Custer, South Dakota 57730

CHARBRAY
American–International Charolais
  Association
1610 Old Spanish Trail
Houston, Texas 77054

CHAROLAIS
American–International Charolais
  Association
P.O. Box 20247
Kansas City, Missouri 64195

CHAR-SWISS
Char-Swiss Breeders Association
407 Chambers
Marlin, Texas 76661

CHIANINA
American Chianina Association
P.O. Box 890, State Highway 92
Platte City, Missouri 64079

CORRIENTE
North American Corriente
  Association
P.O. Box 9390
Casper, Wyoming 82609

DEVON
Devon Cattle Association, Inc.
Drawer 628
Uvalde, Texas 78801

DEVON, SOUTH
North American South Devon
  Association
Box 68
Lynnville, Iowa 50153

DEXTER
American Dexter Cattle
  Association
707 West Water Street
Decorah, Iowa 52101

GALLOWAY
American Galloway Breeders
  Association
Rt. 1, Box 342
Sanger, Texas 76266

GELBVIEH
American Gelbvieh Association
Livestock Exchange Building
Denver, Colorado 80216

HEREFORD
American Hereford Association
P.O. Box 4059
Kansas City, Missouri 64101

LIMOUSIN
North American Limousin
  Foundation
100 Livestock Exchange Building
Denver, Colorado 80216

LONGHORN
Texas Longhorn Breeders
  Association
3701 Airport Freeway
Fort Worth, Texas 76111

MAINE ANJOU
American Maine Anjou Association
654 Livestock Exchange Building
Kansas City, Missouri 64102

MAINE ANJOU, BLACK
American Black Maine Anjou
  Association
6301 Gaston Avenue
Allied Lakewood Bank Center,
  Suite #645
Dallas, Texas 76214

MARCHIGIANA
American International
  Marchigiana Society
Box 342
Atlanta, Texas 75551

MARKY
Marky Cattle Association
Box M
Ellsworth, Kansas 67439

MURRAY GREY
American Murray Grey
  Association
Box 30085
Billings, Montana 59107

NORMANDE
American Normande Association
Rt. 1
Verdon, Nebraska 68457

NORWEGIAN RED
North American Norwegian Red
  Association
Box 5606
Kansas City, Missouri 64102

PINZGAUER
American Pinzgauer Association
123 Airport Road
Ames, Iowa 50010

POLLED HEREFORD
American Polled Hereford
  Association
4700 East 63rd Street
Kansas City, Missouri 64130

RED POLL
American Red Poll Association
3275 Holdredge Street
Lincoln, Nebraska 68503

ROMANGNOLA
American Romagnola Association
6800 Shingle Creek Parkway
Minneapolis, Minnesota 55430

SALERS
American Salers Association
101 Livestock Exchange Building
Denver, Colorado 80216

SALORN
International Salorn Association
P.O. Box 52608
Houston, Texas 77052

SANTA GERTRUDIS
Santa Gertrudis Breeders
  International
P.O. Box 1257
Kingsville, Texas 78363

SANTA GERTRUDIS, POLLED
Polled Santa Gertrudis Breeders
P.O. Box 248
Decatur, Arkansas 72722

SCOTCH HIGHLAND
American Scotch Highland
  Breeders Association
508 Main Street
Box 249
Walsenburg, Colorado 81089

SHORTHORN
American Milking Shorthorn
  Society
800 Pleasant Street
P.O. Box 449
Beliot, Wisconson 53511

American Shorthorn Association
8288 Hascall Street
Omaha, Nebraska 68124

SIMBRAH
American Simbrah Association
1 Simmental Way
Bozeman, Montana 59715

SIMMENTAL
American Simmental Association
1 Simmental Way
Bozeman, Montana 59715

TARENTAISE
American Tarentaise Association
Box 1844
Fort Collins, Colorado 80522

WELSH BLACK
Welsh Black Cattle Association
Route 1
Wahkon, Minnesota 56386

WHITE PARK
White Park Cattle Association of
  America
820 Sixth Street
Nevada, Iowa 50201

ZEBU
Pan American Zebu Association
Registry Office
9039 Katy Freeway, Suite 433
Houston, Texas 77034

Chapter Three

BEEF CATTLE
REPRODUCTION

Understanding the reproductive process is a complicated physiological pathway to travel. So many theories and truths are cloaked in the often difficult language of the physiology involved that it is not an easy task to determine where to start the exploratory journey. Using as much simplicity as practical, the intent of this chapter is to cover reproduction in a logical manner, from one generation to the next. Using a young bull and heifer as a starting point, the changes from age of mating through parenthood will be followed.

ANATOMY

The first step in understanding reproduction is to understand anatomy; thus the artist's conception of the female reproductive system is illustrated in Figure 3-1. The actual appearance of a dissected tract is shown in Figure 3-2. The male reproductive system is shown in Figure 3-3. Further detail of the testis, desirable for explanatory purposes, is given in Figure 3-4.

The male and female reproductive systems do not become fully functional until puberty (the age of sexual maturity). The age at which puberty occurs will vary considerably with breeds, with a range of 8 to 18 months of age. In many breeds, puberty is more a function of weight than age, and many breed-

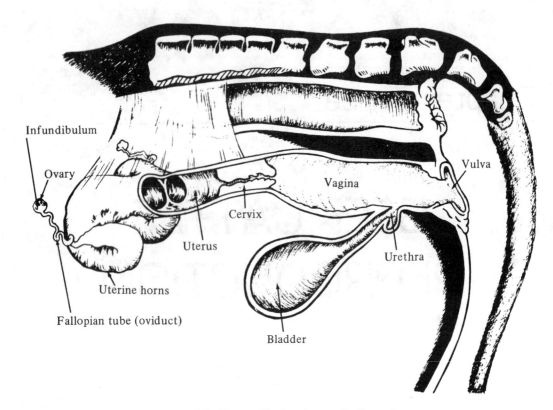

FIGURE 3-1.   The female reproductive system.

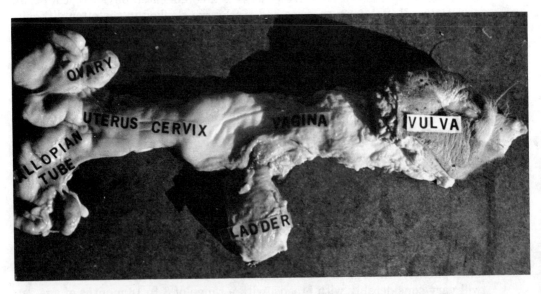

FIGURE 3-2.   The female reproductive tract. Ovaries and fallopian tubes come in pairs.

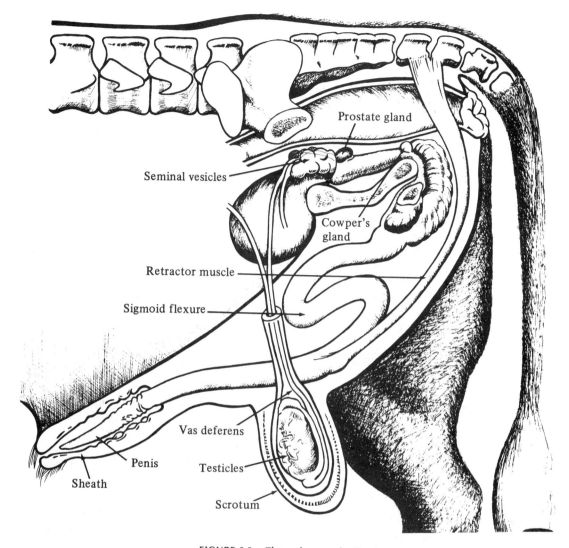

FIGURE 3-3. The male reproductive tract.

ers prefer to use a weight of 600 to 800 lb as a value for heifers. Bulls are generally somewhat heavier, approximately 800 to 1000 lb.

The Male

Because the emphasis is on the male as the foundation of most herds, the anatomy of the bull and changes related to puberty will be covered first. The reproductive system of the bull can be divided into three parts: the *testes*, also called *gonads*, *testicles*, or *primary organs*; the *accessory*, or *secondary sex glands*; and the *external copulatory organ*, the *penis* (Figure 3-3).

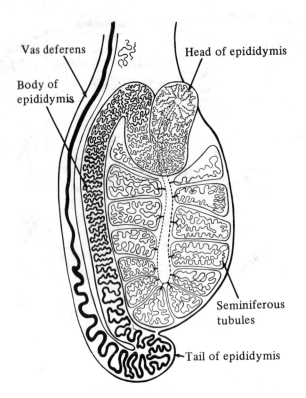

Vas deferens

Head of epididymis

Body of
epididymis

Seminiferous
tubules

Tail of epididymis

FIGURE 3-4. The testes produce both
sperm and testosterone.

Testes. Located inside the scrotum, which is a heat-regulating structure, viable sperm from the testes will not develop at body temperatures. Therefore, the lowering of the testes from the body when environmental temperatures are warm or contracting on colder days in order to provide the testes with the narrow range of temperature required is necessary. The testes descend from the body cavity into the scrotum at birth through a small opening known as the *inguinal canal.* Sometimes this does not take place with one or both testes. This is a condition known as *cryptorchidism.* If only one testis descends, it is called *unilateral cryptorchidism,* and if both fail to descend, it is known as *bilateral cryptorchidism.* A unilateral cryptorchid may have its reproductive efficiency impaired, and a bilateral cryptorchid is sterile.

Under normal development, the testes function by producing *sperm* in the very small, convoluted tubules that make up a large part of the testes structure. If linked together to form one tube, these *seminiferous tubules* from one pair of bull testes have been estimated to stretch the length of 50 football fields. *Interstitial cells,* located in the spaces between the seminiferous tubules in the testes, produce the male hormone *testosterone,* which is responsible for secondary sex characteristics of the male (heavy chest and muscling, masculine appearance). Upon receiving the proper message from the brain, testosterone is released, which causes sexual excitation in the male.

Epididymis. The *epididymis* (Figure 3-4) serves four functions: transport, storage, maturation, and concentration of sperm. This structure, esti-

Chap. 3 BEEF CATTLE REPRODUCTION

mated to be 120 ft in total tubular length, serves to transport the sperm from the testes to the accessory sex glands. Water is resorbed here to increase concentration; maturation is achieved because of cell excretions, and sperm is stored primarily in the tail of the epididymis.

Vas Deferens and Accessory or Secondary Sex Glands.   The *vas deferens* transport the mature semen from the tail of the epididymis past the accessory sex glands, the *seminal vesicles,* the *Cowper's gland,* and the *prostate gland* (commonly called secondary sex glands). These glands are responsible for producing the bulk of the fluid that we commonly refer to as *semen.* This seminal fluid may range from 5 to 10 cc and is ejaculated through the penis into the reproductive tract of the female. Sexual excitation causes blood to be pumped into the chambers of the penis, causing an erection by a straightening of the *sigmoid flexure,* allowing copulation. After copulation, the sigmoid flexure contracts because of a *retractor penis muscle* that retracts the penis into a protective sheath.

The Female

Although the emphasis in most breeding programs is on the bull, the female reproductive system is much more complicated and important. It is therefore necessary to go into more detail to have a full understanding of the anatomy of the cow and the function of each organ or part once puberty is reached.

The Ovaries.   The *ovaries,* which are homologous with the male testes, remain in the body cavity near the kidneys and do not undergo descent. The *ova* (eggs), which when fertilized by the male spermatazoa begin the embryo, are present at birth. Although available ova estimates run as high as 75,000 for both ovaries, relatively few—perhaps 20 to 30—are shed during an average cow's lifetime under natural conditions.

The ovary of a cow is almond shaped, averaging 10 to 20 grams (g) in weight. In contrast to the male, where the "seed" is developed deep within the seminiferous tubules, the tissue that produces the *ovum* (egg) lies very near the surface of the ovary. "Potential ova," called *primary follicles,* are generally believed to be present at birth. Successive stages of maturity follow until a mature ovum called a *graafian follicle* is produced. This protruding "blister" (Figure 3-5) on the surface of the ovary is produced under the influence of the hormone *FSH (follicle stimulating hormone)* from the anterior pituitary gland. The same gland produces *LH (luteinizing hormone),* which causes a rupture of this follicle and a release of an ovum (egg).

Immediately after ovulation, the cavity left by the vacated egg is acted upon and follicular cells increase to produce a scar-like structure, the *corpus luteum,* or *CL* (Figure 3-5). If fertilization of the ovum does not occur, eventually the CL gradually regresses to be replaced by another graafian follicle. If fertilization does occur, the CL retains its size under the influence of the anterior pituitary hormone *prolactin* and itself produces the hormone *progesterone,* responsible for suppressing further heats and maintaining pregnancy.

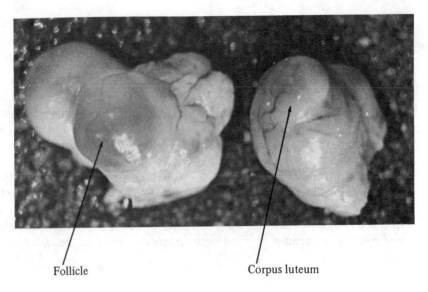

Follicle   Corpus luteum

FIGURE 3-5. The ovarian follicle and corpus luteum.

Fallopian Tube (Oviduct).   The ovary is stimulated to release the ovum into the *infundibulum* of the *fallopian tube* or *oviduct* (Figure 3-6). This action is actually delayed until 12 hours after the end of *estrus* (heat). The egg, swept into the infundibulum of the fallopian tube by ciliated action and muscular contractions, then makes its way into the horn of the uterus. Fertilization (union of egg and sperm) actually takes place in the upper third of the fallopian tube. This sequence of events can occur on either side of the paired system, initiated by either ovary.

The Uterus.   The *uterus* (Figure 3-6) consists of two horns that curve like a ram's horn and a common body. The *cervix*, considered an integral part of the uterus, is discussed separately to simplify study.

In cattle, the horns of the uterus make a complete spiralling turn before connecting with the fallopian tubes. These horns are usually well developed, with one horn or the other as the place where fetal development takes place.

Blood and nerve supply to the uterus are provided through the supporting *broad ligament*. This ligament may stretch in older animals, allowing a lower carriage of the uterus and fetus.

Inside the uterus, the mucosa layer contains *caruncles*. These small projections, which enlarge to about the size of a U.S. half-dollar during pregnancy, are nonglandular and rich in blood supply. Arranged in rows extending into both horns, they have been estimated to number 70 to 120. They have a sponge-like appearance because of small cavities that serve as attachment points for the opposite structure, the *cotyledons* from the *placenta* (membrane enclosing the fetus). The cotyledons and caruncles together, called the *placetome*, might be thought of as two buttons that snap together.

Functions of the uterus are many. For example, it serves as a pathway for sperm at copulation, and motility of sperm to the fallopian tubes is aided

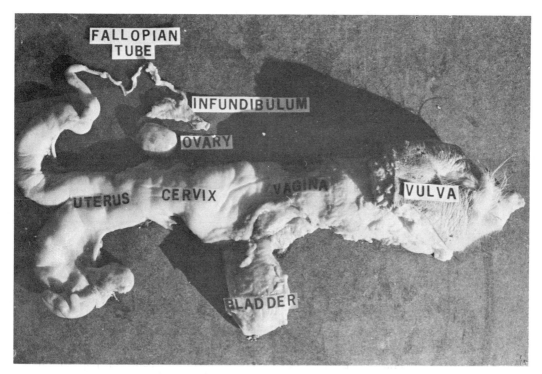

FIGURE 3-6. The female reproductive tract, illustrating dissected ovary and fallopian tubes.

through contractile actions. In the early weeks of *gestation* (pregnancy), it is the uterus that is thought to sustain the embryo by secretions from the uterine glands and blood plasma (uterine milk). The uterus, capable of undergoing great changes in size and shape, serves as an attachment point through the placetome for the growing embryo during gestation. It plays a major role in expulsion of the fetus and membranes at *parturition* (birth) and is capable of regaining its posture quickly after parturition through involution.

The Cervix. A sphincter-like structure that separates the uterine cavity from the vaginal cavity is the *cervix* (Figure 3-6). The basic function of the cervix is to seal off the uterus protecting it against bacterial and other foreign invasions. The sphincter remains closed at all times except during birth.

During estrus and copulation, the cervix serves as a passageway for sperm. If pregnancy results, a cervical plug develops, completely sealing off the uterine canal, protecting the fetus. Shortly before birth, this plug liquefies, the cervix dilates, and passage of the fetus and membranes is allowed at parturition (birth).

The Vagina. Lowermost of the internal reproductive structures, the *vagina* serves as the copulatory organ of the female. It is here that semen is deposited by the male. Like the cervix, the vagina dilates to allow passage of the fetus and membranes.

TABLE 3-1. PUBERTY, ESTRUS, OVULATION, AND GESTATION IN THE COW

Phenomena	Range	Average
Age at puberty (months)	8–18	12
Weight at puberty (lb)	500–800	600
Duration of estrus (hr)	4–30	16
Length of the estrous cycle (days)	14–24	21
Parturition to estrus (days)	16–90	35
Time of ovulation (hours after end of estrus)	2–26	12
Best time to breed	Late estrus	
Gestation period (days)	240–330	283

PUBERTY, ESTRUS, AND BREEDING

Puberty in the male is marked by production of viable sperm and a desire for mating influenced by the hormone testosterone. A similar desire for mating also takes place in the female. Puberty in the female is exhibited by the first *estrus* (heat). This condition is generated through the effect of a hormone called *estrogen*, which is produced by the ovaries (Figure 3-1). The heifer will accept the bull only during the heat period, which lasts on the average 16 hours, and if she does not conceive, this condition will cycle every 21 days (Table 3-1). Thus the term *estrous cycle* is used quite commonly in animal science.

Visual signs of approaching estrus are a swelling and redness of the vulva and a restlessness or nervousness indicating a desire for company, but the most obvious sign is riding and in turn being ridden. The key to determining which heifer is in heat is identifying the one that will stand still when mounted.

Breeding occurs only during estrus, although the bull is capable of breeding at any time. Armed with the anatomical and physiological knowledge of estrus and ovulation, it is easier to select the optimum time to breed for conception. Ovulation is actually delayed until 12 hours after the end of the estrus.

UNION OF THE EGG AND SPERM

The average life of the egg is 6 to 12 hours, and the average life of the sperm is 30 hours. Thus mating must take place at the latter part of estrus for fertilization to occur (Figure 3-7). The deposited sperm makes its way up the vagina through the cervix into the uterus and meets the egg at the upper third of the fallopian tube. The sperm owes its motility to a flagellum (tail) that propels it toward the egg, but it is moved mostly by muscular activity of the uterus and fallopian tubes. The actual head of the sperm causes fertilization with the egg. The first sperm that reaches the egg causes a reaction preventing all other sperm from uniting with the egg.

Chap. 3   BEEF CATTLE REPRODUCTION

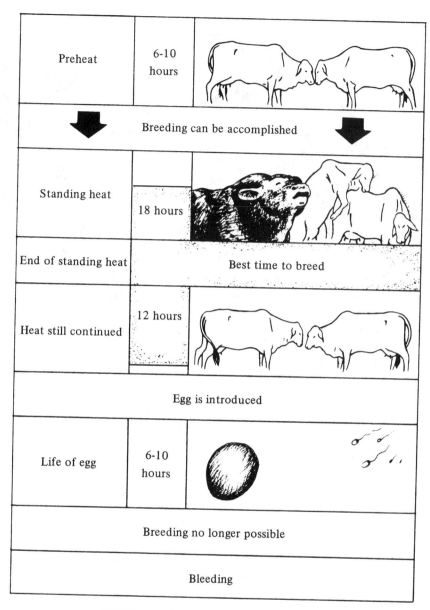

Preheat	6-10 hours	
	Breeding can be accomplished	
Standing heat	18 hours	
End of standing heat	Best time to breed	
Heat still continued	12 hours	
	Egg is introduced	
Life of egg	6-10 hours	
	Breeding no longer possible	
	Bleeding	

FIGURE 3-7. The breeding sequence of events.

GESTATION

Remember the egg that was ruptured from the ovary? The cavity left by the rupture developed into an endocrine gland referred to as the *corpus luteum* or *CL* (Figure 3-5), which produces a hormone called *progesterone*. Progesterone is responsible for maintaining pregnancy. The fertilized ovum floats free before attaching at about 35 days of age.

Caruncles Cotyledons Fetus

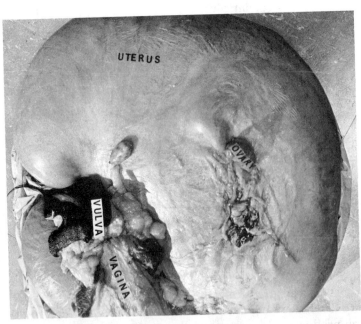

UTERUS

OVARY

VULVA

VAGINA

FIGURE 3-8. A detached placenta, illustrating cotyledons. Note the fetus floating in amniotic fluid.

Chap. 3 BEEF CATTLE REPRODUCTION

The developing *ovum* (egg) now begins development of a *placenta* (membrane), which encloses the developing *fetus* (calf) and protects it in a fluid known as *amniotic fluid* (Figure 3-8). The placenta is attached to the caruncles of the uterus by cotyledons on the placenta. This allows an exchange of nutrients between the dam (cow) and the developing fetus (calf). The primary function of this system is to serve as a pathway for nutrients and oxygen to go into the fetus and waste products to be excreted. This gestation (pregnancy) period normally takes 283 days (Table 3-1). The period of gestation is from the time of fertilization to parturition (calving).

RECTAL PALPATION

The principle involved in rectal palpation (Figure 3-9) is to insert an arm into the rectum of a cow and feel through the wall of the rectum to detect signs of pregnancy. Care should be taken to use a sterile, lubricated glove to protect man and animal from infection. Rectal detection can be as early as 60 days for the average breeder and somewhat less for the experienced veterinarian.

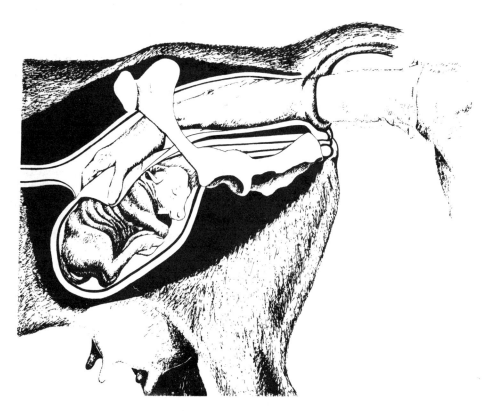

FIGURE 3-9. Simulated palpation of the swollen uterus with a 90-day fetus.

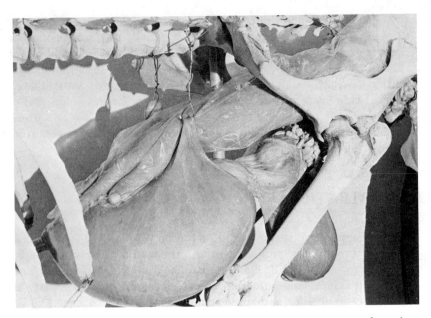

FIGURE 3-9B.   Simulated palpation of a swollen uterus with fetus. Note the position in relation to the skeleton.

The question most often asked in rectal palpation is: "What are you feeling for?" There are certain landmarks that help to orient the palpator to make a valid evaluation of pregnancy or stage of pregnancy. The first point of reference could be the forward slope of the pelvis. Immediately forward of the pelvis one should be able to pick up the cervix (Figure 3-1). To the inexperienced breeder, it may help to describe the cervix as something approaching the size and feel of the neck of a soda bottle. Follow the cervix to the swollen horn of the uterus. This will usually be to the right of center regardless of which horn contains the fertilized egg since the rumen (paunch) occupies most of the area to the left of the center line. Recall that the developing fetus is floating in an amniotic fluid; therefore, one should not expect to feel anything resembling an animal but rather something floating inside a protective membrane. It might be helpful to describe this sensation as the dribbling of a basketball. By gently pushing against the membrane with a dribbling motion, the bouncing ball (fetus) will return because of the tendency of the fetus to float to the top (Figure 3-9). With proper training, one can detect this sensation at 60 days, and as the gestation (pregnancy) continues and the fetus develops, the task becomes easier (Table 3-2).

PARTURITION (CALVING)

Signs of approaching parturition begin with development of the udder. This may occur as early as 6 weeks prior to calving date. A sign that is visible within a week of parturition is a swelling and reddening of the vulva and a

Chap. 3   BEEF CATTLE REPRODUCTION

TABLE 3-2. SIZE OF FETUS AT STAGES OF GESTATION

Stage of Gestation (day)	Length of Fetus (in.)
30	½
45	¾
60	2½
90	10
120	15

relaxation of the pelvis. Signs of very close parturition are distention (swelling) of the teats and a mucous discharge from the vulva. A cow at this point may actually be seen leaking milk from the teats. The hormone *relaxin* causes the relaxation of the pelvis, and *estrogen* opens the birth canal, allowing the calf to pass. Estrogen also causes labor (contraction of the uterus) that helps in parturition. *Oxytocin* from the posterior pituitary of placenta may act in synergism with estrogen in labor contractions. A cow that has labor pains in the early stages will become uneasy, restless, and will move away from the herd. It will frequently be seen lying down and getting up. Normally, no assistance need be rendered and the cow should be left alone unless it is obvious that it has been down too long and is in a weakened condition.

Normal Calf Presentation

The normal presentation of the calf is front feet first with the head lying between them (Figure 3-10). The base of the hooves should be in a downward position. Contractions by the uterus cause the feet to be thrust through the placenta thus releasing the amniotic fluid which serves as a lubricant for passage of the calf through the birth canal. Normal birth may vary considerably, but the average time approximates 30 minutes without assistance. Figure 3-10 also shows the various malpresentations and illustrated corrective procedures.

Rendering Assistance

No aid should be given unless it is absolutely necessary (Figure 3-11). However, if a delivery does not occur within approximately 2 hours from the beginning of labor pains, a veterinarian should render an examination to determine if an abnormal presentation is the problem. Some common causes for concern are breech birth, indicated by hooves of the feet pointed upward (this is characteristic of hind legs coming first); one leg protruding; no legs protruding; head turned to one side making normal delivery difficult or impossible; abnormally large calf. In all these cases, only a qualified individual should render assistance, and extreme precautions should be taken to prevent injury to the dam and/or calf.

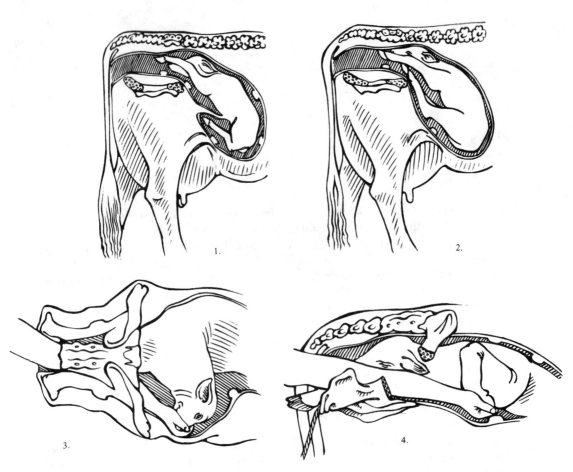

FIGURE 3-10. Normal and malpresentation. (1) Normal anterior presentation of the calf in the proper position for delivery. (2) A "dog-sitting posture." A very serious malpresentation. The rear legs must be retracted before normal birth can occur. Early professional assistance is needed in this case. (3) A "head back" condition that may require pulling the head and neck around with the hand. The calf is pushed back and quickly released in order to grasp the calf's muzzle. The head is then pulled around in line with the birth canal. In some cases an obstetrical chain is looped around the poll, under the ears, and through the mouth in a "war bridle" manner. This allows for greater traction to line up the head. Care should be taken not to cut the birth canal with the gaping jaws of the calf in traction. (4) Correction of a "leg back" condition. The calf is pushed forward and the retained foot grasped in the cupped hand. The foot is carried outward and then forward in an arc over the pelvic brim. More difficult cases may require the use of obstetrical chains on the foot.

Expulsion of the placenta (afterbirth) usually occurs 2 to 6 hours after parturition (calving). Normally, the cotyledons (buttons) that are attached to the uterus (womb) separate allowing the unattached membrane to pass through the birth canal (Figure 3-8). It is quite common for a cow to eat this afterbirth as part of its cleanup ritual. It is thought that this is a throwback to a wild instinct to destroy any evidence of birth to protect the young from

Chap. 3   BEEF CATTLE REPRODUCTION

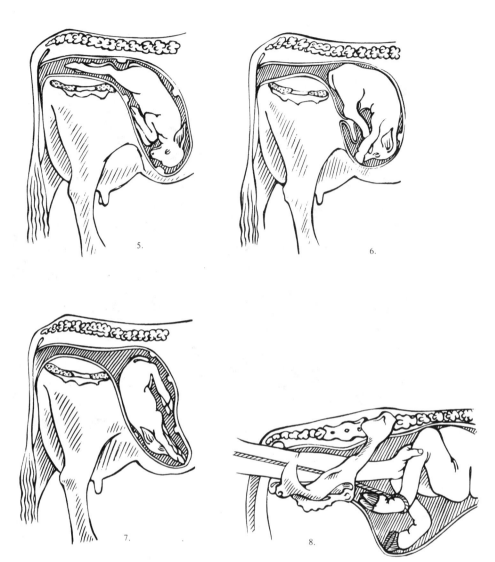

FIGURE 3-10. Normal and malpresentations (continued). (5) A backward presentation with rear legs extended. Birth in this presentation often occurs without incident; however, assistance may be needed if birth is delayed because of the danger of breaking the navel cord and subsequent suffocation of the calf. (6) The breech position is a serious malpresentation. It may be corrected by pushing the calf forward and pulling the rear legs into the birth canal (see number 8). (7) The upside down and backward position. A serious type of malpresentation, this is often caused by twisting of the uterus or rotation of the calf. Delivery should never be attempted in this position. Professional assistance is required. (8) Correction of a breech position. The calf is first pushed forward. Then one foot at a time is drawn back with the hand, as the hock is flexed. The foot is lifted over the pelvic brim into the birth canal. An alternative method is to place a snare around the pastern joint, close to the foot. The snare is then pulled by an assistant while the calf is pushed forward and its foot is guided over the pelvic brim.

FIGURE 3-11. Care of a cow and calf at parturition is so important that the Litton Charolais Ranch uses closed-circuit television to observe calving stalls. (Courtesy of the Litton Charolais Ranch, Chillicothe, Missouri)

predators. There is an old wrangler's tale that a cow eats this afterbirth in order to have a normal lactation (milk production). However, there is no evidence to indicate that this is so, and no harm could be expected to come to a cow not eating the afterbirth. Under pasture conditions, it is not a situation that should be worried about one way or the other.

If the afterbirth is still retained 24 hours after birth, an abnormal condition exists and a veterinarian should be consulted. Retained placenta (afterbirth) should concern the breeder because serious infections could occur.

Care of the Newborn Calf

The newborn calf should be licked clean by the dam. If this is not done to facilitate breathing, the breeder or attendant should make sure that there are no membranes covering the mouth and nostrils. If birth occurs in a corral, barn, or confined pasture where there are other animals, it is advised that

Chap. 3   BEEF CATTLE REPRODUCTION

the navel cord be treated with iodine to prevent tetanus or other illness. This recommendation seldom holds true for cattle on the open range. The calf should nurse within the first few hours after birth, and if it does not, it should be given assistance in finding the udder. *Colostrum* (special milk produced during the first 3 days after calving) is necessary for the newborn calf to live. This colostrum not only contains high levels of energy, vitamins, and minerals to get the calf off to a good start but also carries antibodies (protectors against infection and disease). From this point on, the development of the calf is up to the cow. After weaning, the young calf goes through the normal cycle described earlier in the chapter and thus we are back to puberty.

ARTIFICIAL INSEMINATION

The introduction of sperm into the female reproductive tract by other than natural methods is termed *artificial insemination*. This method first received scientific interest in application to mammals in 1780 when Lazzaro Spallanzoni, the Italian physiologist, used it on dogs. The Russians began its wide-scale use in 1900, and it has since spread to every country in the world. Although only 5% of the beef cows in the United States are artificially inseminated, 50% of U.S. dairy cows are estimated to be so bred and in Denmark the rate is 100%. The opportunity for increased spread of the method in beef cattle is bright.

Semen Collection

Although several methods for collection are available, the artificial vagina and electroejaculator are used most. The *artificial vagina* varies in shape and size but is basically a heavy rubber outer cylinder with an inner rubber lining that simulates natural copulation. The space between the cylinder and liner is filled with temperature-controlled water (40 to 45 °C) that also simulates proper pressure. A lubricant applied to the inner lining allows penetration of the bull's penis and collection of a complete, clean ejaculate from the glass tube connected opposite the end of penetration. A head gate restraining a cow in heat is most effective in eliciting a mount from the donor. Upon mounting, the bull's penis is deflected to one side into the artificial vagina to make the collection. Some bulls will even mount an artificial or "dummy" cow.

The *electroejaculator* is a method used by veterinarians and other skilled physiologists. A small electrical charge is passed through stimulator rings (Figure 3-12) or through an instrument inserted in the rectum of the bull (Figure 3-13). This charge stimulates the reproductive organs lying just under the ventral wall of the rectum, causing an ejaculation. Collection under this system may be as simple as a funnel apparatus leading to a glass tube.

FIGURE 3-12. Semen collection equipment. Note stimulator rings on fingers and collection apparatus.

FIGURE 3-13. The electroejaculator is used in semen collection.

Chap. 3   BEEF CATTLE REPRODUCTION

FIGURE 3-14. Storage tanks hold 600 ampoules for 1200 straws of semen.

Semen Storage

Processing of semen normally involves dilution with egg yolk–citrate with modifications and slow freezing in either 1-ml ampoules or 0.5-ml plastic straws. Plastic straws are the preferred method of semen storage today because of their conservation of storage space and ease of use in insemination. Straws or ampoules are stored in containers maintained at a low temperature of −195°C by use of liquid nitrogen. The field storage tank (Figure 3-14) will hold approximately 600 ampoules or 1200 plastic straws of diluted, processed semen. When storage containers are regularly recharged with liquid nitrogen every 60 to 90 days, this method of storage will preserve semen indefinitely. One ampoule or plastic straw is used per insemination each containing at least 10,000 live sperm cells.

Heat Detection and Insemination

Estrus detection methods can range from the standing mounts previously discussed (Figure 3-7) to more elaborate procedures such as the use of a vasectomized bull and dye-released mount detectors. This sterilized "spotter bull" is often a different color and/or larger than the cows it is detecting for the obvious ease of sighting by the herd manager. Knowing that the most likely time of ovulation is the last 12 hours of estrus, the cow is restrained toward the end of estrus, and the inseminator uses stored semen properly thawed. In insemination, semen from the plastic straw or ampoule is deposited into the reproductive tract of the female (Figure 3-15). A catheter is used if semen is stored in ampoules; a straw gun is used if semen is stored in plastic straws. The advantage of the plastic straw is that semen can be directly deposited in the female reproductive tract from the straw without transfer of the semen from ampoule to catheter. This makes the use of the straw simpler and ensures the maximum number of live sperm being inseminated by the process. Some advantage may be obtained in depositing the semen directly into the uterus,

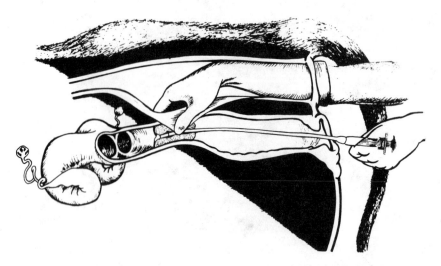

FIGURE 3-15.   The technique of artificial insemination.

passing through the cervix. However, there is the risk of infection, contamination, or abortion if the cow is pregnant. Pregnant cows have been inseminated because of "false heat," an unusual occurrence, or mistaken identity. Therefore, some studies suggest deposits in the vagina are safer and conception rates lowered little, if any.

Advantages of artificial insemination for beef cattle are:

1. Elimination of the cost of owning bulls.
2. The use of outstanding sires at a reduced cost, even in small herds.
3. More accurate evaluation of a bull's breeding ability by observing more of his offspring.
4. Better herd health.

Disadvantages of artificial insemination of beef cattle are:

1. Problems in detection of cows in heat and sorting and restraining beef cows for breeding.
2. Unless properly managed, low conception rates compared with natural breeding.
3. The method requires skilled inseminators.

FERTILITY TESTING OF BULLS

Fertility potential, the breeding power of the bull, should be determined before the breeding season starts and preferably every year thereafter because a fertile bull is susceptible to infertility at any stage of its reproductive life. A fertility examination should include both a physical and a semen examination. From a physical point, the bull should not appear lame, have any imperfections in the sheath that would limit breeding, or have any other critical defects that would prevent it from properly serving (breeding) the cow herd. Semen evaluation is done in order to determine the motility and viability of ejaculated sperm even though the physical examination reveals nothing to be concerned about.

Most veterinarians are well equipped to test fertility in bulls (Figure 3-16). The process is quick, simple, and may be done in the field or laboratory. Usually, the bull is restrained in a squeeze chute, and a semen sample is collected by electroejaculation. Motility, indicated by a mass swirling motion, is viewed first under low power on a microscope. High-power examination reveals individual spermatozoa that are observed by the trained eye for movement, an indication of vigor. A stained slide smear is then prepared, which kills the spermatozoa and allows a detailed physical observation for unwanted defects such as coiled tail, protoplasmic droplets, and other abnormalities. Sperm cells that were alive when the slide was made will not absorb the stain; dead cells will (Figure 3-17). Semen is evaluated on the basis of motility, concentration, percent of live sperm, and structural normality of sperm.

FIGURE 3-16. Tools used in the collection and evaluation of semen.

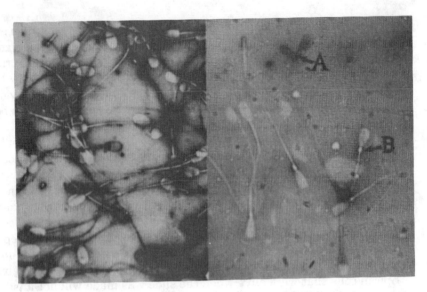

FIGURE 3-17. Abnormal sperm of a bull are shown on the right: (A) sperm with a coiled tail; (B) sperm with a protoplasmic droplet near the base of the head. Normal sperm are shown on the left. The stained cells were dead and the nonstained cells alive. [Reprinted by permission from John F. Lasley, *Genetics of Livestock Improvement* (Englewood Cliffs, N.J.: Prentice-Hall, Inc., 1963)]

RECENT ADVANCEMENTS IN REPRODUCTION

Estrus Synchronization

Synchronization of the estrous cycle is essentially controlling it, treating the animals with drugs to reset their reproductive "clocks" at a common point—no matter where they happen to be in the cycle prior to treatment. The more tightly the estrous cycle can be controlled, the more animals will be in heat at the same time. Although it is unreasonable to assume that all animals will come into heat at exactly the same time, it is possible to synchronize their cycles so that heat occurs within 2 to 6 days with some products and others even claim a more predictable time frame of 24 to 36 hours.

The advantages of a properly managed synchronization program are:

1. Artificial insemination is made more practical.
2. Shorter, earlier breeding season is realized.
3. Uniform calf crop is achieved.
4. The calving season is shortened.
5. The need for the traditional number of bulls for natural mating is reduced and semen purchase of proven, tested, expensive sires for most efficient, economical production is allowed.

To better understand the process of estrus synchronization, it is helpful to briefly review what happens in a typical animal during an estrous cycle.

The hormone estrogen acts on the brain, causing the female to be sexually active, usually for 9 to 18 hours. Estrogen, as discussed previously, is produced by a structure on the ovary called a follicle (Figure 3-5). The blister-like follicle is full of fluid and contains the egg. The pituitary gland, located at the base of the brain, periodically releases FSH and LH to regulate the ovarian cycle.

Near the time of heat, a surge of LH causes the follicle to rupture and the egg to be released about 30 hours postheat. The spot on the ovary where the follicle ruptures develops into a structure called a corpus luteum, "CL," or "yellow body" about 7 days later. Without pregnancy, the CL regresses in the cow due to the action of a substance believed to be *prostaglandin*.

At the time of ovulation, the CL begins producing progesterone, which inhibits heat and the release of FSH and LH from the pituitary, and maintains the uterus in a condition conducive to accepting a fertilized egg. But let us assume, at this point, that mating is prevented so fertilization does not occur. Because high progesterone levels are maintained by the CL, follicles do not mature, eggs are not released, and no heat is shown.

About 15 days after the egg is shed, prostaglandin produced by the uterus causes the CL to shrink. Progesterone levels fall, FSH and LH increase, follicles mature, estrogen is again produced; and heat is the result. This completes a typical cycle in which no pregnancy resulted.

The problem with managing a breeding program has always been that this cycle varies so much with individual cows that artificial insemination had to take place daily in large herds and depended upon time-consuming visual heat detection. Estrus synchronization was researched and developed in order to "reset" all the reproductive clocks to approximately the same time. Progesterone or progestins administered for 10 to 14 days and then withdrawn was found to produce heat in the majority of cows within a few days following withdrawal of treatment.

In more recent research, prostaglandin, or an analog, administered to cows with a functional CL produced estrus within 3 to 7 days and became the preferred treatment. Most recently, a combination of progestin and estrogen (Syncro-Mate-B, CEVA Laboratories, Inc., 10551 Barkley, Suite 500, Overland Park, Kansas 66212) promises heat occurrence in an even more predictable time frame (24 to 36 hours following removal of an ear implant) without the need for heat observation, making timed insemination a practical reality.

Superovulation

Taking the theory of estrus synchronization one step further is the process of superovulating an animal to produce the maximum number of eggs for fertilization. This is especially necessary for the practical application of embryo transfer to be discussed later. Although some variations may exist, the process basically involves administration of prostaglandin to induce heat and a hormone such as pregnant mare serum gonadotropin to achieve a high number of ova.

Embryo Transfer

The use of superior sires through artificial insemination resulting in improved weight, quality, and disease resistance of animals is well known. This method, of course, involves the freezing of semen to be later thawed and used to impregnate the female. The obvious advantage is to make the most extensive use of a superior male's genes. A similar method to take advantage of a superior female's genes is called embryo transfer. Embryo transfer is not new. It dates back to 1890, when Walter Heape, an English biologist, was successful in making a rabbit produce a litter containing purebred young of two nonsimilar breeds. It has since been achieved in cattle, swine, horses, sheep, and zoo animals. There is even some possibility of one species carrying the young of another when compatible conditions exist. For example, there are records of a Mouflon born to a sheep at Utah State University and the Bronx Zoo delivered a rare Indian guar carried by a Holstein cow.

Embryo transfer, in a sense, takes artificial insemination a step further. A cow, for instance, with superior genetic characteristics is injected with a hormone that causes her to superovulate. Instead of releasing one mature egg into the reproductive tract, she produces more typically 6 to 8, but as many as 20.

Through artificial insemination, the eggs can then be fertilized by semen from a genetically superior bull. The superovulated cow then develops a number of fertilized embryos in 7 days. By inserting a long, thin, flexible, latex catheter into the uterus (which seals off the cervix by an expandable bulb at the tip of the catheter), a saline solution can be passed into the uterus. Then, through a second channel in the same catheter, the microscopic embryos are floated out to collection cylinders. This technique, usually called "embryo flushing" (Figure 3-18), can produce many fertilized embryos (Figure 3-19), which, if used fresh, can be transplanted within the next 24 to 48 hours into surrogate (substitute) mother cows.

In about 60% of cases, the transfer "takes" and surrogate mothers give birth to numerous, genetically superior animals unrelated to the surrogate mother. Instead of getting one calf a year out of a gentically superior cow, it might be possible to get 100. One superior "donor" cow is on record at Rio Vista Genetics, San Antonio, Texas, for producing 110 calves in one year, about 150 years of normal production.

This nonsurgical technique has been used in about 1% of the cattle bred in the United States but has spawned approximately 140 companies doing over $20 million worth of business annually dealing in cattle-embryo transfers. Although the cost per embryo transfer still ranges from several hundred to several thousand dollars per embryo transfer, new technologies will undoubtedly bring the price down to benefit the average dairy farmer and beef rancher.

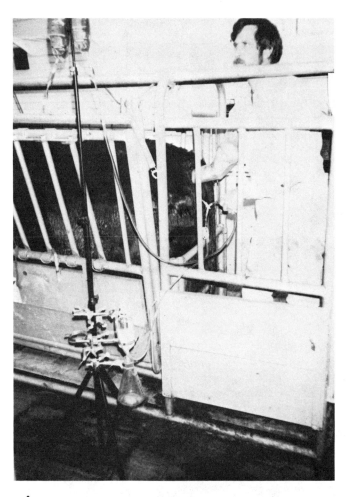

A

FIGURE 3-18. Embryo transfer begins by nonsurgical flushing of fertilized embryos from the uterus of the donor cow. (A) By inserting a dual-channeled catheter with an inflatable head into the cervix, 200 to 300 ml of saline solution can be introduced into the fallopian tubes washing out the embryos into a collecting vial or filter. A technician, through rectal palpation, guides the catheter to its proper position and also manipulates the tract for maximum embryo collection. (Rio Vista Genetics, photo by author) (B) Close-up view of tube arrangement for nonsurgical flushing.

B

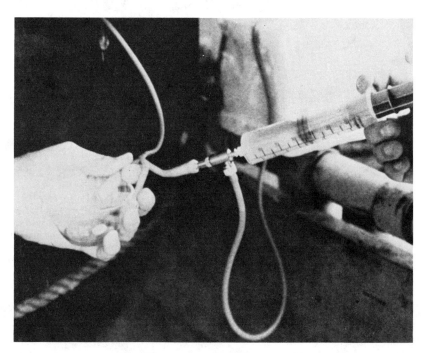

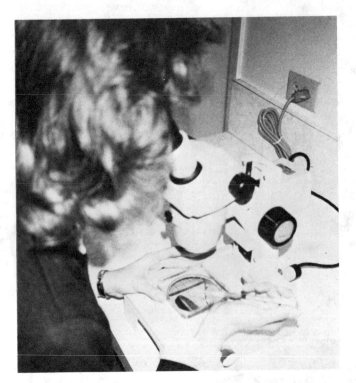

A

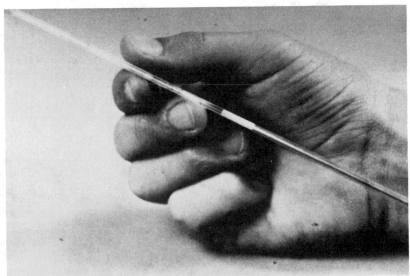

B

FIGURE 3-19. Fertilized embryos, either washed from a filter, or simply discovered in the original flushing solution, are spotted in a grid plate
and individually collected in a simple syringe. Embryos may then
be transferred fresh into recipient cows or frozen for later use in
a "French straw." (Courtesy of Guy Stone, Jr., and D. C. Kraemer,
Texas A&M University)

Chap. 3   BEEF CATTLE REPRODUCTION

One such exciting technique recently developed by Rio Vista Genetics is a new transfer procedure for freezing and storing the embryos much in the same manner that semen has been stored. The technique involves a "French straw" (similar to a plastic soda straw) containing a few drops of protective fluid and the microscopic embryo. The straw is frozen in liquid nitrogen and can be stored indefinitely. From thawing to implantation in the surrogate mother cows takes only about 10 minutes. Although the success rate has been about 50%, the exciting possibility of having embryo transfers in such a simplified and reliable state, no more complicated than artificial insemination, makes it possible for any practicing veterinarian and many skilled laypeople to perform the thawing and implantation successfully (Figure 3-20).

FIGURE 3-20. Mitch West (left), Director of Genetics for Rio Vista International, San Antonio, hands a test tube containing 25 embryos to Tom Risinger, Risinger Ranch, Cody, Wyoming. Collections were made from nine Simmental cows in Wyoming and shipped by air to Gagnon Farms, Cheneville, Quebec, Canada, where they were implanted into recipient cows the same day. This was the first such export from the United States to Canada. (Courtesy of Independent Cattlemen)

Embryo Splitting

Research scientists, taking a lesson from nature, long ago observed that identical twins were produced in a natural splitting of one fertilized embryo that produced two genetically identical offspring. Although the equipment necessary to create this natural occurrence artifically is very expensive, the process of microscopic surgery is relatively simple.

Basically, a microscopic surgical blade is used to puncture the zona pellucida (the embryo's protective shell). The blade is then used to cut in half the 7-day-old embryo, consisting of about 64 cells, and one half of that embryo is then repackaged in a separate but natural shell (a zona pellucida from an unfertilized or inferior embryo). The tough but pliable covering (zona pellucida) that surrounds the developing ball of cells allows the newly twinned embryos to be placed, generally one at a time, in the uteri of surrogate mongrel cows to gestate for the next 9 months (see Figure 3-21).

Chimeras

In the staggeringly complex world of reproductive manipulation, if calf embryos too small to be seen with the naked eye can be split and combined, it is reasonable to assume that cells from numerous embryos might be recombined within one zona pellucida to produce an animal that is a combination of several animals in a single breeding cycle of just over 9 months. Such cattle have

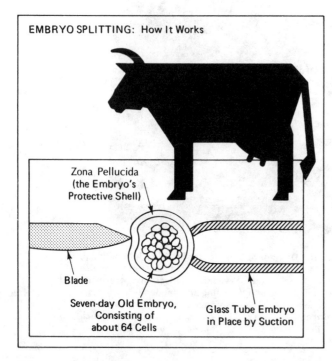

EMBRYO SPLITTING: How It Works

Zona Pellucida (the Embryo's Protective Shell)

Blade

Seven-day Old Embryo, Consisting of about 64 Cells

Glass Tube Embryo in Place by Suction

FIGURE 3-21. The calf embryo, held under a microscope on the tip of a tube, is split with a blade 7 days after fertilization. Both halves of the embryo will develop into identical calves. (Courtesy of Colorado State University)

been produced in Australia using two, three, or even four individual breeds to produce offspring, joining the genetic material together surgically under a microscope. These creatures are called Chimeras, after the mythical Greek monster that was part lion, part goat, and part dragon.

Although the process is still experimental, expensive, and time consuming, the procedure is not much more difficult than embryo splitting because it involves basically the same microscopic surgical technique. Two or more embryos are taken at an early stage of development, the zona pellucida is removed surgically, and the two or more embryos are placed together, fusing a mixture of genetic material inside a common zona pellucida. The Chimera can then be transferred into a recipient (surrogate) cow.

The reason for experimentation with Chimeras is that it may be possible to transfer the milking ability of one cow, the muscular conformation of a second, the disease resistance of another, and numerous other qualities through the Chimeric process. For instance, cows could be made to order for a certain function, to produce at maximum efficiency in a harsh climate, and to resist diseases that may have made cattle production impossible under natural conditions.

The drawback to the Chimera is that it is still very experimental, very expensive, and, at this point, any animal produced as a Chimera does not pass on these outstanding qualities to natural offspring. Chimeras would have to be continually reproduced in the laboratory for replacements. Still, certain aspects of the process make it an exciting possibility.

Sex Control

The option to control the sex of animals within a species would have great impact and has received some successful manipulative research. The obvious positive examples of being able to control sex would be the production of only females in a laying-hen operation, only males in a broiler production, only females for the dairy trade, mostly males for the beef feedlot operator, and so on.

Sex-specific embryo transfer may soon be a reality. Identifying the sex of a fertilized embryo before it is transferred could yield a significant advantage. Encouraging experimental research currently indicates that that service will soon be an option available to cattle breeders throughout the world. A dairy farmer in South Africa, a sheep breeder in Australia, or a beef producer in the United States may soon be able to order from a catalog a fertilized embryo carrying the desired genetic material and of the correct sex and have it delivered to the doorstep. In this way, recipient animals would be used to their maximum potential.

Not only is identification of embryos for sex being investigated, but separation of spermatozoa into the X and Y factions to produce either a male or a female has been attempted by several different methods. Although the research is encouraging, repeatability in sperm cell separation is questionable for commercial application.

One such research project for sex control that holds promise is the immunological approach. The female is sensitized against the X or Y spermatozoa, causing the unwanted sex to be rejected prior to fertilization. In the same way, the unwanted fertilized embryo could be rejected by a sensitized female.

The obvious impact of sex control would be to improve efficiency of animal production beyond all imaginable techniques using natural breeding methods. To be able to control the exact numbers and sex of animals for production of meat, milk, work, sport, and replacements, and to create these individuals made to order, is lending new credibility to the field of animal science through genetic engineering.

Cloning

Much has been written in the popular media about cloning and many misconceptions have arisen as a result. The process is actually not that complicated or that new. Cloning actually means taking the cells from an individual and creating an environment in which those cells can reproduce an individual identical to the one from which the cells came.

Embryo splitting (previously discussed) is, in fact, cloning. This process, and even further splitting into numerous individuals, has been successfully researched at Colorado State University, Cambridge University in Great Britain, and some private research facilities. In the Colorado State University process, a donor cow is inseminated artifically. About a week later, embryos are nonsurgically flushed from the cow. After a few days, the embryo is cut in half inside the zona pellucida (covering) using a microscope and a micro-manipulator equipped with tiny surgical tools. One half is sucked out of its covering, using a pipette, and placed into another covering from which an unfertilized egg has been removed. The other half remains in the original covering. Transferring the two eggs into recipient cows and the ensuing births completes the cloning procedure (Figure 3-22). Colorado State has successfully completed the process over 200 times in Herefords and Holsteins.

With the technique of freezing embryos, a female embryo can even be cut in half, one be put in a recipient cow and the other be frozen. After the cow gives birth and the "cloned" half grows into a heifer, the frozen embryo twin could be transferred into that heifer giving rise to identical twins born several years apart, one of which is surrogate mother to the other. The practical application of such imaginative manipulation would be to get lifetime production values, discover a superior cow, and, when her productive years are ended, go back to the "spare" frozen embryo and grow an exact duplicate.

It is not unreasonable to assume from the research currently available that six to eight identical individuals could result from microscopic embryo splitting of the original cells from a fertilized embryo. To visualize and oversimplify the process, imagine taking a certain amount of the yolk from an egg using a hypodermic syringe and placing it inside empty shells to produce six to eight identical eggs. This illustrates the idea of cloning through embryo

FIGURE 3-22. Timothy Williams holds "Chris" and "Becky," two identical twin heifers born at Colorado State University. Technically clones, the calves are the first in the world produced by nonsurgically recovering embryos from a donor cow, splitting them, and transferring them to a recipient mother. Williams, a postdoctoral fellow at CSU's Animal Reproduction Laboratory embryo transfer unit, developed the splitting technique. (Courtesy of Colorado State University)

splitting. The yolk represents the cells previously discussed and the zona pellucida (covering) is represented by the shells.

Identical clones (embryos) transplanted into different recipient mothers would give invaluable information on maternal effects. When placed in different environments or on experimental nutritional regimens, the clones could yield information not possible to obtain under natural breeding conditions.

On the commercial side, multiple clones of animals could have a tremendous effect on production of milk, meat, fiber, and other animal products. The uniformity that could be produced is only one of the exciting features that commercial producers could look forward to in the future.

Imagine, for instance, producing an entire calf crop at the time of year desired, all born within a few days of each other, to be fattened and marketed at the same time, and for all practical purposes, with all reaching the same weight, yield, and grade.

Spin-off research from this idea prompted University of Minnesota scientists to go a step beyond this technique. It is possible to take an egg from a heifer just born, mature the egg in the laboratory, and fertilize it in four days. If this procedure can be consistently duplicated, it could lead to reduction of the generation interval by as much as three years.

Nuclear Transfer

Taking cloning one step further, researchers have been experimenting with nuclear transfer. Through microsurgery, several hundred cells are removed from a fertilized embryo (see Figure 3-19, embryo transfer). By removing the nucleus of each cell (which contains the chromosomes) and injecting it inside an unfertilized egg that has had its own nucleus microsurgically removed, many hundreds of identical individuals might be possible as a result of embryo and nuclear transfer.

Furthermore, scientists are currently experimenting with *nuclear fusion,* a step beyond nuclear transfer. To greatly oversimplify the process, nuclei from two eggs or two spermatozoa are fused together to create an animal that has two mothers and no father, or two fathers and no mother, and so on. This process would give the maximum amount of in-breeding and concentration of desired genes.

Recombinant DNA

DNA from one species is isolated and transferred to another species, usually a bacterium. For instance, DNA from a cow is transferred to *Escherichia coli* to culture and collect a large amount of a given substance under the direction of genetic instructions from DNA. The DNA can be chemically modified to produce only one such substance, such as a growth hormone. Therefore, natural hormones can be produced in very large quantities, using bacteria as the manufacturing agent. Hormones produced in this way can then be injected back into the cattle from which they came to increase production above what would be possible under normal conditions. Successful application of this technique at Cornell Univeristy has increased milk production in dairy cows and meat production in feedlot animals. Other products produced by recombinant DNA include Interferon, used to combat shipping fever in cattle and a vaccine against hoof-and-mouth disease. These are but a few examples of products that can be produced in large quantities by bacteria incubated in large vats.

Chap. 3   BEEF CATTLE REPRODUCTION

STUDY QUESTIONS

1. The age of sexual maturity is called _____ .
2. The age of puberty ranges from _____ to _____ months or _____ to _____ lb.
3. At puberty, the male is about _____ lb heavier than the female.
4. The heat-regulating structure that allows temperatures compatible with live sperm production is the _____ .
5. Viable sperm will not develop at _____ temperature.
6. A bilateral cryptorchid is _____ .
7. The male hormone produced by the testes is _____ .
8. Estrus is caused by the hormone _____ produced by the _____ .
9. The epididymis is the storehouse for _____ .
10. Most of the semen volume is produced by the _____ .
11. The average estrous cycle is _____ days.
12. The average estrous period is _____ hours.
13. The egg will live about _____ hours and the sperm about _____ hours.
14. The sperm fertilizes the egg at the upper third of the _____ .
15. If the testes fail to descend through the _____ canal, the bull is known as a _____ .
16. The male hormone _____ is responsible for the male _____ characteristics.
17. A heifer experiencing estrus will remain _____ when mounted by another.
18. Estrus is caused by the hormone _____ .
19. The egg is produced from the _____ .
20. The cavity from which the egg was released on the ovary becomes the _____ which produces _____ .
21. The membrane enclosing the fetus is attached to the uterus by _____ , and after parturition, it becomes the _____ .
22. Length of gestation is approximately _____ days.
23. In rectal palpation, the fetus will usually be to the _____ side.
24. Normal presentation at parturition is with feet (hooves) pointing _____ .
25. Colostrum is _____ produced the first few days after parturition.
26. In artificial insemination, semen is placed in the _____ for best conception.

27. Electroejaculation is used for collecting _____ for bull evaluation.

28. The breeding of all cows at one time is made possible through an _____ hormone.

29. A recent advance in animal breeding is producing several calves per year from a donor cow using the method of _____ .

DISCUSSION QUESTIONS

1. Pretend that you are explaining the female reproductive system to someone with no biological training. Using pencil and paper or blackboard, or even tracing with a stick in the dirt, point out important features and explain how estrus and conception occur.

2. Explain what happens when conception does not occur.

3. Using the same system, give a preliminary briefing on rectal palpation as if you were going to instruct the person and actually follow through with live subjects.

4. Suppose that you have complete artificial insemination facilities at your disposal. Describe your program of observing your cattle, signs you would look for, age, weight, and when and how you would artificially inseminate them. What percent conception would you be willing to guarantee on the first service? (Your instructor may have to give you some additional information on conception rates.)

5. Discuss the mechanics of ova transplants.

6. Suppose that you are serving as a livestock specialist. The phone rings and a young, concerned voice informs you that a cow gave birth to a healthy calf, but the cow died an hour later. This was 3 days ago. In spite of supplementary feeding with powdered milk, the calf is doing poorly, losing weight, running a fever, and has scours. What advice would you give for possibly correcting the condition and saving the calf?

Chapter Four

FEEDING BEEF CATTLE

Feed is a substance eaten and digested by an animal that provides the essential nutrients for maintenance (body repair), growth, fattening, reproduction (estrus, conception, gestation), and lactation (milk production). Feeds can be divided into concentrates and roughages. *Concentrates* (grain-type products) and *roughages* (hay or grass-type products) make up the basic structure of a ration. A combination of the feeds illustrated in Figure 4-1 contains essential elements (nutrients) for the well-being of an animal.

ESSENTIAL NUTRIENTS

All classes of livestock require the following six essential nutrients: water, protein, carbohydrates, fats, minerals, and vitamins.

Water

The cheapest nutrient, water, is a necessary part of adequate growth, fattening, or lactation-type rations. A cow will consume about 12 gallons of water per day. She can live for weeks without food but only days without the sub-

FIGURE 4-1. A combination of roughages and concentrates can supply most or all essential nutrients.

stance that regulates body temperature, dissolves and carries other nutrients, eliminates wastes, and constitutes up to 80% of the body. Water is necessary for the digestive processes both as a medium and a participant in body chemical reaction. Some enzymes are more effective when diluted in water, and the ions OH^- and H^+ (which come from water) are involved in the process of *hydrolysis,* the primary method of protein, fat, and carbohydrate digestion.

Circulating throughout the body, water carries dissolved nutrients to the cellular level and waste products through the excretory system. As a vital part of the blood, water has a temperature regulating function because of this circulation, just as temperatures are regulated in water-cooled engines by a circulatory system.

Protein

Protein is the major component of tissues such as muscle and is a fundamental component of all living tissues. It contains carbon, hydrogen, oxygen, nitrogen, sulfur, and sometimes phosphorus. The protein molecule is composed of a number of smaller units linked together. These "building blocks" are called *amino acids.* The name stems from the characteristic chemical structure that

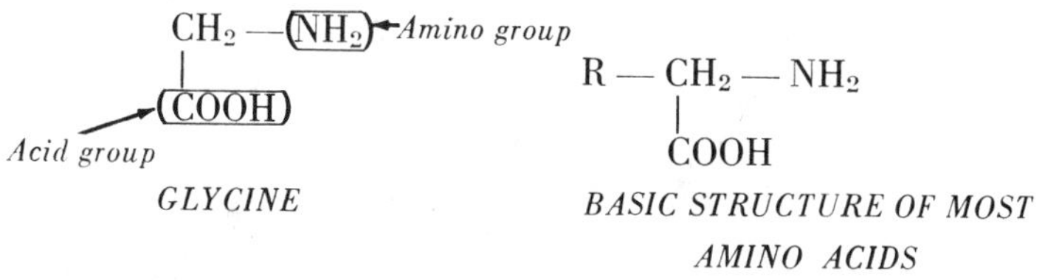

FIGURE 4-2. The simplest "building block" or amino acid is glycine. The presence of amino and acid groups gives rise to the name.

combines an amino group (a base) with an acid. The simplest amino acid, *glycine*, illustrates the structure in Figure 4-2. More complex amino acids are also illustrated by a chemical group represented by R. Amino acids are linked together at the junction CO—NH by a *peptide linkage* as represented by Figure 4-3. Glycine and another simple amino acid, *alanine*, are combined to form *alanylglycine*, liberating a molecule of water. The reverse process, *hydrolysis*, effects separation by taking up a molecule of water.

Twenty-five different amino acids have been identified as constituents of the protein molecule, although most proteins contain from as low as 3 or 4 to as high as 14 or 15 different ones. The average protein contains 100 or more amino acid molecules linked together by peptide bonds.

Nonruminants (swine, poultry, humans) need specific amino acids. Therefore, a high-quality protein containing a large variety of amino acids is desired. Ruminants do not need a wide variety because of their ability to synthesize needed amino acids through the actions of microorganisms in the rumen. The concern in cattle feeding is more for total protein than quality protein.

Amino acids vary slightly in composition but average about 16% nitrogen. Because this figure is fairly constant, protein estimates are most often made by analyzing for nitrogen chemically and dividing by 16% or multiplying by 6.25. Feed samples are treated chemically to release ammonia, a form of nitrogen. Trapped ammonia is titrated for a nitrogen calculation expressed in percent. This is multiplied by 6.25. The resulting figure is called *crude protein* (Figure 4-4).

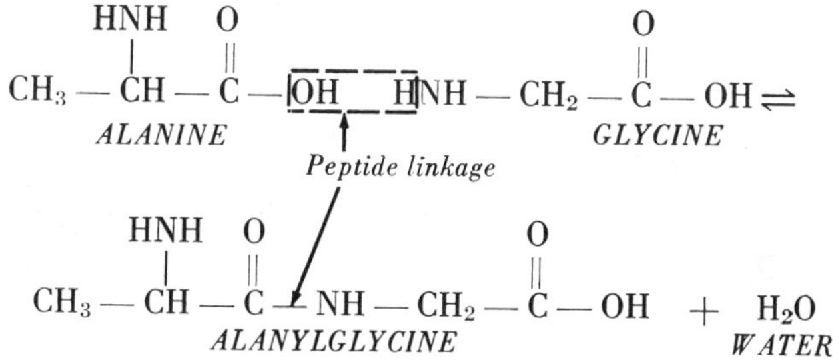

FIGURE 4-3. The combination of amino acids by peptide linkage is the basis of the protein molecule formation. The reverse process, hydrolysis, is the basis of separation.

FIGURE 4-4. The Kjeldahl apparatus is used to determine nitrogen for protein estimates.

Plants bind amino acids in a designated pattern to make plant protein, which, through digestion, provides the amino acids that serve as building blocks for animal protein. Some plant protein sources are soybean meal, cottonseed meal, linseed meal, and peanut meal (concentrates), and legume hays (roughage).

Plant protein is required to build animal protein, which comprises the bulk of muscle tissue and milk solids. Any growth or production that occurs requires this nutrient. Just as manufacturers convert huge quantities of raw materials into desirable products, a cow converts lush, green grass into steaks and fresh milk. Younger, faster-growing animals require large amounts, as do pregnant and lactating cows. Individual amino acids function as a necessary part of enzymes that aid in digestion, hormones that regulate body functions, hair and skin pigmentation, and metabolic body cell reactions, just to name a few examples.

Carbohydrates

The building blocks of carbohydrates are *sugars,* which provide most of the animal's energy requirements. The sugar in a candy bar gives quick energy and may be compared to the carbohydrates in concentrates (grain) and roughages. This energy is used for maintenance, growth, fattening, reproduction, and lactation. It is the fuel that is used to drive the reactions necessary to maintain the fire of life.

Carbohydrates are so called because these compounds are composed of carbon, hydrogen, and oxygen, with the hydrogen and oxygen being in the same proportion as in water. The formula for water is H_2O. In both water and carbohydrates, there are two atoms of hydrogen for each atom of oxygen. This group of compounds includes sugars, starch, cellulose, and other, more complex substances.

The sugars are the simplest of the carbohydrates. Without a doubt, *glucose* is the most important sugar in carbohydrate metabolism, for it is the sugar in blood, and all organisms seem to be able to utilize it. Glucose occurs in combination with other substances in the body. Just to give one example, Figure 4-5 illustrates the combination of two simple sugars, *glucose* and *galactose*, to form a compound sugar, *lactose*. Note that both the glucose and galactose units have six carbon atoms. They have the same general formula $C_6H_{12}O_6$ but differ in arrangement and grouping of the atoms. The simplest sugars contain either six carbon atoms (*hexoses*) or five carbon atoms (*pentoses*). The most common hexoses are *glucose* (blood sugar), *fructose* (found in ripe fruit), and *galactose* (found mostly in milk). The pentoses are seldom found free in

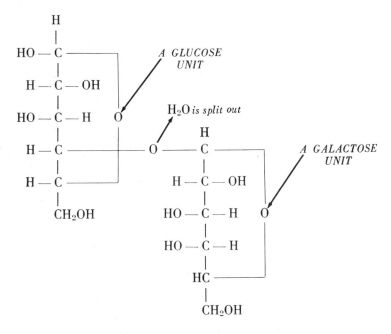

LACTOSE

FIGURE 4-5.   The building blocks of carbohydrates are sugars. The linkage of two simple sugars — glucose and galactose — forms the compound sugar lactose.

nature, but they are found as part of the complex carbohydrates.

A union of two molecules of various simple hexose sugars gives rise to three important compound sugars: *sucrose* (cane sugar), *maltose* (malt sugar), and *lactose* (milk sugar). Most of the carbohydrates in plants and feeds consist of the more complex carbohydrates formed by the union of great numbers of simple sugar molecules, with the splitting out of water. Called *polysaccharides* (many sugars), they include starch and cellulose.

Starch, composed of many molecules of glucose, is the principal form of stored energy in grain (especially corn), wheat, and grain sorghum. Digestion of starch by animals reverses the formation process to yield glucose and is therefore very important in livestock rations because of ease of digestion and availability. The feeding value is therefore relatively high.

Cellulose and related compounds, which comprise the bulk of plant cell walls and form the woody fiber parts, are less completely digested. However, the end product is glucose. This yields energy the same as starch but not as efficiently because some energy must be expended (lost) in the work of digestion. Hay and related feeds typify cellulose-containing products.

For ease of analysis, discussion, and communication, the chemists have divided the often confusing array of carbohydrates into *fiber* and *NFE*. A crude fiber digestion apparatus (Figure 4-6) simulates digestion by alternately washing a feed sample with dilute acid and dilute alkali (base). The material left behind roughly indicates the more poorly digested carbohydrates, fiber,

FIGURE 4-6. Fiber is determined as part of a carbohydrate analysis by a crude fiber digestion apparatus.

Chap. 4   FEEDING BEEF CATTLE

or crude fiber. The chief fibers are represented by cellulose and other carbohydrates not easily dissolved.

NFE stands for *nitrogen-free extract* and represents the more soluble carbohydrates. It is extracted during the fiber determination by the weak acid and weak alkali. Starch, the sugars, the more soluble pentosans, and the other complex carbohydrates are included.

Fats

The building blocks for this nutrient are *fatty acids*, which are also used as a source of energy for cattle. Fats are concentrated, having 2.25 times as much energy as carbohydrates. The customary uses are to raise the energy level and improve flavor, texture, and palatability (animal acceptance) in a ration. Fat choices fall almost entirely in the concentrate category such as by-products of the oilseed crops (vegetable oils) and rendered animal fats. *Fat-soluble vitamins*, to be discussed under "Vitamins," are also associated with this nutrient but are not a part of the molecular structure.

The term *oil* is often used interchangeably with fat. They are alike in composition except that fats are solids at ordinary room temperature while oils remain liquid. Fats and oils are soluble in ether and certain other solvents. Because of this characteristic (analysis is effected by chemically extracting with ether), all the substances dissolved are included under the classification *fats* or *ether extract*. The term *lipids*, often used by the chemist, also means fats.

Like carbohydrates, fats are also made up of carbon, hydrogen, and oxygen, but the oxygen proportion is much lower, making the energy value greater. It is generally accepted that a pound of fat has 2.25 times the oxidation energy value of a pound of carbohydrate. Fat is composed of three fatty acids linked chemically to glycerol (Figure 4-7).

The fatty acids that are linked to glycerol vary with the type of fat from which they originate. Generally, two types are referred to for simplification:

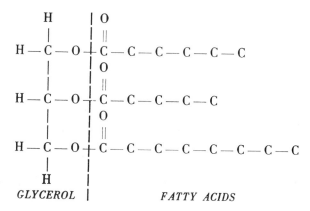

FIGURE 4-7.   Fats are composed of three fatty acids linked chemically to glycerol.

saturated and unsaturated. The term *unsaturated fatty acids* means they are able to absorb oxygen or other chemical elements. It is thought that some unsaturated fatty acids are necessary for normal life. Three fatty acids (*linoleic, arachidonic,* and *linolenic*) are referred to as *essential fatty acids,* indicating the need for some fat in a feed or ration.

Digestion of fat liberates the fatty acids from the glycerol molecule allowing movement through the bloodstream to muscle tissue or adipose (fatty) tissue to be united again to form fat. This fat deposit in muscle tissue is called *marbling.* In other forms it is a storehouse of energy.

Minerals

What are minerals? Simply stated and easily visualized, minerals are the ashes of a burned tree or other organic matter, including feed. To analyze for minerals, feed is burned at 600 °C until a constant weight is reached. All that is left is *ash,* often seen on feed tag information. This ash residue is the inorganic portion of feed, minerals. There are 15 minerals required for normal body function. Table 4-1 gives a condensed appraisal of functions, deficiency signs, and sources of supply.

When feeds are grown on fertile land with no deficiencies, they will supply sufficient mineral elements. Some deficiencies do exist (Figure 4-8); therefore, a common practice is to supply a supplement of one part trace mineralized salt and two parts dicalcium phosphate or bonemeal. This would ensure against the more common deficiencies. The information in Table 4-1 does not include all the functions of minerals, signs of deficiencies, or sources of supply; however, it does cover the more common information and indicates the general roles that minerals play.

TABLE 4-1. MINERALS REQUIRED BY BEEF CATTLE

Mineral	Function	Deficiency Sign	Source
Salt (Na, Cl— sodium and chlorine	Used in gastric juices, maintains body water percentage	Lack of appetite, coarse coat	Salt
Calcium (Ca)	Role in blood coagulation, bone formation, and other vital functions	Fragile bones, rickets in young animals, osteomalacia in mature animals	Dicalcium phosphate; steamed bone meal
Phosphorus (P)	Formation of bones; aids in absorption of simple sugars and fatty acids	Depraved appetite, fragile bones, stiffness of joints, chewing bones and wood	Steamed bone meal; dicalcium phosphate
Potassium (K)	Muscle control and bone formation	Vague, seldom seen	Quality roughages; potassium chloride[a]

TABLE 4-1. (Continued)

Mineral	Function	Deficiency Sign	Source
Magnesium (Mg)	Bone, teeth formation, and muscle coordination	Convulsions, grass tetany disease (grass staggers)	Magnesium sulfate[a] (epsom salt)
Sulfur (S)	Synthesis of sulfur containing amino acids	Vague, seldom seen	Quality roughage; elemental sulfur[a]
Iodine (I)	Necessary for thyroid function	Goiter, poor growth, listlessness	Iodized salt[a]
Cobalt (Co)	Needed by bacteria to synthesize vitamin B_{12}	Loss of appetite, weakness	Cobalt sulfate[a]
Copper (Cu)	Hair development, hemoglobin (oxygenated blood) formation	Severe diarrhea; weight loss; appetite loss; rough, coarse, bleached coat; anemia	Copper sulfate[a]
Manganese (Mn)	Bone formation	Deformed or "crooked calves"	Quality roughages; trace mineral salt
Iron (Fe)	A part of hemoglobin (oxygen carrier in blood)	Anemia (paleness of blood vessels of eyelid)	Quality roughages; iron phosphates[a]
Zinc (Zn)	Skin and hair; a component of insulin	Rough skin	Quality roughages; zinc sulfate[a]
Selenium (Se)	Associated with vitamin E	White muscle disease in calves (muscular dystrophy)	Linseed meal; some soils have toxic levels that lead to poisoning
Molybdenum (Mo)	Stimulates fiber digestion; low level is required	Toxicity occurs more often than deficiency	Quality roughages

[a]Trace mineralized salt is a convenient source.

Vitamins

There are two divisions of vitamins: *fat soluble* and *water soluble*. Table 4-2 illustrates that fat-soluble vitamins are obtained from the ration or from synthetic supplements. The water-soluble group is synthesized by the cow, and vitamin C, which is required by humans, monkeys, and guinea pigs, is not needed by cattle.

Vitamins are organic in nature and function mainly as a catalyst. Only very small amounts are needed (as little as 1 gram or less per ton of feed in the case of some B vitamins), but they are vital to life. The name is derived

FIGURE 4-8. Phosphorus deficiency in feedlot cattle. The steer at the bottom received the same ration as the one at the top plus ½ lb of steamed bone meal daily to correct the phosphorus deficiency. [Reprinted by permission from Committee on Animal Nutrition, National Academy of Sciences–National Research Council, *Nutrient Requirements of Beef Cattle*, 4th revised edition, Publication ISBN 0-309-01754-8 (Washington, D.C.: NAS–NRC, 1970)]

TABLE 4-2. VITAMINS REQUIRED BY BEEF CATTLE

Vitamin	Function	Deficiency Sign	Source
Fat Soluble			
A	Strong calves at birth, normal vision, normal joint function	Weak calves at birth, night blindness, and anasarca (swelling of joints and brisket)	Alfalfa and synthetic vitamin A
D (the sunshine vitamin)	Calcium and phosphorus metabolism	Rickets in young animals, osteomalacia in mature animals	Sun-cured hay (sunshine plus certain animal and plant sterols produce vitamin D); synthetic vitamin D not needed except when animals are housed out of sunlight
E	Thought to be associated in some way with reproduction; associated with selenium	White muscle disease in calves (muscular dystrophy)	Any feed containing natural oil; synthetic vitamin E
K	Blood clotting	Excessive bleeding from cuts or bruises (often caused from consumption of dicumarol, product of grass smut)	Quality forages; synthetic vitamin K
Water Soluble			
C	Not required by ruminants		
B complex	Aids in nutrient building block metabolism	No deficiencies noted in functioning ruminants	Synthesized in adequate amounts by rumen bacteria

from the French, who in the early part of the twentieth century thought these small quantities of nutrients were amines. The term *vital amine* was reduced to *vitamin* and is used to this day.

The fat-soluble vitamins (A, D, E, and K) are associated with and found along with fat, as the name implies. They are not exclusively found there; it just happens that solvents that will extract fat will also extract these vitamins. Table 4-2 gives some common functions, deficiency signs, and sources. Actually, only vitamins A and D are of much concern in ruminant feeding (Figure 4-9).

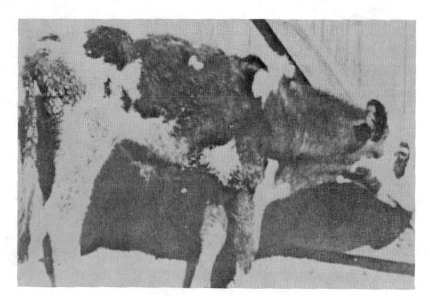

FIGURE 4-9. Vitamin A deficiency in a calf. The animal is suffering from night blindness and scours. [Reprinted by permission from Committee on Animal Nutrition, National Academy of Sciences–National Research Council, *Nutrient Requirements of Beef Cattle*, 4th revised edition, Publication ISBN 0-309-01754-8 (Washington, D.C.: NAS–NRC, 1970)]

Vitamin C, often referred to as the citrate-soluble vitamin, is not needed by cattle. It is occasionally used in cattle feed rations for nonnutritional purposes, generally as an antioxidant to stabilize other, more volatile vitamins. The B-complex vitamins (niacin, pantothenic acid, pyridoxine, riboflavin, thiamine, B_{12}, etc.) are generally not needed by mature ruminants and are discussed in Chapter 24.

All the essential nutrients found in plant tissue are broken down by the cow in digestion and rebuilt for its purposes. Amino acids, the building blocks of protein, for example, are lined up by a plant in a specific sequence much as cogs on a zipper and "zipped" up to make plant protein. Animals digest plant protein down to amino acids by "unzipping" them, realigning them in another sequence, and "zipping" up animal protein.

THE USES OF FEED NUTRIENTS

Although each of the six essential nutrients has a specific function, they are required by the animal in specific combinations to perform a desired body function. Although all six are needed, the amount of each needed will vary depending on the use of the feed by the animal. In general, nutrients from feed are utilized by the animal in two or more of the following body functions.

Maintenance of the Body

Maintenance requires those nutrients necessary to maintain life without a gain or loss in body weight, and without any production (milk, fattening, reproduction). This includes heat to warm the body, energy for normal body work (heartbeat, breathing), and protein and minerals for repair and replacement of worn-out body tissues. Since maintenance requirements are based on body size, the larger the animal the more nutrients will be needed to maintain the body. On the average, from one-third to one-half of the nutrients in a ration go toward maintaining the body. This requirement must be supplied first, with the excess of nutrients above maintenance used for desired production.

Production Functions

Production body functions use those nutrients left after maintenance requirements have been satisfied. Production functions include one or more of the following: growth, fattening, milk production, reproduction, or work. As seen in Table 4-3, the main nutrients required for each will vary depending on the type of production desired. The amount of nutrients required will also depend on the amount of production desired. For example, more nutrients are required for the growth of 1.5 lb per day than 0.5 lb per day. The total requirement of an animal is the sum total of all the desired body functions. For example, the requirement of a pregnant heifer growing 1.0 lb per day is the total of the requirements of body maintenance, growth, and fetal development.

The amount of each of the six nutrients in a feed is important in selecting and mixing the right feeds in feeding. Essential feed nutrients are obtained by a chemical analysis of the feed. In general, the main system used is the Wende System of Proximate Analysis. This system, devised over 100 years ago, is still used. By the chemical processes of the Wende system (discussed briefly

TABLE 4-3. MAIN NUTRIENTS REQUIRED FOR SPECIFIC BODY FUNCTIONS

Body Function	Main Nutrients Required[a]
Maintenance	Energy, protein, minerals
Growth	Protein, energy, minerals, vitamin D
Fattening	Energy
Reproduction (fetal development)	Protein, energy, minerals
Rebreeding	Energy, protein, vitamins
Milk production	Energy, protein, minerals
Work	Energy

[a]The other nutrients are needed also but at minimal amounts. Water is essential for all functions.

for each class of nutrient) the nutrients in a feed are separated into the following fractions:

Wende Analysis	Corresponding Nutrient
Dry Matter	Feed minus dry matter equals water
Ether extract	Fats
Crude protein	Total proteins
Crude fiber	Hard-to-digest or undigested carbohydrates
Nitrogen-free extract	Easy-to-digest carbohydrates
Ash	Minerals

Note: Vitamins are not included since they are contained in very small amounts and must be analyzed separately.

The Wende analysis works very well for concentrate feeds but has limitations in forages, which vary greatly in digestibility of the fiber content.

To better analyze forages, P. J. Van Soest developed a detergent system of analysis which breaks down the forage into the following: *Neutral detergent solubles* (NDS) consist of plant cell contents (proteins, fats, easy-to-digest carbohydrates such as starch and sugars). These are all digested and used by the animal. *Neutral detergent residue* or *fiber* consists of the plant cell walls (cellulose and undigested carbohydrates). This group varies in digestibility and use to the animal. A further breakdown of the neutral detergent fiber by acid detergents separates those carbohydrates which have some digestibility and use from those which are totally undigestible.

Van Soest Fraction	Percent Digested or Used by Animal
Neutral detergent solubles	100
Neutral detergent fiber	62
Acid detergent soluble (hemicellulose)	79
Acid detergent fiber	30
Cellulose	50
Lignin	0

A comparison of the two systems is shown in Table 4-4. The student of animal science may be confronted with either or both systems and should recognize what each stands for.

Feed Digestibility

Although the methods discussed above indicate the total amount of nutrients present, it does not tell about the amount of the nutrient that will be available to the cow for use. For a feed nutrient to be utilized by an animal it must be consumed (eaten) and digested. Undigested parts of the consumed feed are not

TABLE 4-4. NUTRIENT SYSTEM COMPARISON

Wende System	Nutrient Component	Van Soest Analysis	
Ash	Minerals		
Ether extract	Fats	Neutral detergent solubles (cell contents)	
Crude protein	Proteins		
Nitrogen-free extract	Carbohydrates Starch Sugar		
	Hemicellulose		
Crude fiber	Lignin	Acid detergent fiber	Neutral detergent fiber
	Cellulose		

used by the animal but excreted as feces (manure). Thus the term *digestible nutrient* is a better value than total nutrient, that is, digestible protein versus total or crude protein.

Feed Energy

The energy in a feed usually comes from carbohydrates and fats. Its importance in animal feeding has resulted in expression of the energy of a feed in one of two methods.

The *total digestible nutrient* (TDN) system is the oldest and is based on the digested fractions of the Wende analysis and their energy contribution. It is still widely used by many persons in animal science.

The *calorie system* is based on the actual energy (calorie) content of the feed as measured by the heat given off in burning a small pellet of feed, usually in a measuring device such as the bomb calorimeter. Based on this system the energy content of a feed will be expressed in one of the following terms:

1. *Gross energy:* The total amount of energy in a feed.
2. *Digestible energy (DE):* The amount of energy in a feed digested by the animal, expressed in megacalories (1,000,000 calories of energy) or kilocalories (1000 calories of energy).
3. *Metabolizable energy (ME):* A measure of the energy in a feed used by the animal.
4. *Net energy (NE):* The amount of energy in a feed used for specific body functions.

There are two types of digestive systems discussed in this book. The *monogastric system* (simple-stomached animals) is discussed in later chapters. The *ruminant digestive system* is unique. In order to understand the mechanism of ruminant anatomy, it may be helpful to trace feed through the system. A ruminating animal (in this case a cow) uses its tongue to wrap around grass and twist it off. The animal chews this feed very little before swallowing, but does mix it with saliva in the mouth and lubricates it. Movement continues down the *esophagus* to the *rumen (paunch* or *fermentation vat), reticulum (honeycomb, water bag, pace setter), omasum (manyplies), abomasum (true stomach), small intestine, cecum, large intestine,* and *anus* (Figure 4-10). Swallowing feed arrives first at the rumen, which may reach 50 gallons in capacity. The rumen contains microorganisms, bacteria, and protozoa that break down fibrous materials, digesting them for their own benefit, forming volatile fatty acids, and synthesizing B vitamins and amino acids. The organisms, amounting to 200 billion per teaspoon, are short lived and, upon dying, are digested, releasing nutrients (fats, carbohydrates, proteins, minerals, vitamins) from their bodies to be absorbed by the ruminant.

It should be noted that a calf is a potential ruminant at birth but is not a true ruminant until its digestive system becomes inoculated—by eating or drinking with more mature animals—with microorganisms that gradually develop in the rumen. Calves do not start ruminating until about three months of age, and even then, only slight amounts of roughage are digested. The system continues to increase in size and efficiency so that by weaning, it should be functioning completely.

When microorganisms act upon feed, a by-product called *volatile fatty acids* (VFA) is produced. These VFA are absorbed through the wall of the rumen via finger-like projections called *villi* (Figure 4-11) and yield energy. This is the only nutrient absorbed from the rumen.

Feed floating on the rumen fluid moves in a circular pattern becoming heavier and slowly sinking. This circular, tide-like motion becomes most active after grazing is completed. The resting cow then begins *rumination (cud chewing),* which is a common sign that a cow is feeling well. Early cattlemen observed that cows not regurgitating were often ill and explained it as having "lost her cud." Not understanding the function of ruminants, they forced artificial cuds (balls of grass, even rags) in the mouth, trying to simulate nature and initiate a return to health. A *bolus (cud)* is formed through muscular action of the *reticulum (pace setter)* from material delivered to it by the tide. This cud is regurgitated (forced up the esophagus) back to the mouth, where it is more thoroughly chewed, reswallowed, and this time goes through the reticulum that due to its honeycomb structure (Figure 4-12) prevents most foreign objects like wire from further travel. The reticulum is often punctured by these sharp objects, creating a condition known as *hardware disease.* Because the heart is nearby, the condition is easily fatal.

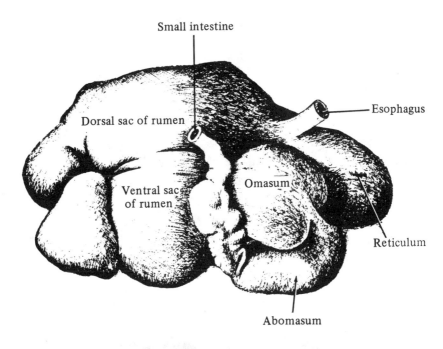

Small intestine

Esophagus

Dorsal sac of rumen

Ventral sac of rumen

Omasum

Reticulum

Abomasum

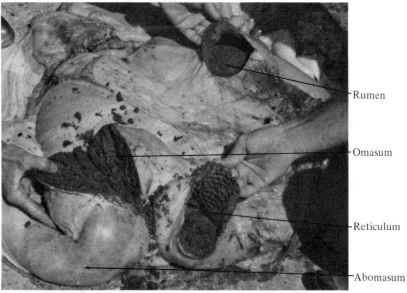

Rumen

Omasum

Reticulum

Abomasum

FIGURE 4-10.   The ruminant's digestive system.

FIGURE 4-11.   Villi line the wall of the rumen.

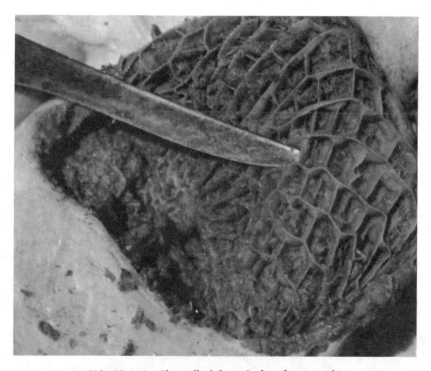

FIGURE 4-12.   The wall of the reticulum (honeycomb).

FIGURE 4-13.   Folds of tissue make up the omasum (manyplies).

The *omasum (manyplies)* receives the slurry mixture and removes most of the water through the large absorptive area of the manyplies (Figure 4-13). Most digestion is completed by the *abomasum,* referred to as the *true stomach* because it relates closely to simple stomach functions. The various nutrient building blocks (amino acids, sugars, fatty acids, etc.) are produced here through action of digestive juices on the bacteria and protozoa and absorbed through the wall of the small intestine. The undigested material is moved through the *cecum* and *large intestine,* and is excreted as *feces (manure)* through the *anus.*

FEEDING WITH A PURPOSE

The type of ration selected will depend on the goal intended—fattening a steer for slaughter, feeding show animals, feeding dairy cows for maximum lactation, or wintering cattle. Regardless of the type of feeding desired, there are some rough rules of thumb that can be used for developing basic rations. The Bible tells us that "the eye of the master fattens his cattle." This is an excellent idea to keep in mind when developing rations. No matter what type of feeding system is used, any signs of unthriftiness in cattle should indicate a change in the ration if parasitism and illness can be ruled out. Basic rules of thumb can be used as a guideline. For example, 2% of the body weight of cattle may be fed as hay (roughage) or three times this amount for silage.

Feeding 1½ to 2 lb of oil meals (protein) or three times this amount for alfalfa or legume hay will meet protein requirements. For fattening, 3% of the body weight is generally accepted as a guideline with most rations containing 60% concentrates (grains) and 40% roughage (hay or silage). Some basic rations suggestions are contained in Table 4-5.

For the more particular feeder or student, rough rules may not suffice. Researchers in the United States have accumulated a massive amount of data on nutritional requirements to provide information wherever needed. The National Academy of Sciences–National Research Council has made available a summary of this information. Using these requirements, a ration may be scientifically balanced for the specific purpose desired.

To balance a ration, the content of the various feed ingredients should be analyzed. Because this is expensive and not always practical, average values for numerous feeds have been compiled by the National Research Council. Because the balancing of rations is time consuming, the extension service in most states offers assistance through the extension animal scientist, in the form of prepared table suggestions. Such a publication is offered in Table 4-6.

Because of the ever-changing costs and availabilities of feed ingredients, no fixed ration appears likely to fill nutritional and economical needs. Combinations will continue to shift with supply and demand making the business of cattle nutrition worthy of study and, when necessary, rations will change to keep pace with the times.

TABLE 4-5. SUGGESTED RATIONS FOR BEEF CATTLE[a]

	Supplementing a 1000-lb Cow (lb/day)	Fattening a 600-lb Calf (lb/day)
1. Nonlegume hay	20	10
Oil meal	2	2
Grain (shelled corn or ground grain sorghum)	—	6
2. Silage	60	20
Oil meal	2	1.5
Grain (shelled corn or ground grain sorghum	—	6
Nonlegume hay		3
3. Nonlegume hay	16	8
Legume hay	6	2
Grain (shelled corn or ground grain sorghum	—	6
Oil meal		1.5
4. Silage	48	—
Legume hay	6	10
Oil meal	—	1
Grain (shelled corn or ground grain sorghum)	—	7

[a]With any ration, provide suitable mineral supplements.

STUDY QUESTIONS

1. Grain-type products are _____ (roughages, concentrates).
2. A sample of alfalfa hay containing 3% nitrogen would mean a protein content of _____ %.
3. If 1 lb of corn yields eight units of energy, 1 lb of rendered animal fat would yield _____ units.
4. _____ is a mineral required by bacteria to synthesize vitamin B_{12}.
5. Amino acids contain _____ .
6. Corn is considered to be a source of _____ or _____ .
7. The two building blocks for carbohydrates and protein are _____ and _____ .
8. The most concentrated, highest sources of energy are _____ .
9. A roughage protein feed is any _____ .
10. A deficiency of calcium in the diet of a young animal may cause _____ .
11. Elemental _____ is not needed by beef cattle but can be used by bacteria in the rumen to synthesize some amino acids.
12. The two elements most routinely supplied to livestock are _____ and _____ that come from _____ .
13. _____ vitamins are synthesized by the cow and not needed in the ration.
14. Cows that constantly chew on old bones, paper, or rags are quite likely showing symptoms of a _____ deficiency.
15. A common condition known as *grass staggers* is caused from a lack of _____ .
16. A feed analysis indicates one percent ash in a ration. This refers to its _____ content.
17. An element necessary for bone growth and found primarily in forages is _____ .
18. A deficiency of _____ shows up as a paleness of the blood vessels under the eyelid.
19. The cow requires _____ (fat-soluble, water-soluble) vitamins from the ration.
20. Sun-cured hay, green in color, supplies vitamins _____ and _____ .
21. The disease rickets is associated with vitamin _____ .
22. Night blindness (inability to see in dim light) results from a vitamin _____ deficiency.
23. The regurgitation of a bolus (cud) of a cow is termed _____ .

TABLE 4-6. RATIONS FOR FATTENING CATTLE[a] (Choice of any one of six rations—read down)

	First 30 Days						Second 30 Days					
	1	2	3	4	5	6	1	2	3	4	5	6
Calves—Initial Weight 400 lb												
Ground: shelled corn, grain, sorghum grain, or barley	5		5	4	6		7		7	5	7	
Ground: ear corn or grain sorghum heads		6				6		8				7
Hay: prairie, sorghum, ground grain sorghum fodder, etc.	7	5	2				5	4	2			
Cottonseed meal or cake	1.5	1.5	1.5	1.5	1	1	2	2	2	2	1	1
Alfalfa hay	1	1		2	7	7	1	1		2	6	6
Cottonseed hulls				6						6		
Silage: corn or sorghum			14						12			
Limestone or oyster shell flour	0.1	0.1	0.1	0.1			0.1	0.1	0.1	0.1		
Yearlings—Initial Weight 600 lb												
Ground: shelled corn, grain, grain sorghum, or barley	7		6	6	7		9		9	8	9	
Ground: ear corn or grain sorghum heads		8				8		10				10
Hay: prairie, sorghum, ground grain sorghum fodder, etc.	11	10	3				9	7	3			
Cottonseed meal or cake	1.5	1.5	1.5	1.5	1	1	1.5	2	1.5	2	2	2
Alfalfa hay	2	2		2	10	10	2	2		2	9	9
Cottonseed hulls				8					15	7		
Silage: corn or sorghum			20						15			
Limestone or oyster shell flour	0.1	0.1	0.1	0.1			0.1	0.1	0.1	0.1		
Two-Year Olds—Initial Weight 800 lb												
Ground: shelled corn, grain sorghum, grain, or barley	8		8	8	10		10		10	10	12	
Ground: ear corn or grain sorghum heads		9				10		11				14
Hay: prairie, sorghum, ground grain sorghum fodder, etc.	12	12	4		3	3	10	10	4		3	3
Cottonseed meal or cake	2	2	2	2.5	1.5	1.5	2	2.5	2.5	2.5	2	2
Alfalfa hay	1	1		1	12	12	1	1	1	1	11	9
Cottonseed hulls				12						10		
Silage: corn or sorghum			25						20			
Limestone or oyster shell flour	0.1	0.1	0.1	0.1			0.1	0.1	0.1	0.1		

Source: Courtesy of Texas A&M Agricultural Extension Service.

[a]When barley is fed, it should not make up over one-third of the grain ration.

24. A cow will normally consume _____ % of its body weight as forage.

25. Protein requirements are met by _____ to _____ lb of oil meals fed per day.

26. The pig with its one-compartment stomach has a _____ digestive system.

TABLE 4-6. (Continued)

Third 30 Days

Block 1

1	2	3	4	5	6
8		8	6	9	
	9				9
4	3	2			
2	2	2	2	1	1
1	1		2	5	5
			6		
		11			
0.1	0.1	0.1	0.1		

Block 2

1	2	3	4	5	6
11		12	10	11	
	12				12
7	5	3			
2	2	2	2	2	2
2	2		2	8	8
			7		
		12			
0.1	0.1	0.1	0.1		

Block 3

1	2	3	4	5	6
13		12	12	14	
	14				16
9	8	4		3	3
2.5	2.5	2.5	3	2.5	2.5
1	1	1	1	10	8
			8		
		20			
0.1	0.1	0.1	0.1		

Fourth 30 Days

Block 1

1	2	3	4	5	6
10		9	8	10	
	11				11
4	3	2			
2.5	2.5	2.5	2.5	2	2
1	1		2	5	5
			6		
		10			
0.1	0.1	0.1	0.1		

Block 2

1	2	3	4	5	6
13		14	12	13	
	14				14
6	5	3			
2.5	2.5	2.5	2.5	2	2
2	1		2	7	7
			5		
		10			
0.1	0.1	0.1	0.1		

Block 3

1	2	3	4	5	6
15		14	14	16	
	16				18
8	5	3		3	3
2.5	3	3	3	2.5	2.5
1	1		1	9	8
			8		
		15			
0.1	0.1	0.1	0.1		

Fifth 30 Days

Block 1

1	2	3	4	5	6
11		12	10	12	
	12				12
4	3	2			
2.5	2.5	2.5	2.5	2	2
1	1		2	5	5
			5		
		8			
0.1	0.1	0.1	0.1		

Block 2

1	2	3	4	5	6
15		16	14	14	
	16				16
5	4	3			
2.5	3	3	2.5	2	2
2	1		2	7	7
			5		
		8			
0.1	0.1	0.1	0.1		

Sixth 30 Days

Block 1

1	2	3	4	5	6
13		14	12	14	
	14				15
4	3	2			
3	3	3	3	2	2
1	1		2	5	5
			5		
		6			
0.1	0.1	0.1	0.1		

27. Hardware disease results when sharp objects puncture the walls of the _____ .

28. Many microorganisms are located in the _____ , which enables the cow to digest fibrous foods.

29. Actual digestion of fats is accomplished in the _____ .

30. The function of the small intestine is to _____ the nutrient building blocks into the bloodstream.

31. A 600-lb steer being full fed a fattening ration would eat about _____ lb of feed per day.

32. One ton of feed for fattening cattle in dry lot should contain about _____ lb of roughage.

33. A ruminant is able to digest roughage because of _____ found in the _____ .

34. The ruminant's digestive system has four components: _____ .

DISCUSSION QUESTIONS

1. What are the basic elements and/or building blocks of the six essential nutrients?

2. Name every possible nutritional deficiency that could result in rickets.

3. Give examples of a mineral–vitamin interaction; that is, how could an excess or a deficiency of one create signs of imbalance of either? Example: Even though cattle do not normally require B vitamins, a deficiency of cobalt could create deficiency signs similar to B-complex deficiencies.

4. Discuss the movement of feed through the ruminant digestive system and the function of each portion from the esophagus to the anus.

5. Calculate the total roughage, grain, and supplement you would recommend stocking as the new manager of 30,000-acre ranch in a southwestern range environment to winter 900 cows and 35 bulls.

Chap. 4   FEEDING BEEF CATTLE

Chapter Five

BEEF CATTLE SELECTION

Selection criteria of beef cattle will vary some with individual techniques. However, most selection principles are based on (1) *eye appraisal*—judging, as commonly seen in the show ring, (2) *pedigree*—selection based on reputation of ancestors, (3) *animal performance*, and (4) *production testing*.

EYE APPRAISAL—JUDGING

Perhaps there is no more exciting event than the tension that is built up just prior to the judge's slapping the rump of the Grand Champion at a leading livestock show. The selection is more than just the judge's personal preference. Based on a systematic approach to judging and observation, the professional arrives at a decision.

The basic categories in types of animals are breeding and slaughter classes. Judging will be somewhat different for each of these, but the same principles apply in breeding stock as in slaughter steers because the characteristics produced in steers must be inherited from the breeding herd. The main differences between the two are the degree of finish and secondary sex characteristics. Breeding stock are often judged on the basis of pedigree as well as the characteristics of the slaughtered offspring. Visual characteristics as well as pedigree are used in breeding stock selection. For purposes of simplicity, the selection of breeding cattle will be skipped to concentrate on slaughter steers. Breeding cattle selection de-emphasizes finish and emphasizes sex characteristics. The selection principles are quite similar, although breeding cattle certainly look different from slaughter cattle.

Cattle Terminology

In order to intelligently discuss observations of the superiority of one animal over the other, it is necessary to know cattle terminology. Figure 5-1 shows the major parts of an animal that should be known before attempting an evaluation of observations of cattle. It is more descriptive to say that an animal is thicker through the quarter with a wider, more bulging round that lets down into a deeper twist, than to say that an animal is broader across the rear. Thus familiarity with parts of an animal indicates interest, knowledge of the field of animal science, and especially a critical eye in livestock judging.

Judging the Steer

The steer or a class of steers should always be judged in a logical manner. It is quite common to view all animals first from the *side*, then from the *rear*, and lastly, from the *front*. Usually, in the show ring, animals are led around in a circle so that the judge can get an overall view of them. Then the judge starts the logical sequence of mentally judging all animals against each other.

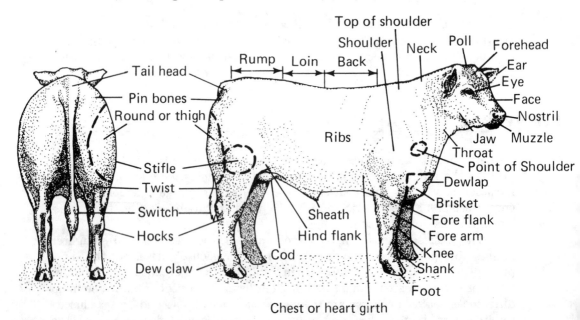

FIGURE 5-1. The body parts of a steer.

FIGURE 5-2.   The side view shows length, trimness, muscle, and bone.

The judge looks for *size for age from the side view* (Figure 5-2). Perhaps the most important overall observation here is *length*. The modern trend is toward cattle that are showing more stretchiness because the more expensive cuts are in the area of the backbone. An animal should be straight across the topline, long from its hooks to its pins (lots of round steak), fairly deep in the flank (more round steak), and trim about the middle (less waste, more meat). A tight middle means a higher dressing percentage, 60 to 65% in better-quality steers. The steer should stand on sound bone (an indication of muscling) and should have uniformity in appearance that indicates a good quality animal.

Figure 5-3 indicates the mental observation that should be kept in mind

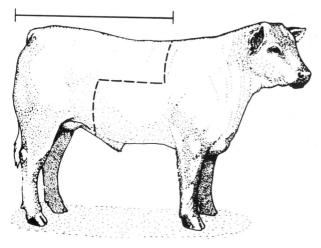

FIGURE 5-3.   Seventy percent of the retail value is found in the top rear.

regarding the influence of economics on selection. Although 49% of the weight of the average steer is in the rear portion illustrated, 70% of the value is represented by it. Look for an animal with much length, strength in the topline, and heaviness in the quarter.

After judging from the side and making comparisons between them, most judges will have the animals turn for a rear view. A steer that is wide and muscular through the center of the quarter is illustrated in Figure 5-4. The top of the round should be straight and full in the tailhead area. Adequate depth from the top of the tailhead to the twist and a wide, bulging round are desired, indicating a large volume of choice steaks. Also from the rear, one should observe the placement of the rear feet. An animal that is heavy boned (large circumference) and "tracks wide" indicates a muscular loin eye (backstrap) as in Figure 5-5. This muscle is often measured in carcass shows. One square inch or more per hundred pounds of live body weight is the production goal. Heavy bones in the live steer indicate heavy muscling in the carcass. Moving close enough to look over the top from the rear (Figure 5-6), the uniformity of width from pins to hooks to crops is noted.

From the front (Figure 5-7), one should observe the thickness through the forequarter and the fullness in the dewlap and brisket. A full, pendulous brisket will indicate excessive fattening. A thin, rather loose dewlap indicates an animal that is carrying less finish, which is becoming more important in today's market as the consumer demands lean, muscular cuts. Bone is also

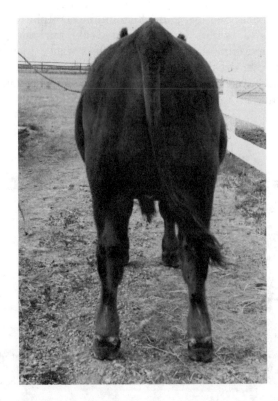

FIGURE 5-4. Width, muscle, and bone from the rear view.

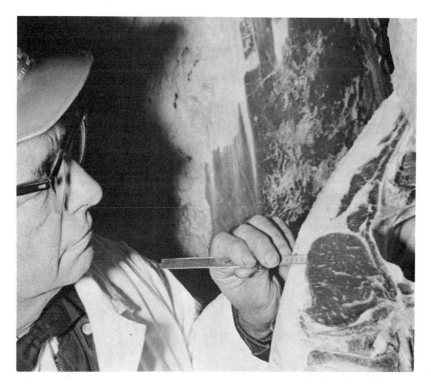

FIGURE 5-5. The top view shows uniformity. (Courtesy of Texas A&M Agricultural Extension Service)

observed in the front legs similar to the way it is from behind. That is, the desirable animals are thick boned and stand on straight, heavy legs, indicating thickness of muscling in other areas that are correlated to this observation.

Handling the steer is the final touch in estimating carcass traits and grade. A thin, firm, mellow covering of fat (finish) is expected to cover the ribs, loin, and back of superior steers. A loose, pliable skin is desired because it indicates carcass quality.

Judging is never more than the comparison of two animals, regardless of the size of the class. The judge always compares one with another, then once a decision has been made between them, the winner of these preliminary thoughts is compared with the winner of other comparisons. Therefore, it should be mentioned to the beginning student that judging is a very simple process if taken in this systematic manner. To illustrate the thought process involved in a typical livestock judge's selecting a grand champion for a show, the following reasons might be given for placing this animal number one.

I place this steer Grand Champion, because he was the longest steer in the class, being neater and trimmer about the middle, indicating a higher dressing percent than any other steer in the class. He was straight in his topline and firm in his quarter, showing thickness of round, great length from his hooks to his pins, deep in the flank and standing on straight, firm bone. He is thicker through the

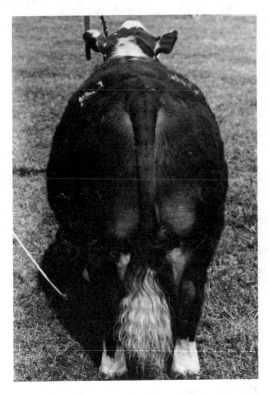

FIGURE 5-6. Note the dewlap and brisket from the front.

FIGURE 5-7. Carcass measurements include loin-eye development and backfat thickness. (Courtesy of the U.S. Department of Agriculture)

quarter, shows greater depth in his twist and more uniformity in width throughout, carrying the broadness from the loin up through the crops and over his back than any of the steers standing below him. He is carrying a desirable amount of finish over his back, loin, and ribs but does not indicate an excessive degree of fat, due to the uniformity of his brisket and the thinness of his dewlap. He is a steer that will hang up a higher percentage of primal cuts and the highest grading carcass of any steer in his class.

Judging is the process of seeing through the live animal and evaluating the final product, the carcass. Figure 5-7 shows a muscled, well-finished carcass with large loin eye. This is what any judge really sees when the final slap on the rump is made.

PEDIGREE

Selection of cattle based on the supposed merit of their ancestors has been used for hundreds of years. This system is still used heavily in some purebred herds, not always with happy endings. A *bloodline,* or being a descendant of an outstanding individual, does not always mean that outstanding traits will result from selection and mating to that descendant.

Pedigrees are most useful if they are employed regarding the most recent individuals; each generation may reduce the potential by one half because half the genes of an offspring come from each parent. Generally, the parents and grandparents are about as far back as it is recommended to go for genotypic material.

Fads and promotional advertising are possible fallacies in pedigree selection. Breeders must be careful that names are backed by the right kind of performance desired by the breeder. Some undesirable inherited traits (such as dwarfism) have been disadvantages that have accompanied seemingly progressive selections based on pedigree.

The highest and best use of pedigree information is to use it as the basis for selection of young animals before their performance or that of their progeny is known. Traits expressed late in life, such as longevity or continued soundness, are arguments for pedigree surveillance. Also, traits expressed in only one sex, such as milk production, have been effectively selected for in numerous instances. For example, a sire that consistently produces daughters and granddaughters with high milk production and mothering ability is more likely to contribute strength to a pedigree than a bull with descendants weak in those traits. Young bulls selected from the strong pedigree will more likely produce daughters with improved milk production and mothering ability.

Pedigrees serve as a useful tool once their limitations are realized. However, once an individual animal's performance and that of its progeny are measured, the pedigree should receive less weight in an individual's evaluation.

ANIMAL PERFORMANCE

Measuring the individual animal's ability to perform efficiently and economically involves some simple and some complex techniques. Prolificacy is often measured in females by observation of regular heat periods and the ability to settle quickly—both are noted as positive for this trait. The males are observed for masculinity, aggressiveness, and good sperm count. A long life means longer production and is a trait observed and selected for by memory in many commercial herds, by pedigree in registered herds. Efficient growth from birth to weaning to production is measured in terms of rate of gain, amount and type of feed consumed, and quality (which determines price) of the end product.

Simple Measuring Techniques

Pencil and Pad. A pocket notepad is helpful to record data such as date of birth, birth weight, date of first heat, breeding dates, and so on. These data are most helpful if later transferred to individual records and kept for the life of that individual (Figure 5-8).

FIGURE 5-8. A notepad is a handy item for making on-the-spot recorded notes.

Scales.   Perhaps the most widely used tool of selection in the United States is the livestock scale. Large or small, stationary or portable, it has become an indispensable part of measuring techniques. Calving weight, weaning weight, and weight at calving are the three most accepted, useful observations (Figure 5-9).

After the initial starting weight at birth, growth to weaning is largely a reflection of the dam's milk production and the calf's inherited ability to gain. Heavier weaning calves are usually grouped together for selection of the best for breeding replacements or "keepers." It should be noted that weights, for comparison purposes, should be adjusted for sex differences and age variation. At weaning, bulls are about 15 lb heavier than steers and steers are about 20 lb heavier than heifers.

Yearling weights indicate growing ability, mostly a function of inherited qualities provided that proper feed, health, and other factors are normal. Because half the inheritance comes from each parent, this weight measurement also gives an indication of the sire and dam's potential for contributing further genetic material.

Weight of the calf at first calving completes the cycle of weight observations from birth to reproduction. Those cows having lighter calves are culled first even though their past records have been good. In a few years, by culling out the normal 20%, a herd of cows may be selected that were superior calves, grew well after weaning, and are reproducing the heavier, higher grading kind of calves.

Bull Testing Stations.   Most stations now have one or more centers designed to measure economically important traits such as rate of gain and feed efficiency. Other measurements may be added but these two are common to all. The purpose is to gain information on young sire prospects after wean-

FIGURE 5-9.   The scale has become an indispensable tool in livestock selection techniques.

ing and before breeding. A controlled, identical environment over a specified length of time (usually 140 days minimum) makes more accurate comparisons possible. The higher-gaining, more efficient bulls are more likely to produce calves with these characteristics than the low gainers and poor converters (Figure 5-10).

FIGURE 5-10.   A bull testing station compares prospective sires under controlled, identical environments. (Courtesy of J. Custom Fitters, Wharton, Texas)

Complex Measuring Techniques

Ultrasonic Device. A device similar to sonar, which is used by submarines to detect objects under water, has been produced on a small, portable scale for use in livestock research. The device emits high frequency sound waves and records the time it takes for an "echo" to bounce back from the junction between two tissues (fat and lean for instance). These echoes are displayed on a readout instrument that allows the plotting of the shape and size of some muscle. The loin-eye muscle is plotted with accuracy using this method and is sometimes used to supplement bull testing data. The obvious advantage is that important data on a high-priced muscle can be obtained without surgery or slaughter (see Figure 5-11).

Cut-Out Data. This is another method used by some bull testing stations equipped with slaughter facilities to add feed efficiency and even ultrasonic measurements to rate of gain. A steer, preferably a full brother to the bull being tested, completes the same program. The steer is then sacrificed to obtain actual data. The most useful guides are percent lean cuts and square inches in area of loin eye. Additional data often include grade, dressing percent, and marbling scores. This information coupled with the performance of the bull gives a more valid projection of the potential improvement based on this line of selected genetic material.

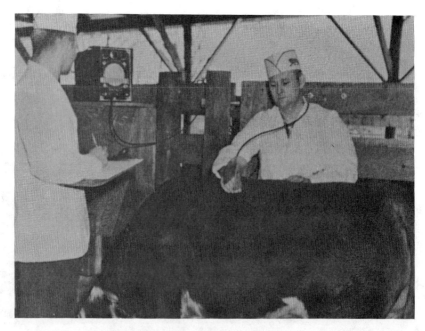

FIGURE 5-11. Ultrasonic sound is used to measure loin-eye area in live animals. [Reprinted by permission from John F. Lasley, *Genetics of Livestock Improvement* (Englewood Cliffs, N.J.: Prentice-Hall, Inc., 1963)]

Chap. 5   BEEF CATTLE SELECTION

PRODUCTION TESTING

Production testing encompasses *performance testing* (individual merit) and *progeny testing* (merit of offspring). It is used most effectively when based mainly on characteristics of economic importance that are highly inherited. For example, these traits are often considered of highest economic value: (1) reproductive efficiency, (2) mothering ability, (3) rate of gain, (4) economy of gain, (5) longevity, and (6) carcass merit.

Production testing is not a new idea, having been used by the Romans over 2000 years ago. Of course, techniques have advanced significantly. The basis, in the beginning as now, is that individuals will differ in their performance and that traits are inherited.

Thus production testing as a selection technique is a systematic measurement of economically important traits, the recording of these traits, and the use of the recorded information for selecting superior and culling inferior producers. Production testing is the most infallible of all the methods of selection, particularly for selecting young herd sires.

To be effective, the superior characteristics to be selected for must be of economic importance and high heritability. Objective measurements such as pounds, inches, and so on, should be used. Minimum standards are set forth and any animal falling below these standards is culled.

Improvement in an economically important trait may be attributed to heredity and environment. If traits are selected for that are influenced mainly by environment, the heritability of that characteristic will be low and little progress can be expected. Therefore, traits that are on the upper end of the scale of probability should be used. (Table 5-1).

Developing a Production Testing Program

Generally speaking, five basic steps are necessary to embark on a production testing project. It should be remembered that both performance testing (merit of the individual) and progeny testing (merit of the offspring) make up the broader procedure of production testing.

TABLE 5-1 HERITABILITY OF SOME ECONOMICALLY IMPORTANT CHARACTERISTICS IN BEEF CATTLE

Characteristic	Heritability
Birth weight	Medium to high
Weaning weight	Low to medium
Daily rate of gain	Medium to high
Feed efficiency	Medium
Live and carcass grade	Medium
Area of loin eye	Very high
Milking and mothering ability	Medium
Reproductive efficiency	Low to medium

Make Permanent Markings.   Each cow and sire should be marked by branding, tattooing, and so on. Furthermore, for ease of identification from a distance, these marks should be easily read on foot, horseback, or from the cab of a pickup. Brands work well in this regard. However, some breeders prefer to supplement brands or tattoos with large plastic ear tags that are easily read and may be color coded to quickly differentiate between groups or bloodlines (Figure 5-12).

Record Ages.   All cows, sires, and calves in the herd should have their age recorded. If records do not exist for cows, they should be mouthed by a qualified person for an estimate. This is necessary because comparison between families must be adjusted for age influence. Fair comparisons between young, old, and cows at their peak of performance are not possible unless age influence is considered. Weaning weights of the calves will be the adjusting factor because very young and old cows wean lighter calves than those cows in their prime (Figure 5-13).

Observe from Birth.   Each calf should be weighed at birth and records made concerning the date, weight, sex, dam, sire (if known), and any other information that may be important to the breeder. A permanent identification (ear tattoo or brand is most common) should link the pair. Again, for ease of identification, ear tags are suggested to save needless rechecking.

Record Information.   Weaning weights and grades at 6 to 8 months should be recorded and information updated for each calf, each dam, and each sire. An example of one form of headings is shown in Figure 5-14. Adjust and work the data where required. These forms can be added to for further rate of gain, feed efficiency, and other measurements.

FIGURE 5-12. Cows and sires are permanently marked and made easy to identify for production testing purposes.

Chap. 5   BEEF CATTLE SELECTION

FIGURE 5-13.   Weaning weights are measured by the scale and calves graded by the eye.

Make Decisions Based on Records.   Production testing really pays off when the records are put to work and sentiment is retired. The cows that produce the lightest, poor-doing kind of calf crop should be culled regardless of personal preference for minor things like markings, provided they do not weigh heavily on the economic market. The sires that produce the lighter calves should also be culled.

The heavier, faster-growing, more efficient heifers and bulls should be kept for replacements. For best results, the heifers are fed out and yearling weights and grades are taken. The bottom (20% as an example) is culled out and the superior heifers bred to the superior sires. After first calving, these heifers are again culled, based on the weaning weight of their first calves.

In this manner, the best animals are tested for individual performance (performance testing) and offspring performance (progeny testing), and a strict culling program endorses selection of the best to be bred to the best. Thus production testing removes the grandeur of reputation, the magnificence of ancestors, and the glory of visual excellence to pace the contest with the inevitable—documented results.

A SYSTEM OF SELECTION

Finally, using a combination or all of the techniques (eye appraisal, pedigree, animal performance, and production testing) discussed, the breeder needs to follow a system of selection that will result in maximum progress (Figure 5-

a. CALF & YEARLING WEIGHTS & GRADES RECORD

Sire	Dam		Calving Record				Weaning Record					Yearling Record				
Herd Identity Number	Herd Identity Number	Age of Dam	Calf Number	Sex	Date Calved	Birth Weight	Weaning Weight	Age in Days	*Adjusted 205 Day Weight	Weight Per Day of Age	Weight Ratio or Grade	Yearling Weight	Age in Days	**Adjusted 365 Day Weight	Weight Per Day of Age	Weight Ratio or Grade

b. SIRE'S ANNUAL PRODUCTION RECORD

Sire's Herd Identity Number	Calf Number	Sex	205 Day Adjusted Weight	Weight Ratio	Grade	365 Day Adjusted Weight	Weight Ratio	Grade	Replacements	Remarks

c. LIFETIME PRODUCTION RECORD OF DAM

Herd Identity Number	Year of Birth	Average Total Production at Weaning				Production Index	Replacements	Remarks
		Number of Calves	Adjusted Weaning wt.	Weight Ratio	Grade			

*To compute the adjusted 205 day weaning weight, apply the formula: $\dfrac{\text{actual weight} - 70}{\text{age in days}} \times 205 + 70 = 205$ day weight

then correct the adjusted weight for the age of dam according to the following table:

Percent to be added to calf weights after adjusting for age

Age of dam	Percent to be added
2	15
3	10
4	5
5-10	none
11 and older	5

**To compute the adjusted 365 day weight, apply the formula: $\dfrac{\text{final weight} - \text{actual weaning weight}}{\text{number of days between weights}} \times 160 + 205$ day adjusted weaning weight = adjusted 365 day weight

FIGURE 5-14. Example of a useful production record form. (Courtesy of Vocational Instructional Services, Texas A&M University)

Chap. 5 BEEF CATTLE SELECTION

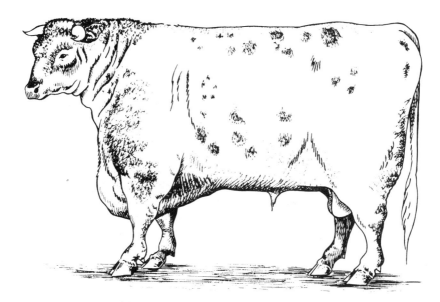

FIGURE 5-15. Success of selection of stockmen through the years can be seen when you compare the ideal animal of the nineteenth century with modern Grand Champion steers.

15). Three common systems follow: tandem selection, independent culling level, and selection index.

Tandem Selection

One trait is selected for at a time until maximum progress has been achieved; then another trait becomes the target for improvement. *Tandem* means one after another. The process sounds simple. For instance, cattle are selected for the polled character and once the herd breeds true for this trait,the same group is selected for weaning weight, and so on.

The disadvantages to this system are that progress is slow, many characteristics are not suited for selection alone because of linkage to another characteristic, and income is generally dependent on several traits. Therefore, if economics enters the picture, this system may have strong limitations.

Independent Culling Level

Production testing fits well with this systematic procedure. Minimum standards are set for several traits. A failure by any animal to meet the minimum standard for any one trait results in removal from the herd to be sold or slaughtered. The obvious disadvantage is that if standards are too high and too many traits are involved, the level culled could be too high to leave sufficient animals to work with. With common sense and some adjustments in the standards from time to time, this system becomes very effective.

Selection Index

Generally accepted as the most efficient of the systems, an index evaluates all important traits and combines them into one figure or score. Higher scores mean a more valuable animal for breeding purposes. The weight assigned to each trait to be included in the index will depend on its economic importance, its heritability, and its genetic linkage to other traits. This method is accepted by many breeders because slightly substandard performance in one trait can be offset by excellence in another trait. The only disadvantages appear to occur when too many traits are included in an index or progress is attempted on characteristics of low heritability or little economic importance. When used with a sensible number of traits of relatively good heritability, the selection index gives perhaps the fairest appraisal of merit.

STUDY QUESTIONS

1. Selection of breeding stock is similar to slaughter cattle except for _____ .

2. A trim middle indicates a higher _____ percent.

3. A 1000-lb steer may be expected to yield a carcass that weighs as much as _____ lb.

4. The rear half of a carcass worth $500 would represent about $_____ .

5. A steer that stands wide should have a large _____ muscle.

6. A carcass weighing 800 lb should have at least _____ square inches of loin-eye area.

7. The rear quarter shows depth in the _____ and _____ .

8. A full, wasty brisket means too much _____ .

9. A thin dewlap is related to a _____ carcass.

10. Perhaps the most important view from the side is _____ of body.

11. Pedigrees are most useful when the traits of the most _____ generations are considered.

12. For traits that are expressed by one sex (as milk production), _____ and _____ testing are used as selection measurements.

13. The use of an animal's performance is valuable, providing accurate _____ are kept for the traits upon which selection is based.

14. Ultrasonic sound devices are used to measure size of muscle on _____ (live, dead) animals.

15. Production and progeny testing is used more effectively when characteristics are of _____ importance or of _____ heritability.

16. In the _____ selection system, one trait is selected for at a time until maximum progress has been made.

17. Minimum standards are set for several traits in an _____ culling system.

18. A _____ evaluates many important traits by weighing each according to economic importance, heredity, and genetic linkage.

19. To make the fastest progress possible for just one trait, you would use the _____ system of selection.

20. The most efficient way to select for several different traits is the _____ .

21. the most widely used tool or piece of equipment of selection is the _____ .

22. The three most widely used weights in an animal's record are _____ , _____ , and _____ .

23. All weights except birth weight should be adjusted for _____ and _____ .

DISCUSSION QUESTIONS

1. Explain the advantages and disadvantages of traditional judging, pedigree selection, production testing, and progeny testing.

2. How can performance testing be used to evaluate breeding stock for both carcass data and breeding purposes?

3. Using the form in Figure 5-14, develop a mock set of records and have them evaluated for realism by your classmates or instructor.

4. Discuss the facilities, equipment, and methods needed to set up a production testing station.

5. Explain exactly what a livestock judge looks for and why in a breeding class; in a slaughter class.

Chapter Six

THE BEEF CARCASS

Although indications of quality can be noted in the live animal, the final test of a beef animal is slaughtering to yield the product demanded by the consumer. The selection techniques used in judging (Chapter 5) are utilized by the packer buyers who determine the price they are willing to pay for an animal. This price reflects what the buyer considers to be a reasonable gamble that will allow the buyer's company to make a profit. If the buying decisions are right most of the time, the buyer remains a buyer. A sharp buyer is often described as one with keen observation, able to size up a prospect in an instant and observe the present and future—these predictions to materialize or perish on the kill floor.

SLAUGHTERING PROCESS

A premortem (before death) inspection by a veterinarian precedes slaughter. If an animal is determined for any reason not to be fit for human consumption, it is tagged and must be disposed of for nonedible products. Occasionally, an animal may be questionable in the mind of the examiner and it is tagged as "suspect," the final decision resting with the veterinarians doing the postmortem (after death) inspections, which all must pass. Animals passing the pre-mortem inspection proceed directly to the kill floor.

Stunning

For humane reasons, animals to be slaughtered are rendered unconscious as quickly and painlessly as possible without causing the heart to stop pumping. Because of safety factors, most stunning (Figure 6-1) is done by a *captive bolt*, a rod that works inside a cylinder and is activated by an explosive charge similar to a blank cartridge triggered by pressure of the device against the animal's head. Calves are electrically stunned in some plants. Except for kosher-slaughtered animals (which are not stunned), this procedure is used almost nationwide.

Sticking

The stunned body is hoisted by the rear feet using special shackles (Figure 6-2). Because the heart is still beating, the blood is actively pumped from the body upon sticking with a sharp knife to sever the jugular veins and carotid

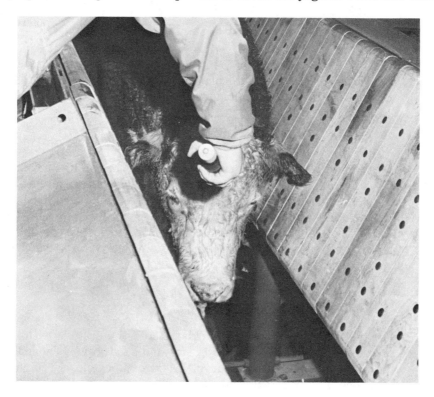

FIGURE 6-1. Animals are rendered unconscious or "stunned" by a device known as a captive bolt. (Courtesy of MBPXL)

FIGURE 6-2. The stunned body is hoisted, the jugular veins and carotid arteries are severed by "sticking" with a sharp knife, and blood is drained with the aid of the still-beating heart. (Courtesy of MBPXL)

arteries in the neck. After the heart ceases to beat, the carcass continues to drain blood because of the hanging position. Good drainage is important to the appearance of the retail cuts.

Skinning

After drainage is complete, the animal is lowered on its back so that the hide may be partially removed. The carcass is rehoisted (Figure 6-3) and the skinning process completed by removal of the feet, hide, and head. The carcass is then eviscerated (intestinal organs removed) as in Figure 6-4. The kidneys are left in the body. The carcass and viscera are inspected and parts (such as the liver) may be rejected for human consumption or, in some cases, the entire carcass condemned.

Halving

The tail is removed and the backbone split (Figure 6-5) with a power-driven saw to create two halves—a *tight side* and a *loose side*. The left half is always the tight side, so called because the fat adheres closely to the kidneys and backbone, giving a more desirable appearance to retail cuts, such as T-bone steaks.

Chap. 6 THE BEEF CARCASS

FIGURE 6-3. Skinning involves removal of the hide, feet, and head. (Courtesy of MBPXL)

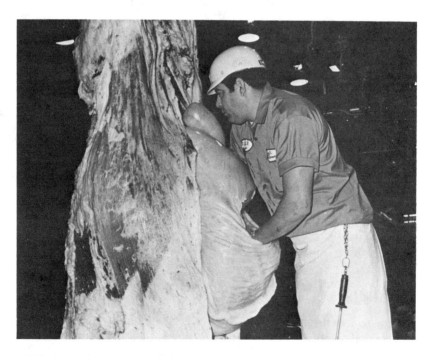

FIGURE 6-4. Evisceration allows inspection of the carcass and viscera to determine suitability for human consumption. (Courtesy of MBPXL)

FIGURE 6-5. Halving of the carcass is done by removing the tail and splitting the backbone with a power-driven saw. (Courtesy of MBPXL)

Cooling

Prior to further inspection, the hot carcass halves are washed—some may be shrouded—and sent to the coolers where the temperature is held at 34 °F for 24 hours before ribbing. U.S. Department of Agriculture (USDA) inspectors inspect and pass on wholesome carcasses. An inspection stamp on the meat

FIGURE 6-6.   The carcases are washed, cooled, and cut between the twelfth and thirteenth ribs to expose the loin-eye. (Courtesy of Monfort of Colorado, Inc.)

signifies acceptance as fit for human consumption. These inspections are mandatory in the United States on the state and national levels.

Grading

A knife is used to cut between the twelfth and thirteenth ribs *(ribbing)* to the backbone (Figure 6-6). This exposes the loin-eye area and condition that along with other observations are used to determine the quality of beef represented by a value given by a government grader, discussed later in this chapter.

Dressing Percent

Yield, as it is often called, is the percent of the product left in the cooler as compared to live weight. Conformation and finish influence dressing percent the most, provided a normal "fill" is allowed. The more muscular, fatter animals are expected, on the average, to grade higher and dress (yield) higher (Figure 6-7). Because the viscera are part of the weight subtracted from the live weight to arrive at dressing percent, it is easy to see how an extra "fill" of water or feed can add to selling weight but lower yield. This is taken into

FIGURE 6-7. Carcasses may differ in muscling, grade, and other factors. In this famous frozen steer exhibit from Colorado State University, the steer on the right shows an excessive fat trim (note the trim under the belly) compared to the steer on the left.

account by the packer buyer when making a bid. If there is any question, the buyer will bid very conservatively so it is to the seller's advantage not to allow the cattle undue fill just prior to sale.

Dressing percent will vary with each individual but some average figures for the grades involved are: Prime and Choice—63%; Standard and Good—60%; Commercial—56%; Utility, Cutter, and Canner—52%.

Aging

For optimum tenderness, beef carcasses that are fat enough to seal out air can be aged for 2 to 5 weeks at temperatures between 34 and 38 °F. This is higher than freezing and allows the natural enzymes to break down the connective tissue (collagen) surrounding the cells, resulting in a tenderizing process.

KOSHER SLAUGHTER

Beef used for the Jewish market must be slaughtered according to religious customs. A representative of the rabbi uses a special double-edged, razor-sharp knife to sever the jugular veins and carotid arteries. Stunning is not permitted, and only one stroke of the blade is allowed. Jewish law forbids the consumption of the meat of undrained beef, and because the hind quarters are considered to have small veins, they are not utilized in the Jewish trade; only the forequarter is marketed. In some areas largely populated by Jews, all cattle are kosher slaughtered to cater to the total population.

CARCASS GRADES AND GRADING

Beef carcass grades are based upon the quality and palatability of the meat (USDA *quality grades*) and the quantity or cut-out yield of trimmed, boneless, major retail cuts (USDA *yield grades*). Since 1976, both quality and yield grades are assigned to a beef carcass. Grading is not mandatory but a service provided by the USDA, paid for by the processor. Because consumers depend mainly on USDA quality and yield grades in purchasing, most processor feel the service is advantageous.*

U.S. Quality Grades

Consumer demand for beef has long indicated a need for a systematic way to express meat quality and carcass value. This led to grading meat for quality by the USDA as early as 1927. USDA quality grades for beef have been revised since 1927 to ensure that they serve as a reliable guide to the eating quality of beef. USDA quality grades are *prime, choice, good, standard, commercial, utility, cutter,* and *canner.* They are based on carcass characteristics such as maturity, marbling, texture of the lean meat, and color of the lean meat (Figure 6-8).

Maturity is determined by the size, shape, and ossification of the bones and the color and texture of the lean meat, with more mature animals having darker and tougher lean.

Marbling refers to the intermingled fat in the lean muscle as evaluated between the twelfth and thirteenth ribs (place of ribbing). *Texture* or firmness of the lean meat and color of the lean meat are also evaluated at the twelfth and thirteenth ribs.

Carcasses are grouped according to sex classes. Male classes consist of steers, bullock (young bulls), and bulls. Female classes are heifers and cows.

Five different maturity groups (A, B, C, D, and E) are recognized for the approximate age at slaughter of 9 to 30, 30 to 48, 48 to 60, 60 to 96, and 96 months or older. Marbling is broken down into slightly abundant, moderate, modest, small, slight, traces, and practically devoid. The relationship between marbling and maturity with quality grades is shown in Figure 6-9. As seen in Figure 6-9, the higher USDA Quality grades (prime, choice, good, and standard) are reserved for animals in early maturity groups (A and B) with adequate marbling. In 1976, marbling standards are reduced and now allow more animals in the prime, choice, and good grades. This reflects today's consumer

*For a discussion of the Canadian grading system, see Appendix A.

FIGURE 6-8. These frozen steers from Colorado State University are dissected to illustrate the relationship between the live animal and the various internal factors that influence carcass grades and grading.

RELATIONSHIP BETWEEN MARBLING, MATURITY, AND CARCASS QUALITY GRADE*
Maturity**

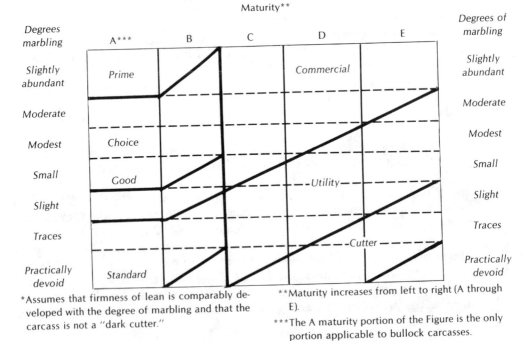

*Assumes that firmness of lean is comparably developed with the degree of marbling and that the carcass is not a "dark cutter."

**Maturity increases from left to right (A through E).

***The A maturity portion of the Figure is the only portion applicable to bullock carcasses.

FIGURE 6-9. Relationship between marbling, maturity, and carcass-quality grade.

preference for lean meat and the cattle feeders' practice of feeding animals for early marketing. Overfeeding or overfinishing animals is inefficient and not practiced in today's feedlots. Examples of the more popular carcass grades of choice, good, and standard are shown in Figures 6-10, 6-11, and 6-12.

U.S. FEEDER CATTLE GRADES

In 1979, feeder cattle grades were revised to reflect the expected weight of the animal when it grades low choice, yield grade 3. Thus feeder cattle grades reflect the feeding value of the animal. Feeder cattle grades are applied to cattle less than 36 months of age and are based on three general characteristics: frame size, muscle thickness, and thriftiness.

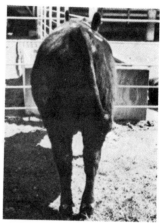

FIGURE 6-10.   Examples of choice carcass grades.

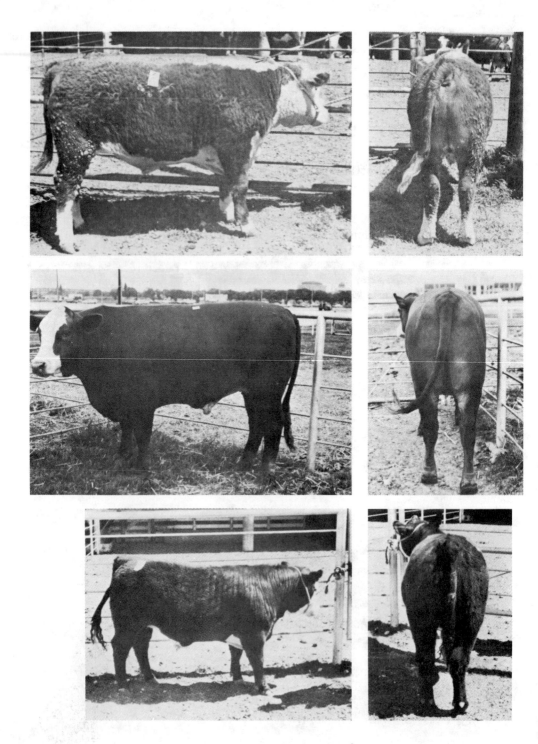

FIGURE 6-10. (Continued)

Chap. 6   THE BEEF CARCASS

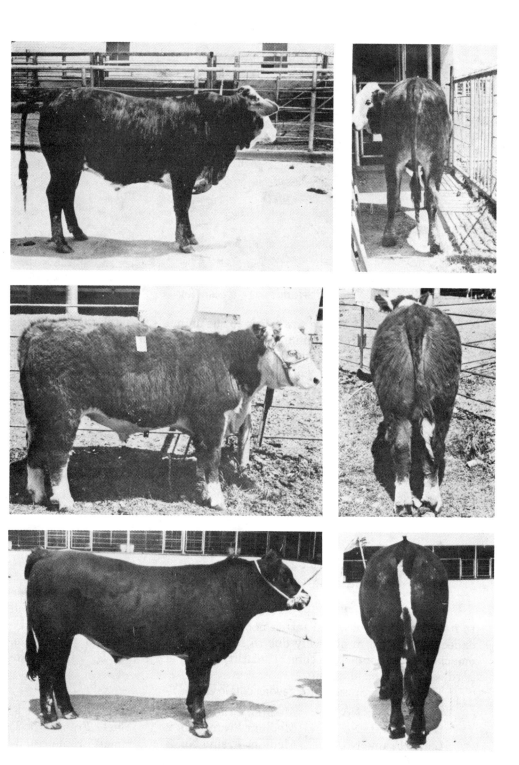

FIGURE 6-11. Examples of good carcass grades.

FIGURE 6-11. (Continued)

Frame Size.   Frame size refers to the animal's skeletal size in relation to its age. Smaller-framed animals will finish out faster than a large-framed animal. There are three frame sizes. Large-frame (L) animals are tall and long bodied for their age. They will be expected to produce a U.S. Choice carcass (0.5 in. fat at the twelfth rib) at 1000 lb for heifers and 1200 lb for steers. Medium-frame (M) animals have slightly larger frames for their age and would be expected to produce a U.S. Choice carcass at 850 to 1000 lb for heifers and 1000 to 1200 lb for steers. Small-frame (S) animals would expect to produce a U.S. Choice carcass at under 850 lb for heifers and under 1000 lb for steers.

Muscle Thickness.   Muscle thickness refers to the relationship between muscle development and skeletal size. Three standards for thickness are designated:

No. 1: thick muscling
No. 2: moderate muscling
No. 3: slight muscling

Thriftiness.   Thriftiness refers to the apparent health of an animal and to its ability to grown and fatten normally. Unthrifty animals would not be expected to perform normally due to such conditions as disease, parasitism, emaciation, or double-muscling. Unthrifty cattle are graded U.S. Inferior regardless of size or muscling.

The resulting grades for feeder cattle are:

Large Frame No. 1	Medium Frame No. 1	Small Frame No. 1
Large Frame No. 2	Medium Frame No. 2	Small Frame No. 2
Large Frame No. 3	Medium Frame No. 3	Small Frame No. 3

FIGURES 6-13, 6-14, and 6-15 are shown as examples of feeder cattle grades.

Chap. 6   THE BEEF CARCASS

FIGURE 6-12. Examples of standard carcass grades.

FIGURE 6-13. Feeder grade M1⁺ (medium frame, very thick muscling).

FIGURE 6-14. Feeder grade M2⁺ (medium frame, moderate muscling).

U.S. Yield Grades

Beef of the same quality grades varies widely in fatness, which directly affects the yield or cutability of beef carcasses for the retail markets. In the 1960s, retailers stressed the importance of controlling cutability or yield in carcasses as well as quality. The USDA in 1965 provided yield grades to identify the quantity or amount of salable meat from beef carcasses. There are five USDA yield grades numbering 1 through 5 with yield grade 1 having the highest yield of retail cuts and yield grade 5 the lowest. Carcasses in each yield grade are expected to yield about 4.6% more retail cuts than carcasses in the lower yield grades. The expected yields of boneless retail cuts for carcasses of yield

Chap. 6   THE BEEF CARCASS

FIGURE 6-15. Feeder grade S1⁻ (small, thick muscling).

grades 1, 2, 3, 4, and 5, respectively, are 52.6% or above, 52.6 to 50.3%, 50.3 to 48.0%, 48.0 to 45.7%, and 45.7 to 43.4%.

Yield grades, expressed as a whole number or as a percent, are determined by an equation based upon four factors: hot carcass weight, ribeye area at the twelfth rib, fat thickness at the twelfth rib, and estimated percent kidney, pelvic, and heart fat. The equation for determining yield grade is as follows: Percent closely trimmed, boneless retail cuts from the round, loin, rib, and chuck = 51.34 − 57.84 (inches fat thickness over ribeye muscle)—0.462 (percent kidney, pelvic, and heart fat) + 0.740 (ribeye area, square inches)—0.0093 (pounds hot carcass weight). Figure 6-16 gives examples of yield grades.

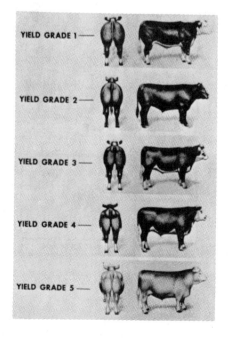

YIELD GRADE 1

YIELD GRADE 2

YIELD GRADE 3

YIELD GRADE 4

YIELD GRADE 5

FIGURE 6-16. USDA yield grades for slaughter cattle. (Courtesy of the U.S. Department of Agriculture)

WHOLESALE AND RETAIL CUTS

The beef halves after inspection and grading are usually divided into fore and hind quarters between the twelfth and thirteenth ribs and further divided into major wholesale and retail cuts as illustrated in Figure 6-17. Most beef carcasses are distributed from packinghouses to markets or restaurants in the form of halves, quarters, or wholesale cuts depending on customer demands.

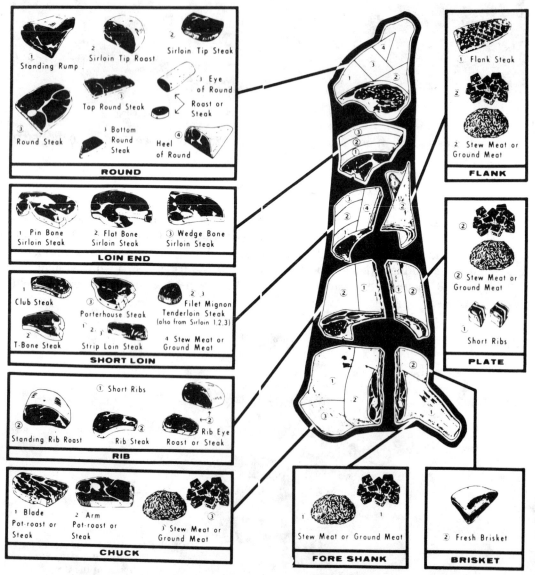

FIGURE 6-17. Major wholesale and retail cuts of beef. (Courtesy of the U.S. Department of Agriculture)

Wholesale cuts are then broken down into the familiar retail cuts found along meat counters.

The very low quality grades such as cutter and canner are usually not sold in halves or cuts. The meat is boned out and sold as boneless cuts or used in a variety of prepared meats such as sausage, weiners, canned meat products, and so on. This meat is completely acceptable, edible, and nutritious. The reason for boning out is simply to make a better appearance in comparison to higher grades.

BEEF BY-PRODUCTS

In slaughtering and marketing beef, only about 60% of the steer is ultimately converted to a beef carcass and less than that converted to retail cuts. Thus, only 400 to 500 lb of retail cuts could be expected from a 1000-lb live steer. Utilization of the entire animal is the goal of every packer. More than 100 by-products from beef slaughter are processed and marketed in products ranging from hides to glue, medicines to candles, soaps to brushes, feeds to fertilizers, cellophane to chewing gum.

The hide is the most important by-product of cattle slaughter. Shoes, belts, purses, furniture, clothing, athletic equipment, even musical instruments such as drumheads, can be made from the hide. The list of uses is limited only by human imagination.

Hides are soaked in a salty brine to cure for a minimum of 24 hours before tanning. Most tanners are specialists and buy hides from the slaughter facilities. The heavy hides are soaked for several weeks in tanning solutions made from wood bark to give maximum pliability; chromic salts are used for lighter, less particular hides.

Fats are rendered to be used as edible products for humans, such as oleomargarine, or in animal feeds to raise the energy content. Tallow is a product of fat that can be used in candlemaking. Fats are also used in chemicals, plastics, lubricating oils, detergents, antifreeze, paints, cellophane, and a host of other items. The by-product gelatin is used in candy, ice cream, pharmaceuticals, and photographic materials and processes.

Blood is dried and used as an animal feed (blood meal). Meat scraps and bone scraps are mixed and ground and provide another livestock supplemental feed. Extracts from hide and bone, mainly a type of connective tissue called collagen, are used to make glue and other adhesives.

Animals or animal products that are condemned as inedible are also treated to make products safe for use as fertilizers. Hair scraped from hides is used in brushes, felt padding, rug pads, upholstering material, insulating material, and so on. Horns and hooves are used in plant foods and fertilizers, combs, buttons, ornamentals, dice, even teething rings. Pulverized and burned, they provide a product used in refining sugar.

Finally, although this discussion is nowhere near complete in mentioning all by-product divisions, over 35 medicinal and pharmaceutical drugs are purified from the organs and glands taken from livestock. One example, insulin, taken from the pancreas, is commonly known to the public as a treatment for diabetes.

STUDY QUESTIONS

1. Veterinarians perform a _____ and _____ examination at the large slaughterhouses.
2. An animal is stunned rather than killed to keep the _____ functioning.
3. Good drainage of blood is important for _____ of the meat.
4. Removal of intestines is called _____ .
5. The left half of the carcass is always the _____ side.
6. Cooling temperatures are _____ .
7. Enzymes break down _____ through aging.
8. Stunning is not permitted in _____ slaughter.
9. The intermingled fat in muscle tissue is called _____ .
10. The _____ is the muscle exposed by cutting between the twelfth and thirteenth ribs.
11. The lowest grade is _____ .
12. Yield grades range from _____ through _____ .
13. The short plate comes from the _____ quarter.
14. T-bones come from the _____ .
15. Round steak and sirloin come from the _____ .
16. A 1000-lb steer will produce _____ lb of retail cuts.

DISCUSSION QUESTIONS

1. Why is beef aged, how is it done, and what kind of carcass is selected for this purpose?
2. Explain the basic divisions for current grades and their relationship to wholesale and retail cuts.
3. Name 25 nonedible by-products of cattle.
4. Discuss at least three ways in which finish (fat) influences grade.
5. How many religious or ethnic requirements influence form of slaughter and choice of cuts?

Chapter Seven

SYSTEMS
OF PRODUCTION

The options available to the cattle breeder for systems of production are limited only by imagination. Some of the more common systems follow.

COW-CALF PROGRAM

The oldest and most established form of animal husbandry in the United States might well be the cow-calf program. The cow-calf business is normally separated into three basic categories.

Registered Program

The purebred cow-calf program started in many parts of Europe and spread to the New World. This system is based on selection of a breed to fit the environment. It is usually on small acreage; the market must be in a condition to absorb or demand the type of purebred animals that are being produced; and quite often the availability of feed is a primary factor in determining establishment and success. The purebred breeder must also make the decision whether to use natural means of breeding or the recent developments allowed in purebred associations such as artificial insemination. Perhaps the most rewarding benefit from a purebred operation is the production of a generally superior animal for which the breeder receives a premium price for the effort involved. This effort does not come easy. The registration procedures, the paperwork involved in keeping breeding records, and the certification of animals require a great deal of time (Figure 7-1).

Generally, the purebred breeder interested in showing must maintain cattle to look their best at all times of the year. Therefore, there is more expense involved in maintaining a herd for show purposes. The more successful breeders also find it necessary to fit and show animals on a show circuit, increasing even further the expensive business of maintaining pure bloodlines. If a breeder must maximize income on small acreage and if he or she realizes the time, efforts, and management problems incurred, then the purebred operation may be worthy of consideration.

F-1 Cattle

F-1 refers to a crossing of two pure breeds of cattle (Figure 7-2). Because cattle are basically divided into the *Bos taurus* and *Bos indicus* species, the first attempt at producing true crossbred cattle was to cross Brahman cattle with European breeds. The first cross from this type of breeding program was referred to as the F-1 generation, meaning the first generation from a cross. More recent developments led to the F-1 Heifer Registry Association that has

FIGURE 7-1. Registration procedures and paperwork involved in breeding and certification of purebred animals require a great deal of time. (Courtesy of the American–International Charolais Association)

Chap. 7   SYSTEMS OF PRODUCTION

FIGURE 7-2. A typical F-1 resulting from crossing a Brahman and a Hereford. (Courtesy of the American Brahman Breeders Association)

established the general rule that both parents must be from registered stock before the heifer can be listed in the association. Production of quality animals with excellent hybrid vigor results because of the crossing of two species.

Hybrid vigor or *heterosis* refers to an added performance advantage of the offspring in comparison to either parent in crossing races, breeds, or species. The greatest vigor normally results from crossing of species. However, it might be pointed out some F-1 cattle are only from different breeds. The advantages in the use of F-1 cattle in a cow–calf operation are improved mothering ability, production of milk, weaning weights, disease and insect resistance in both the dam and the offspring, longevity in the dam, and reproduction rate.

The disadvantage to this program is that a premium price is generally paid for what amounts to a crossbred heifer, and the offspring are sold at market prices. Also there is the problem of hybrid cattle (F-2, F-3 generations) kept as replacements from the herd reverting back to the normal, thus eventually losing the hybrid vigor established by the original cross. This has led to a production technique known as *termination cross* in which no offspring is kept and replacement heifers, perhaps dairy-beef crosses, are commonly purchased from F-1 breeders. Superior sires of a third breed are used to complete the program (Figure 7-3).

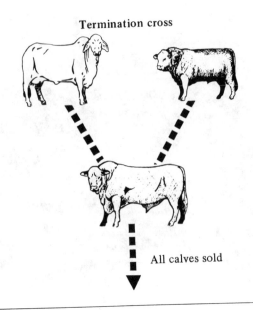

Termination cross

All calves sold

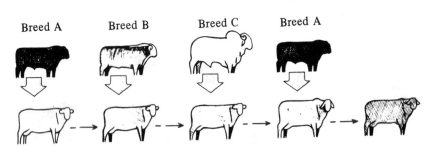

3-breed rotational cross

Breed A   Breed B   Breed C   Breed A

FIGURE 7-3. Examples of termination (top) and rotation (bottom) crossbreeding programs that keep hybrid vigor high for commercial beef breeders.

Another more recent system of production is the *rotational cross*—breeding crossbred heifers to a bull of a breed is not used in the original cross. Heifers are continually retained from the existing herd, rotating a different breed of bull into the herd to maintain hybrid vigor throughout. This system, illustrated in Figure 7-3, maintains approximately 67% of the possible heterosis or hybrid vigor.

Commercial Cow–Calf Program

The commercial cattle herd is an old established method of handling grade cattle of some dubious breed such as "whiteface cattle," "black cattle," or some color pattern that designates major bloodlines but not necessarily a pedigreed animal. The commercial cattle breeder usually takes advantage of regis-

FIGURE 7-4. The commercial cow–calf program varies from confinement to the open range as shown here. (Courtesy of Devonacres, Eagle Point, Oregon)

tered bulls, either to maintain the same bloodlines or to crossbreed. It should be mentioned at this point that crossbred bulls used on purebred cattle do not generally impart the same qualities of hybrid vigor in the offspring that can be expected through the use of crossbred cows and purebred bulls. Because of hybrid vigor, a crossbred bull invariably looks better than it breeds. That is, the inherited vigor exhibits itself in rapid growth, muscular development, feed efficiency, and other desirable measurements, but the bull may not be able to pass on these characteristics. In the same way that seed saved from hybrid corn reverts back to original parent material, the hybrid bull offers this possibility. Therefore, unless a definite system is planned to keep a heterogeneous mixture of blood in offspring, the crossbred bull is of much less importance than the crossbred cow.

The commercial cattle breeder can use a system of confinement on improved pasture (controlling diseases and other problems that come from concentrations in small quarters) or the open range method. Stocking rates vary from 250 square feet per cow in dry lot systems—including 1 to 3 acres per cow on well-managed, improved pastures—to 640 acres or more per cow in open range (Figure 7-4).

DUAL- OR TRIPLE-PURPOSE CATTLE

Some cattle in the United States are classified as *dual-purpose* (meat and milk) or *triple-purpose* (meat, milk, draft). This is primarily a European method of use that has been perfected by such countries as Switzerland, Germany, Italy, and France in order to cope with the problem of heavy human population and maximum production of animal proteins. European countries having little available pasture utilize it for maximum production of meat and milk. For example, the Brown Swiss or European Holstein was developed as a dairy animal, yet retained a great deal of beef type through selection. Therefore, the average European cow produces milk plus an acceptable beef carcass. The problem is partly resolved because Europeans simply eat less beef than Americans. In the United States, we do not have as many dual-purpose cattle, although for some years such breeds as Red Poll and Milking Shorthorn and more recently Devon, Simmental, and Chianina have made inroads. Because of the market demand for beef, this country is also beginning to use dairy animals in a crossbreeding program to produce both milk and beef of a higher quality.

BABY BEEF PRODUCTION

This is a system in which calves are creep-fed (Figure 7-5) up to weaning time and pushed for rapid gains on high-quality feeds so that maximum production (1000 lb live weight or more) is reached in a minimum length of time. The reason for the name is simply that the animal is pushed so that it is never allowed to lose the baby fat deposited in the early growth cycle.

VEAL PRODUCTION

The production of veal refers to a very young calf almost always of a recognized dairy breed that is generally less than 3 months of age (Figure 7-6). These young calves are traditionally bulls that have been culled from the dairy herd because of a lack of need for sires, which is in turn because of the wide use of artifical insemination in dairy cattle.

FIGURE 7-5.   Baby beef production refers to calves creep-fed up to weaning time and pushed for rapid gains in a minimum length of time.

Chap. 7   SYSTEMS OF PRODUCTION

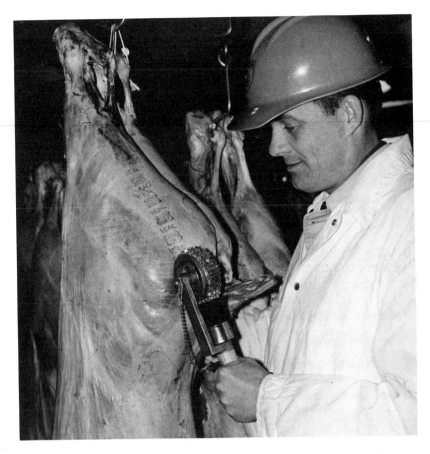

FIGURE 7-6. Veal refers to a very young calf, usually less than 3 months of age, that has been fed only milk. (Courtesy of the U.S. Department of Agriculture)

PRODUCING FAT CATTLE

Three basic ideas generally come to mind in the production of high-quality American-type beef: fattening cattle immediately, roughing them through the winter followed by a full feed, and utilizing a winter pasture followed by full feed. Any one of these techniques may be employed.

Immediate Fattening

The feedlot business is based on finishing cattle quickly (Figure 7-7). Utilizing a full feed program with high concentrate rations, thin calves weighing 350 to 750 lb can be finished in 180 days or less. The feed conversion under this sys-

FIGURE 7-7. Feedlots handle thousands of cattle for fattening. (Courtesy of Monfort of Colorado, Inc.)

tem can be expected to be about 7 lb of feed per pound of gain with an average daily gain of 2 lb or better producing a desirable carcass. It should be kept in mind that younger calves are more efficient but somewhat slower gainers than older cattle. Thin calves are invariably sought by some feeders, especially if they are partially mature because of rapid weight gains from sufficient nutrition.

Another feedlot method is to use fleshy calves that have been on creep feed and are conditioned to eating. These calves continue to gain at a steady rate once full feed is reinstated. This better-quality calf may range in weight from 500 to 800 lb and be finished in 170 days or less. The average daily gain and feed required should be quite similar to the method using thin calves. The shorter feeding period may offer a distinct advantage when the goal is for a specific marketing time.

Older cattle, generally termed *medium-grade yearlings,* are quite often selected when a ration containing maximum roughage is necessary. Some cattle feeders at certain times of the year have coarse roughages that must be or should be utilized and because older cattle have a more fully functioning rumen, they are able to convert poor quality roughage to beef. Research indicates that roughage fed the first half of the period followed by full feed concentrate gives better gains and efficiency than a 50–50 mixture. Mature yearlings weighing only 700 lb often gain up to 2 lb per day on 10 to 12 lb of these coarse feeds per pound of gain.

Conclusions on finishing cattle immediately indicate that large amounts of concentrated feeds are needed with small amounts of roughage and considerable skill. Weanling calves are good gainers because they are growing as well as fattening. Thus grains are economically marketed through cattle (Figure 7-8). Some roughages produced on farms and ranches that might otherwise go unused can be marketed through the use of older cattle with more highly developed digestive systems.

FIGURE 7-8. Smaller farm feedlots are used extensively in the Corn Belt, utilizing the "home-grown" grain crop. (Courtesy of the U.S. Department of Agriculture)

Roughing Cattle through the Winter — Full Feed

This system might be summarized as follows: Buy calves weighing about 400 lb in October or November; winter to gain 1 lb per day or less because cattle gaining these moderate amounts will gain faster once they are put on high concentrate rations; finish out the cattle on spring pasture (Figure 7-9), feeding grain and roughage when necessary or full feeding in dry lot.

The conclusions to be noted here are that costs are reduced by the use of dry roughage during the winter. Moderate gains through the winter are offset by greater gains on the spring pasture in the finishing period. The utilization of a pasture feeding or drylot finishing system depends on individual conditions and feeds available.

Winter Pasture — Full Feed

The use of pastures that will grow in cooler seasons of the year such as wheat, rye grass, barley, oats, and mixtures of these with clovers can be an effective means of producing cheap gains with little or no supplemental feeding. Cattle are purchased in the early fall, grazed through the winter, and may go directly into the feedlot for full feed rations. Or they may be grazed through summer

FIGURE 7-9. Cattle finished out making maximum use of existing pastures. (Courtesy of the U.S. Department of Agriculture)

pastures for additional cheap grass gains and enter the feedlot the following fall. The major advantage to this type of system is maximum utilization of roughages, which are generally much cheaper than concentrate forms of energy.

STOCKER SYSTEMS

Stocker cattle are heifers used to replenish the cow herd or cattle that remain on the farm or ranch for additional gain before entering the normal American system of fattening cattle. A 250- to 600-lb stocker can be handled by roughing through the winter, grazing on winter pasture, grazing on summer pasture, or a combination of these methods. The main objective here is to produce either replacement cattle for the herd or to produce a mature animal for feedlot, resale, or slaughter that has been grown solely on roughage feeds. This is the most common world-wide system of producing beef. Although it is unusual in the United States to actively produce grass-fed cattle, it is the only method in most parts of the world. Grains are more efficiently utilized by humans, but roughages that cannot be utilized by the human population are converted through ruminants into a desirable, edible product—beef.

GROWTH-PROMOTING SUBSTANCES

Because of residues found in the organs of slaughtered animals and carcinogenic (cancer-producing) accusations, DES is no longer used as a growth-promoting substance. It is only used by veterinarians as a drug. For instance, accidentally bred young heifers or a prize cow bred to a scrub bull can be made to abort very soon after conception by DES injections.

A few growth-promoting substances may survive on the market to be implanted or fed. Special preparations are available for heifer and/or steer use. Specific product names will not be discussed because they vary and are subject to replacement or withdrawal.

One natural growth promoter that will likely remain without accusation or harmful effect is the *UGF (unidentified growth factor)* found in some feeds, notably alfalfa. This is one reason that alfalfa is so widely used in feed formulas.

In conclusion, systems for handling cattle vary with the size of property, availability of feed, labor, and financing, and the environmental situation. The successful cattle breeder appears to be the one with knowledge of many different types of production systems and imagination enough to incorporate the desirable factors of many into a program that fits the breeder's needs.

STUDY QUESTIONS

1. The purebred cow–calf operator's expense for _____ is generally higher than the commercial breeder's.
2. The _____ procedures distinguish purebred breeders from commercial cattle breeders.
3. The greatest response to hybrid vigor is found in crossing _____ .
4. The Brahman breed is widely used as the representative of the _____ species.
5. The first generation from a crossbreeding program is called _____ .
6. The F-1 Registry Association requires that both parents be _____ of different breeds.
7. F-1 cows make better _____ .
8. Increased _____ , _____ , _____ , and _____ are four advantages of a F-1 cow.
9. Replacements _____ (should, shouldn't) be kept from F-1 cows.
10. The disadvantage of F-1 heifer programs is that the heifer is bought at _____ prices and calves are sold at _____ prices.
11. A termination cross means that no _____ are kept.
12. The use of three or more breeds in a hybrid program is a _____ system.
13. Beef cattle breeders are tolerating more _____ breeds in hybrid systems.
14. A _____ bull is not recommended for any breeding program.

15. Triple-purpose cattle are used for _____ and also produce lots of _____ and yet retain _____ .
16. A baby beef is a weaning calf. (True, False) _____ .
17. Most vealers are bulls. (True, False) _____ .
18. The feed required to produce 300 lb of gain on light, thin calves is about _____ lb.
19. Calves in feedlots should gain _____ lb or better in 180 days.
20. Younger calves are _____ (more, less) efficient than older cattle.
21. Older cattle can convert poor quality _____ because of a more fully developed _____ .
22. Weanling calves gain so well because they are _____ as well as fattening.
23. Cattle roughed through the winter should gain about _____ pound(s) prior to the feedlot.
24. Winter grass mixed with _____ makes good winter pasture.
25. Cheapest gains come from _____ .
26. The most common system of fattening cattle around the world is the feedlot system. (True, False) _____ .
27. Cattle are important worldwide because they can convert _____ that humans cannot.
28. Estrogenic feed compounds cause the production of more _____ and less _____ .
29. Hormones may improve normal gains of 2 lb per day to _____ .
30. One reason alfalfa is widely included in feed programs is because it contains _____ .
31. DES can cause _____ in pregnant females.

DISCUSSION QUESTIONS

1. Describe in detail the breeds or type of cattle and the system you would use to produce and maintain maximum heterosis.
2. Explain and give examples of single-purpose, dual-purpose, and triple-purpose cattle.
3. Explain the diference between prime beef, baby beef, and veal.
4. Develop and be prepared to defend a plan to buy stocker heifers for development and addition to the herd as replacements. Give age, weights, type of feed, length of time to develop, and so on.
5. What is the likely future of growth-promoting substances—synthetic and natural?

Chapter Eight

USE OF EQUIPMENT

Every occupation has particular "tools of the trade"; yet few have such a variety of historic tools as animal science. In the early days of the Old West, the cowboy's saddle was seat, pillow, luggage, easy chair—a very prized possession. Knowledge of the various tools available to the modern day animal scientist makes even the most complicated tasks of the modern cowboy relatively easy even though the saddle is still around. For example, branding was a detailed task of rounding up, sorting, running down, roping, restraining, and finally branding a wriggling, unwilling steer. Today, it is simply a matter of sorting cattle in well-designed pens and moving the desired steers through a covered corral to squeeze chutes where one or two people can not only brand, but vaccinate, drench, dehorn, and check for other problems in a matter of minutes. Although horses and cowboys are still around, one is more likely to see mechanized vehicles and tools handling the bulk of the work.

SHELTER FOR CATTLE

Beef cattle on pastures with some natural windbreaks may not need buildings for shelter. However, in extreme regions of the nation, cattle need protection

FIGURE 8-1. Loose, open shelters such as this provide enough protection for range cattle in most areas.

from wind, rain, cold, and heat. Extremes in temperatures can cause diseases and a general inability to function properly. Often a cold, rainy condition is the most dangerous, requiring some shelter to block the wind and rain. Loose, open shelters (Figure 8-1) are adequate for range cattle in most areas.

Cattle in the feedlot are often under more stress than range cattle and thus require more attention for shelter needs. Shaded areas of various kinds have been used in feedlots. Open sheds should allow 30 to 50 ft^2 for each animal unit. Housing should be simply constructed, easy to clean, and reasonable in price.

FENCES FOR PASTURES AND CORRALS

Fences throughout the world range in construction from water ditches in Holland to thickly planted hedges or vines in France to split rails and stones used in the United States before the advent of barbed wire. All fences serve the same function of restricting cattle to a specific area. Modern fences for range use consist of either wood or steel posts with usually two to four strands of barbed wire. Post spacing varies according to cattle concentration with 10 to 20 ft in concentrated grazing areas or 50 to 120 ft suspension-type fences in sparsely stocked areas. Figure 8-2 illustrates some of the types of fences. Woven wire is often used instead of barbed wire in areas of concentrated grazing or where sheep also utilize pastures.

Corral fences need to be stronger and more durable than range fences. Materials used for construction of corral fences are wood, metal cables, pipe or sucker rods, or heavy woven wire. Fences are usually 5 ft high or above with post spacing from 4 to 10 ft. Materials used will vary with availability and cost.

Chap. 8   USE OF EQUIPMENT

FIGURE 8-2. Pasture fences vary depending on cattle concentration, materials used, and personal preference.

One of the most creative forms of fencing utilizes that Osage corner post (Figure 8-3). First reported in areas of Oklahoma where large, flat, heavy rock dots the pastureland, this type of corner post is reported to be immovable.

Simplicity is the key to the Osage corner post. A large wire ring of hog netting, several feet across, is held in place by one or two metal fence posts driven into the ground to give stability to the ring in the initial stages of filing. Large rocks are then placed inside the ring without mortar and with only minimal effort in arrangement. When completed, wire strands are attached by completely encircling the stack of rocks. With so much weight and load-bearing surfaces, the pile of rocks takes on the characteristics and resistance to being moved of a boulder several times its size.

FIGURE 8-3.   Osage corner posts are quite effective in areas of the world where rock is readily available.

High-Tensile Wire Fencing

A relatively new innovation in fencing for all classes of livestock is the high-tensile wire fence. In 1973, a New Zealand sheep rancher, John R. Wall, wrote to U.S. Steel in the United States for a few thousand feet of type III galvanized wire for an experimental fence replacing barbed wire. The next year, a huge tree fell across the high-tensile wire and pinned to the ground one of the fences. When the tree was removed, the wires sprang back into place and the only repair needed was to replace staples on the posts.

Manufacturers' tests show that high-tensile wire fencing has nearly twice the breaking strength of two-ply barbed wire. Each wire stretched to the recommended 250 lb of tensile will withstand at least 1200 lb of livestock pressure or extremes in temperature without losing its elasticity. When properly grounded to protect from lightning or the possibility of overhead electrical lines falling on it, this type of fence will stop charging cattle, sheep cannot squeeze between the wires, horses cannot weigh it down, and the cost is about the same as a five-strand, two-ply barbed wire fence.

The estimated length of life is 35 to 50 years and requires a different type of installation, corner posts installed to manufacturer's specifications, and specific tools for the fence to perform as it should. Figure 8-4 shows the recom-

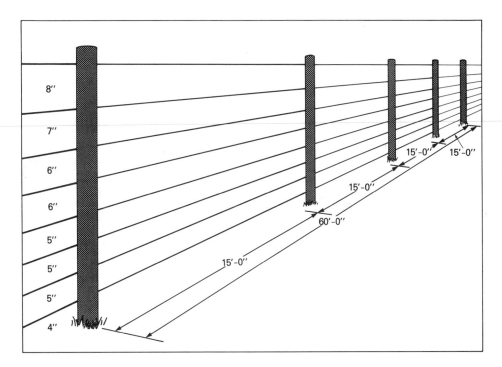

FIGURE 8-4. An eight-wire high-tensile fence with line posts 60 ft apart and drop-pers every 15 ft has replaced barbed wire in many parts of New Zealand and Australia.

mended spacing of line posts and setting of wires. For a small fee a booklet of information is available from England Distributing, Route 3, Bloomfield, Iowa 52537.

PENS AND CORRALS

The arrangement of pens and their size are important in feedlots and confined beef operations. Cattle pens should be easily accessible by livestock and vehicles, afford ease in feeding, cleaning, and caring for the animals, provide comfort and adequate room for all animals, and allow for future expansion. Figure 8-5 illustrates a well-designed pen layout for a modern feedlot. Pens should be well drained and even cemented if possible in wet areas with heavy rainfalls.

Pens usually should allow 40 to 50 ft^2 of space for each mature animal. The number of animals per pen varies from 50 to 200, depending on the management plan of the operation. All pens should have adequate feed and water facilities for the cattle. Feed bunks should be 30 to 36 in. wide and about 20 in. from the ground. Each animal should be allowed 20 to 24 in. at the feed

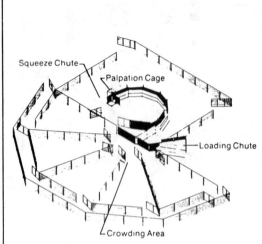

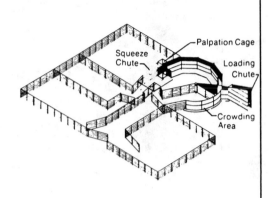

70-400 head or more.

Circular crowding pen and curved working chute. Expand from 1 to 4 pie-shaped holding pens, which require more fence than rectangular ones. Careful construction is needed.

USDA 6229. Expansible corral.

70-400 head or more.

Circular crowding pen and working chute. Good sorting and loading arrangement. This layout can be a hospital area, receiving lot, or combination. Consider roofing over the working area.

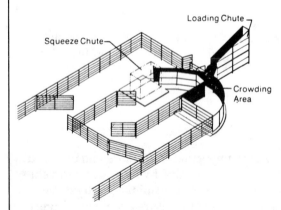

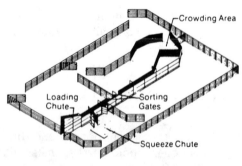

25-75 head.

Circular crowding pen and working chute. Good layout for loading and sorting. The plan is adapted to only limited expansion and has no ideal scale location.

USDA 6230. Corrals with working facilities.

Up to 200 head.

Circular crowding pen. Straight working chute to loading chute or squeeze. Holding pens next to the crowding area. There is poor crowding for loading and restricted expansion.

FIGURE 8-5. Plans like these are available through the U.S. Department of Agriculture.

FIGURE 8-6. Various types of watering systems can be used for beef cattle. The water trough at top supplies water to three different pastures. (Courtesy of the U.S. Department of Agriculture)

bunk. A clean water source is a must. Water facilities vary in size, shape, and design with concrete and metal troughs the most commonly used. Drinking fountains have been used successfully to supply a fresh supply of water at all times. Water troughs should be heated in cold regions to prevent freezing and shaded in hot regions to keep water cool. Troughs should be cleaned regularly for sanitation and disease prevention. Figure 8-6 illustrates the types of watering systems from which a breeder may choose.

FIGURE 8-7. An example of a well-designed corral and working area.

The layout of working corrals is one of the most important points in feed-lot or beef cattle equipment design. Corrals are highly specialized facilities built to assist in the handling, sorting, and restraining of both cows and calves. Good corrals consist of a crowding area or holding area that leads into a working chute. The chute allows for sorting cows and calves into desired groups, into separate pens, or into other chutes. Sorted cattle can then be moved through scales, a restraining chute, loading chute, or holding pens for future care. Figure 8-7 is only one example of a well-designed corral layout. Note both the rounded design of the crowding pen to avoid cornering and in-juring cattle and the working facilities (squeeze chute and scale that are cov-ered and well protected).

RESTRAINING EQUIPMENT

Most procedures such as dehorning, vaccination, castration, weighing, identi-fying, and treating for sickness or disease require restraining of individual animals. Early restraining was accomplished by use of ropes, like roping the animal in a pen using two ropes pulling in opposite ways. The popular rodeo event of steer roping, restraining a steer by roping the head and hind legs, began in this way. Another method was to throw the animal by a rope as shown in Figure 8-8. The principle is to make a loop over the neck, a half hitch around the heart girth and another half hitch around the flank-loin area. An-other rope or halter holds the animal to a post or fence. By pulling straight

182   Chap. 8   USE OF EQUIPMENT

FIGURE 8-8.   Even very large animals can be thrown and restrained by a rope.

back on the rope, the half hitch is tightened, resulting in a paralysis of the hind legs due to pressure on the nerves. A very large animal can be thrown to the ground in this way with surprising efficiency and can be worked on with ease.

Today's facilities for restraining can be a chute with a blocking gate, a chute with head clamp, a squeeze chute, or the modern chute that performs the functions of head clamp, squeeze chute, tilting table, and scale all in one. Hydraulically controlled levers are operated by one person. All chutes should be concreted for heavy use, be just wide enough to allow one animal to move through at a time, and be designed to keep slippage at a minimum. Figure 8-9 illustrates some of the different types of restraining devices used today.

EQUIPMENT FOR SPECIFIC FUNCTIONS

Castration

Castration by a sharp knife has long been practiced by cattle breeders. In this method, the scrotum is washed with an antiseptic solution and the bottom third of the scrotum is removed by the knife. The testicles are pushed out of the scrotum, and the cords are either severed or crushed. The wound is dusted with a powder to prevent infection and aid in healing. This system allows drainage and fairly fast healing. An alternate method is to slit the scrotum vertically on both sides and remove the testicles through the slits. Care should

FIGURE 8-9. Top—Modern restraining devices. Bottom—Cattle are secured to the tilt-table by a canvas and metal brace. This hydraulic table was custom built at a cost of about $14,000. (Courtesy of *Farm and Ranch Weekly*, Mexia, Texas)

Chap. 8 USE OF EQUIPMENT

be taken to slit the scrotum low enough to allow for drainage of fluids during healing.

Another method of castration is by using either an elastrator or Burdizzo pincer (Figure 8-10). The elastrator stretches a rubber ring that is released around the scrotum right above the testicles. Blood circulation is stopped below the ring, and the testicles and that part of the scrotum actually die and slough off. The Burdizzo method is similar, with the cords crushed by the pincer, cutting off circulation to the testicles only. The lower part of the scrotum is not affected. Cords from each testicle are pushed to the side of the scrotum where they are clamped.

In addition to the use of the knife and Burdizzo, there is now an effective method of chemical castration of bull calves. An injectable castrating agent with the trade name Chem-Cast (Bio-Ceutic Laboratories, Inc., P.O. Box 999, St. Joseph, Missouri 64502) is a patented chemical solution formulated for injection directly into the testicles of young bulls weighing up to 150 lb.

Injection of 1 to $1\frac{1}{2}$ ml through a small, 18- or 20-gauge hypodermic needle leaves no open wound and results in little or no bleeding, eliminating most castration complications. The solution also desensitizes the treated area once it is injected, reducing or eliminating shock or stress from irritation or swelling.

Although the product does create some swelling, research indicates that it is painless in destroying the testicles and spermatic cords, leaving calves totally castrated within 60 to 90 days following injection. The remaining cod will be similar to that found on a surgically castrated calf.

The advantage to a chemical method of castration is that it can be done any time of the year because of the painless, bloodless, simple operation and

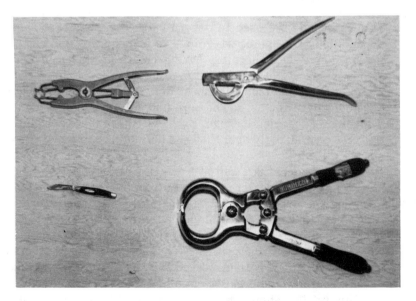

FIGURE 8-10. Castration equipment.

also fits in well with other management programs, such as vaccinating, dehorning, branding, and parasite control, without adding to the stress already created.

Branding and Identification

Any accurate recordkeeping method requires some method of identifying cows and calves. Many options are available for identification including ear tags, ear tattoos, photographs of color markings, and the most popular—branding. Branding allows for identification of ownership as well as a permanent number for individual identification. Two methods of branding are now used, the traditional hot iron brand and the more recent freeze brand. In *hot iron branding,* the iron should be hot but not quite red hot and be held firmly on the animal for 5 seconds. Electric brands are available and can be used effectively when an electrical source is available. *Freeze branding* uses copper irons that are kept in dry ice or liquid nitrogen. The area to be branded is closely clipped and washed with alcohol before applying the brand. The brand is held firmly on the animal for about 30 seconds. White hair grows on the branded area about 3 months after branding. Although it can be used on white breeds because of a different shade of white that grows in the brand, this method is especially suited for dark-colored cattle because of the ease of reading the white brand. Figure 8-11 illustrates just some of the equipment that is used for animal identification.

Dehorning

Many methods are used to dehorn cattle. The one that should be used will depend upon the age of the animal to be dehorned and the experience of the person doing the dehorning. Young calves are often dehorned by use of a strong chemical paste (potassium or sodium hydroxide). The paste is placed around the horn button during the first week of age, destroying the growth or development of the horn. The horn button can also be taken off by the use of a spoon or a scoop and a tube. Both instruments are designed to cut the horn button out at an early age. If all the horn button is removed, a clean dehorned animal will result.

Electric dehorners can also be used at an early age. An electric heated steel ring is pressed over the horn button, burning the surrounding tissue and arresting the horn growth. The experienced administrator finds this method easy to use and efficient, while many inexperienced operators wind up with a partly killed horn button that later grows into a deformed horn, called a *scur.*

Older animals must either have their horns sawed off or cut off by use of a Barnes dehorner. Either method may result in bleeding and should be followed with the usual practice of pulling the arteries and veins to reduce blood loss. Figure 8-12 shows the various equipment that can be used in dehorning cattle.

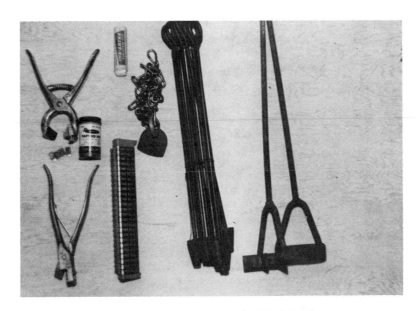

FIGURE 8-11. Equipment used in identification.

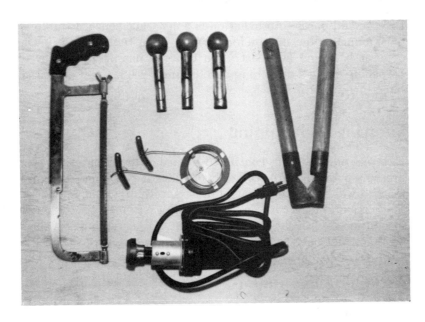

FIGURE 8-12. Dehorning equipment.

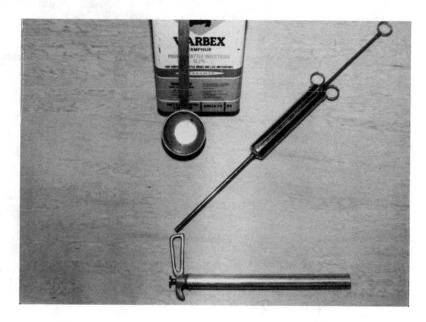

FIGURE 8-13. Drenching guns and systemic pour-on insecticides help in combatting parasites and diseases.

Vaccination and Drenching

Disease and parasite prevention is a constant concern throughout the life of most animals. Vaccination for the more important diseases in the area is easily done at branding and castration time by use of various syringes. Periodic drenching using a drenching gun (Figure 8-13) helps fight parasites and other intestinal disturbances.

Spraying and Dipping

External parasites are controlled by use of insecticides sprayed with a high-pressure pump (Figure 8-14) or by the use of a dipping pit (Figure 8-15). Either method will effectively control most external parasites. A routine procedure for range and feeder cattle of the southwest is the use of a systemic pour-on insecticide to prevent the development, under the hide, of the heel fly larvae (grub, ox warble).

Grooming Equipment

Equipment used to groom cattle from head to tail is important to prevent diseases and infections (foot rot, for example), to control external parasites, and to add beauty to the animal. Various equipment used in grooming is illustrated in Figure 8-16.

Chap. 8   USE OF EQUIPMENT

FIGURE 8-14. High-pressure sprayers can be used to control insects and external parasites.

FIGURE 8-15. The use of dipping vats ensures good coverage of insecticide over the animal's body.

FIGURE 8-16. Equipment used in grooming animals.

STUDY QUESTIONS

1. Open sheds when used should allow _____ square feet for each animal unit.
2. Although post spacings vary in range and pastures, corrals require posts _____ to _____ ft apart.
3. Pens should allow _____ ft² per animal.
4. Arrangement of pens and working _____ is very important in beef equipment design.
5. Most operations in beef management require _____ of animals.
6. Knife, elastrator, and Burdizzo are instruments used in _____ .
7. Branding can be done by hot brands or by _____ branding.
8. The method of _____ depends on the age of the animal, size of its horn, and experience of the handler.
9. Many diseases and internal parasites are controlled by _____ and _____ .
10. _____ parasites can be controlled by spraying or dipping.

DISCUSSION QUESTIONS

1. A feedlot shed measuring 25 ft by 100 ft would be sufficient protection for how many steers? Give a low to high optimum range.

2. How much feed trough space would be needed for the example above?
3. Name and describe the advantages and disadvantages of three types of castration.
4. Discuss dehorning methods for different ages of cattle. When and why should dehorning be considered?
5. Describe every practical method of ownership and individual identification for livestock.

Chapter Nine

DISEASES OF CATTLE

Cattle breeders from every century since we domesticated the cow have noticed a variety of diseases and concocted every treatment imaginable to cure them—astrology, witchcraft, liniments, bleeding, and surgery included. An old remedy in the United States was to determine if the cow was suffering from "hollow horn" or "hollow tail" by sawing off the horns, splitting the tail, and filling the space with salt. Because hollow spaces always exist, the suspicion was always confirmed. If the cow lived, the operation was a success. If not, treatment was started too late. This kind of historic background gave rise to the cow's point of view as expressed in the following definition of disease, namely anything that makes an animal ill at ease.

Ill at Ease

Winter winds bemoan my fate
Feeling bad and I'm losing weight.

My temperature must be one hundred and five
Skin and bones, just barely alive.

They say I have the hollow horn
Because I look so frail.

I'm roped, tied, dehorned
And then they split my tail.

I'm kicked and prodded to get on my feet
My wounds filled with salt, my body filled with heat.

If they'd just do something to ease the pain
I wouldn't mind the hunger, the cold and the rain.

I'd like to live but it would serve them right
If I just lay down and died for spite.

Ruling out the hollow horn, hollow tail, and nutritional (hollow belly) theories, major diseases of cattle contain little mystery, and treatment is now based on the best knowledge available from researchers.

The rectal temperature of normal cattle should fall within the range 100.4 to 102.8 °F. Most, but not all, diseases exhibit animal temperature change above and sometimes below this variation. This temperature measurement is usually one of the first observations made before proceeding to specific diagnosis.

ANAPLASMOSIS

Anaplasmosis is a protozoan disease of cattle produced by a destruction of red blood cells, causing anemia and death. The most common signs of anaplasmosis do not develop very quickly. Body temperature rises slowly, there is poor appetite, and weight loss progresses. The anemic victim shows a slight jaundice or yellowing of the pink membranes. This anemia is one of the main clinical diagnostic characteristics of anaplasmosis. Progression to recovery or death may last anywhere from 2 days to 2 weeks. Only mild discomfort and few deaths occur in calves; yearlings appear sicker but usually recover. The most severely affected are cattle 2 years of age or older, exhibiting 20 to 50% death losses. Abortion is the common result of pregnant cows infected with anaplasmosis. All affected cattle appear hyperexcitable just before they die and many attack attendants. Cattle that recover will experience symptoms of anemia on and off for several months thereafter.

The worst form, acute or peracute, is deadly, characterized by high temperature, anemia, difficult calving, and death within 24 hours. Anemia makes the heart beat so hard that a "jugular pulse" can be seen at the jugular vein. A microscopic organism, *Anaplasma marginale* (Figure 9-1), is the cause of the disorder. These small protozoa attach themselves to red blood cells, causing them to rupture. Oxygen cannot be carried through the blood stream and death occurs through internal suffocation.

Transmission of this disease is brought about by insect vectors or unsanitary needles and dehorning instruments. These instruments should be disinfected between use on each animal, and insects should be controlled through

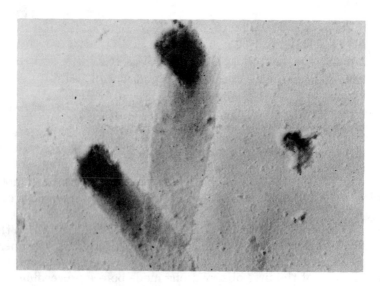

FIGURE 9-1. Anaplasmosis is characterized by anemia caused by a protozoan-like parasite, *anaplasma marginale,* which destroys an animal's red blood cells. The organism is enlarged here 7000 times. (Courtesy of the U.S. Department of Agriculture)

the use of back rubbers and spray to keep down infestation of ticks, flies, and so on.

A curious and distressing situation sometimes arises the year following an outbreak of anaplasmosis. Some immunized cows that were protected against anaplasmosis have healthy calves that die suddenly within days of birth, yet the cow is not affected. The problem (*neonatal isoerythrolysis*) has been defined as a genetic rarity. Anaplasmosis vaccine is made from the red blood cells of donor animals. The problem starts if there is a factor in the genetic makeup of the vaccinated cow that causes her to produce antibodies to those red blood cells in the vaccine. At birth, those antibodies are found in the colostrum milk. If the calf is genetically sensitive to those antibodies when it nurses, the antibodies that are ingested destroy the red blood cells, causing jaundice and death.

The problem almost always appears to affect the most vigorous and healthy calves because they nurse so rapidly after birth and consume the antibodies at the most potent level. If the cow is milked completely of colostrum milk soon after birth and before the calf nurses, the level of antibodies will be sufficiently reduced and will create no reaction.

In vaccinated cattle, a mortality rate as high as 30% might be anticipated. Because the condition involves genetics, changing bulls after an outbreak of anaplasmosis and vaccination has been somewhat effective because of dilution of the genetic makeup that apparently lowers sensitivity.

The condition appears worse in Charolais cattle due to closer inbreeding practices because of fewer animals available as compared to other breeds. However, the situation has occurred in all breeds and in crossbreeds.

For these reasons, field evidence suggests that vaccinating brood cows for anaplasmosis can be a calculated risk. Protective benefits must be weighed against this risk. As always, a qualified veterinarian should be consulted before a final decision is made.

Treatment has been prescribed by veterinarians for anaplasmosis. Broad-spectrum antibiotics such as tetracycline, 3 to 5 mg per pound of body weight daily for 3 days, is often the recommendation. For more valuable animals, blood transfusions may be recommended. Extra excitement aggravates the condition, so cattle should be handled with great care so as to create as little stress as possible. Whole herds treated with injections of tetracycline have recovered from the disease. Tetracycline has also been added to the feed for 30 days, an alternative effective treatment.

ANTHRAX

This is a disease that affects the entire body (septicemia) producing most often sudden death. Cattle appear to drop dead for no apparent reason. The key sign of anthrax is a carcass that bloats very quickly and discharges blood the color of tar from the rectum, nose, and other body openings. Rigor mortis does not set in as quickly as normally expected, blood fails to clot, and the animal assumes a "saw horse on its side" appearance because of the characteristic bloat (Figure 9-2). Extreme care should be taken if anthrax is suspected as the cause of death because it is extremely contagious to both man and animals. Blood samples, taken with extreme caution, can verify the disease by submission to a qualified laboratory for culture and identification under a microscope. If anthrax is confirmed or even suspected, the carcass and any contaminated bedding should be quickly buried at least 6 ft deep and topped with adequate quick lime before complete burial. An alternate disposal method is complete cremation to kill all spores.

The peracute (very fast) form of the disease is most common. Only about 1 to 2 hours elapse from infection to death. Should you happen to be observing the animal at the point of infection and watch through progression (which is very unlikely) you would see muscle tremors, difficult breathing, total collapse, and convulsion ending in death. Discharges at the body openings, previously mentioned, and bloat begin to develop almost immediately.

The acute form, running its course in about 48 hours, is exhibited by obvious depression, a reluctance to move alternating with periods of hyperexcitability, a temperature of about 107 °F, rapid breathing, congested membranes (possibly bleeding), lack of appetite, and abortion by pregnant cows. Milk production declines to almost nothing, and what milk is produced may be blood-tinged or deep yellow in color. Swelling in the throat and tongue is

FIGURE 9-2. Typical appearance of the carcass of an animal that has died of anthrax. Note the bloated condition which occurs soon after death due to rapid decomposition. (Courtesy of the U.S. Department of Agriculture)

common. Death in the peracute form is expected to be 100%, about 90% in the acute form, even with treatment.

Anthrax is often confused with less contagious conditions. Redwater, cattle struck by lightning, several types of poisoning, and acute bloat may be confused and diagnosed by the amateur as anthrax. However, extreme caution should be used in any case because of the possibility of the highly contagious anthrax, which affects both man and animals.

Bacillus anthracis is the microorganism that causes anthrax. Upon exposure to the air the organism forms a spore called anthrax bacillus. It has been known to live in the soil in a viable condition for over 60 years. A preventive program of annual vaccination is good protection against the disease. However, it is recommended only in areas where the disease is a problem.

The most common method of transmission is associated with rough, stemmy feeds that have been grown in areas where anthrax is known to exist. By puncturing membranes of the mouth or digestive system, the organism may be eaten or inhaled, or gain entrance to the body through these injuries. Absorption of the spore through these minor wounds creates onset of the disease.

Treatment is almost without response, although antibiotics such as penicillin and antiserum are prescribed in the very early stages of the disease. Since the very early stages are almost impossible to detect, the only real defense is prevention through vaccination.

Chap. 9   DISEASES OF CATTLE

True blackleg is caused by *Clostridium chauvoei.* It should be mentioned that other blackleg-type diseases are also caused by the Clostridial organism. They are malignant edema and enterotoxemia.

Signs of blackleg are inflamed muscles, severe toxemia (poisoning), and a death rate approaching 100%. Strangely, the youngest, healthiest, fastest-growing animals between 6 months and 2 years are usually first affected (Figure 9-3). The warmer months of spring and autumn account for the highest rate of outbreaks.

The first sign of blackleg is usually a dead calf. High fatalities within 24 hours of onset of signs are common. Since it occurs so quickly, signs are seldom seen; but close observation of younger cattle, once an outbreak is known to have occurred, should reveal some telltale signs of infection during the first 24 hours. These signs are obvious lameness, a swelling of upper parts of the leg, and a slight "dragging" of the hind toe as if the calf were showing signs of exhaustion. Hot, painful swelling is obviously observable. Later the same swelling may be cold and painless. Depressed appetite and temperatures of 105 to 106 °F are characteristic.

The upper part of the leg often produces discoloration and a gassy swelling under the skin. When touched, these swellings have a dried, crackling, tissue paper feeling. Occasionally, lesions are seen at the base of the tongue, heart muscle, diaphragm, brisket, or udder. Blackleg is contagious and may be spread by direct contact from one animal to another. The only reliable means of control is by vaccination. All calves 4 to 6 months of age should be vaccinated. This may be modified somewhat upon the advice of a local veterinarian. The most common vaccine is a blackleg-malignant edema bacterin that

FIGURE 9-3. Young animals (6 months to 4 years old) can be affected with the fatal blackleg disease. (Courtesy of the U.S. Department of Agriculture)

produces a high degree of immunity in 10 to 12 days and lasts for 9 to 12 months or longer. Treatment is usually ineffective once calves are infected but a few cases may respond to tetracycline or penicillin.

BLOAT

Although not considered a true disease, this unhealthy condition causes widespread discomfort and death to ruminants totalling annual losses in excess of $100 million. The sign of bloat is a swelling to abnormal proportions of the left side of the animal. Severe cases cause pressure on the diaphragm and lungs inducing gasping for breath. It is thought that toxic substances produced during this period are inhaled or absorbed through the lungs, causing death.

The exact cause is not clear, but it is well known that legume pastures, alfalfa hay, and high concentrate feeds are most likely to produce bloat. Two types are recognized, a *gaseous* form and a *gas bubble* form. The gaseous type may be relieved by walking, passing a tube down the esophagus, or as a last resort, puncturing the rumen. The bubble type must be burst like a balloon. A *surfactant* (something to prevent or break the tension needed to produce a bubble) is usually quite effective. Vegetable oils and even detergent soap powders have been used in an emergency. However, a prepared surfactant such as poloxalene is recommended. It may be used as a drench or in critical cases injected through the wall into the rumen.

Prevention of bloat is preferred over treatment. Filling cattle on dry hay prior to grazing legume pastures, a heavy stocking rate to prevent rank growth, and use of surfactants are common procedures that have been very effective.

BRUCELLOSIS (BANG'S DISEASE)

Abortion after the fifth month of pregnancy is the key sign of brucellosis. The cow may conceive afterwards but usually has poor or irregular calving records. Occasionally the cow will carry the fetus to full term, but the calf usually dies and the cow has a retained placenta (the bag surrounding the calf) along with metritis (inflammation of the uterus). In the other instances, a cow or heifer may abort once or twice, then calve normally for the rest of her life. Although she is immune to the disease, she remains a "spreader" of the condition.

Bulls are affected by the same disease through the scrotum, which appears swollen and reddish. Known as orchitis, brucellosis infections can result in sterility in bulls. In both cases, the microscopic bacterial organism, *Brucella abortus,* is the cause.

The U.S. government has developed a vaccination program in young heifers to create immunity. *Strain 19* vaccine is given between 2 and 10 months of age for beef replacement heifers (2 to 6 months for dairy) to produce an immunity to brucellosis. This has been a controversial program because the testing program that follows vaccination often mistakenly indicates a reactor. Obviously, cattlemen object to a testing program that needlessly forces animals to be slaughtered that never abort. Supplemental tests are currently under development to differentiate between a reactor and a true carrier. Serum samples must be submitted for a final analysis. Vaccinated heifers receive a tattoo in the ear with the date of the vaccination and an ear tag or a brand for follow-up observation.

Individual herds as well as those passing through regular market channels may also be evaluated through a testing program. Testing involves taking a blood sample and analyzing it by a simple card test or the plate agglutination test. Positive reactors must be branded with a B on the left jaw. All positive reactors must be slaughtered. There is no transmission through the marketing of the dressed carcass or by-products, but the disease is easily spread from one cow to another. All animals known to be in contact with a reactor are quarantined and must be tested at 30-day intervals for two negative tests, at least 120 days after the last reactor is found. The herd may be certified clean if no reactors are found after this quarantine and two negative tests.

Cattle marketed through livestock auctions and other outlets are traced back to the place of origin through this system, and herds may be quarantined until they are tested and certified clean. Government enforcement is difficult and results have been frustrating because of lack of 100% compliance. A few cattlemen may be tempted to ignore these regulations because of the expense and trouble involved, but the program is a must if the disease is ever to be completely eradicated. Replacement heifers definitely are recommended to have had negative tests within the last 30 days prior to entry into the herd.

Brucellosis is spread through contact with unpasteurized milk, causing a condition in man known as *Undulant fever*. It can be spread to animals through direct contact with infected cattle, contaminated grass, ground, or water. A very common method of transmission is through other cows licking the calf of an abortive cow.

No known treatments exist. For this reason, government regulations have tried to eradicate the disease through a search and destroy mission (Figure 9-4). Strict enforcement of a cleanup program has reduced the incidence of brucellosis from 50% of the cattle herds in the United States in the 1930s to less than 2% currently. Complete cooperation between cattlemen and government officials will be required to ever completely remove the threat of brucellosis from this country.

CALF SCOURS (WHITE SCOURS)

Scours or diarrhea is one of the least understood diseases of cattle. Ten to twelve percent of all calves born die from calf scours in the first 30 days of

FIGURE 9-4. Many counties require brucellosis blood testing and disposing of reactors to control the disease. (Courtesy of the U.S. Department of Agriculture)

their lives. Death rates commonly exceed 50%, and animals that do recover often are stunted for the rest of their lives.

Scours is generally thought to be caused by an outbreak of bacterial or viral invasion. However, research has revealed that the condition is much more complicated than this. It may be caused by bacteria, virus, and environmental conditions—high concentrations of cattle, lack of colostrum, overfeeding, vitamin A deficiency, and parasitism. Because of such a wide variety of causes, it is extremely important to determine the specific reason for it. A veterinarian or diagnostic laboratory can isolate the specific organism or other cause. This is the key to successful treatment.

The principal damage caused by scours is loss of water, bicarbonate, sodium, and potassium from the blood and body fluids. Irritation to the intestines by invasion of the microorganisms produces a body reaction to try to flush out the invading organisms. This is why the calf passes watery feces and loses weight. If 15% of the body weight is lost as a result of dehydration, the calf goes into a coma and dies.

The key to treating dehydration is to replace fluids lost from the body through *electrolyte therapy*. Table 9-1 gives recommended mixtures to replace lost ions, energy, and liquids brought about by this dehydration. The calf can be orally drenched at recommended rates while the diagnostic laboratory is determining the specific organism responsible for the diarrhea. These ingredients work somewhat like Gatorade, which professional athletes drink to prevent their bodies from excessive dehydration and loss of electrolytes (ions).

Once the specific cause has been diagnosed, a recommended antibiotic may be given orally or injected to treat the viral or bacterial invasion. If calves respond to antibiotics, treatment should not be stopped too quickly. Twelve

TABLE 9-1. RECOMMENDED ELECTROLYTE MIXES FOR FLUID THERAPY

Mixes That Can Be Made from Household Ingredients

Formula 1 (to be given orally)
White corn syrup (dextrose)	8 tbsp
Salt (sodium chloride)	2 tsp
Baking soda (sodium bicarbonate)	1 tsp
Warm water	1 gal

Feed 2½ pints to a 90-lb calf four times/day (total of 1¼ gal).

Formula 2 (to be given orally)
Condensed beef consommé	1 can
Warm water	3 cans
Baking soda	1 tbsp

Feed this to calf twice a day.

Mixes That Can Be Made from Drugstore Ingredients

Formula 3 (to be given orally)
Sodium chloride	4 oz
Potassium chloride	5 oz
Sodium bicarbonate	5½ oz
Potassium monobasic phosphate	4½ oz

Add 1 oz of above mix plus ½ lb of dextrose
to 1 gal warm water. Feed 2–3 qt of this
solution four times/day (total of 2–3 gal).

Formula 4 (to be given either orally or intravenously)
Sodium bicarbonate	1 tbsp
50% dextrose solution	100 cc
Warm water	900 cc

Give 1–2 liters (quarts) four times/day
(total of 1–2 gal).

Mixes That Can Be Made from Grocery Store Ingredients

Formula 5
MCP Pectin	1 pkg.
Low-sodium salt	1 tsp
Baking soda	2 tsp
Beef consommé	1 can

Plus water sufficient to make 2 qt
Feed this mixture 3 times per day, or more if
dehydration is severe, in place of milk. After 1 or
2 days, gradually introduce milk (i.e., ⅓ milk,
⅔ formula 5, etc.) (*Source*: R. D. Phillips,
Colorado State University)

Source: Michigan State University.

hours is about the maximum effective time for injectable antibiotics. The normal recommendation is to continue treatment for 1 to 2 days after the scours has cleared up.

Other preventive measures are adequate amounts of vitamins A and D for cows before calving, keeping calves in clean environments, disinfection of stalls, isolation of infected calves, dipping the calf's navel in iodine immediately after birth, making sure the newborn calf receives a full feed (about 2 quarts) of colostrum milk within the first hour after birth, and assuring that the calf does not gorge itself, thus creating a food-induced diarrhea.

CALF DIPHTHERIA (SORE MOUTH)

This is a disease of suckling calves. Drooling and the appearance of infected patches of dead, yellow tissue at the edges of the tongue are the most obvious symptoms. The same soil organism that causes foot rot also causes calf diphtheria. Death may occur if treatment is delayed. A veterinarian should direct antibiotic injections, the removal of dead tissue, and the painting of the tongue with iodine.

FOOT-AND-MOUTH DISEASE

This is a dreaded disease around the world. The first signs are watery blisters in the mouth (Figure 9-5) and between the claws of the hooves. The udder may also be affected and high temperatures are common. Mortality is low but the economic losses are catastrophic because of strict quarantine and eradication programs carried out by almost all governmental agencies of developed countries (Figure 9-6). Foot-and-mouth disease is feared because it is so highly contagious. However, the United States has not had an outbreak since 1929. Caused by a virus, foot-and-mouth disease has no cure. Most governmental agencies rely on the quarantine of imported cattle until it is certain no animal is infected or a carrier.

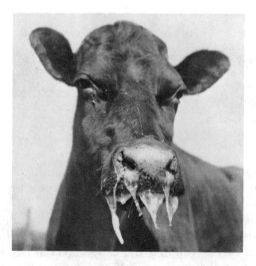

FIGURE 9-5. This steer shows excess saliva flowing from its mouth, one of the symptoms of foot-and-mouth disease. (Courtesy of the U.S. Department of Agriculture)

Chap. 9   DISEASES OF CATTLE

FIGURE 9-6.   One of the cows in a Mexican dairy herd dead from foot-and-mouth
disease. (Courtesy of the U.S. Department of Agriculture)

FOOT ROT

Foot rot is common in feedlots, corrals, or among cattle in muddy, confined areas. The skin between the toes becomes red and swollen, sometimes breaking open (Figure 9-7). Hooves may become deformed and lameness can develop.

A soil organism causes this inconvenient disease. Losses are caused by failure of cattle to eat because of pain and slight fever. A low level of antibiotic in the feed is the best preventative. Treatment by trimming affected parts of hooves and walking animals through a suitable disinfectant (such as iodine solution) is also effective.

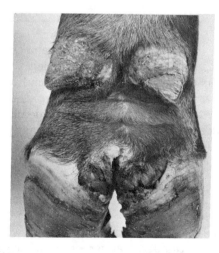

FIGURE 9-7.   Foot of a steer showing
foot rot. (Courtesy of the
U.S. Department of Agri-
culture)

FOUNDER

The condition of founder is more common in cattle and horses, but it may affect most domestic farm animals. An inflammation of the area between the bony part of the foot and the hoof wall is the most obvious clinical sign. In cattle, the chronic form occurs when the hoof wall separates from the bony tissue and the toes grow long and turn up without touching the ground. Toes may grow so long as to cross over, giving a distorted gait when the animal walks. In the acute form, signs are less drastic. Lameness, stiffness in the joints, and a painful stiff-legged gait are common signs.

Founder is often associated with overeating, but is more likely an allergic reaction to grain components when an animal accidentally gets into feed and gorges itself on high-concentrate rations. Cattle that accidentally eat all they want of a high-concentrate feed, for instance, develop symptoms of founder within 24 hours after having gorged themselves. The first signs, of course, are lameness and stiffness. It is much later that the crippling effects of deformation of the hooves takes place.

When increasing grain rations for milk production or fattening, care should be taken that the increase is gradual. A reasonable "warm-up" of 2 to 4 weeks in which slight increases are programmed will normally prevent the occurrence of founder. Storeroom doors and other storage openings should be kept firmly secured to prevent the possibility of a break-in by cattle.

Treatment of the chronic form consists of trimming the abnormal growth of the hooves back to normal. The most practical method of treatment thereafter is through sale or slaughter, because the condition generally recurs and complete cures are rare. If the acute form is in evidence, treatment with steroid therapy and antihistamine injections may provide a good chance of recovery.

GOITER

Goiter is an enlargement of the thyroid gland in the neck causing a swollen, lumpy appearance on both sides of the neck. A deficiency of iodine in the diet brings on goiter. Iodine is needed by the thyroid gland for the manufacture of the hormone thyroxine. When the supply is reduced, the thyroid gland is stimulated to grow larger to produce what the body demands. The feeding of iodized salt is a simple cure in the early stages and an easy preventative. There is no effective cure for advanced cases.

GRASS TETANY

Grass tetany is indicated by a nervousness and tetany (twitching) of the muscles. The animal may breathe rapidly, stagger, or fall when grazing, giving rise to the common name *grass staggers*. The cause is low blood magnesium usually originating from soil or forage deficiencies. Treatment consists of intravenous injection of a magnesium salt by a veterinarian. Preventive measures dictate the feeding of a free-choice supplement containing magnesium.

JOHNE'S DISEASE (PARATUBERCULOSIS, CHRONIC SPECIFIC ENTERITIS)

This is an infectious, incurable disease characterized by reappearing diarrhea and gradual loss of condition (Figure 9-8). Eventual refusal to eat and death may occur. Johne's disease produces one of the most serious forms of diarrhea. This infectious intestinal disorder creates emaciation, thickening or folding of the intestinal wall, a condition of edema characterized by "bottle jaw," a very offensive odor of the diarrhea but no blood, and a general weakened condition. The disease may recur at intervals throughout the life of the animal. It is slow to develop, most animals affected being 2 to 6 years of age before developing signs. The tuberculosis-like disease is caused by a small bacterial organism, *Mycobacterium paratuberculosis*. Although vaccines are available, they are not normally recommended and treatment is not very successful. Culling of the affected animal is the most practical recommendation.

Two types of tests are used to identify cattle suffering from Johne's disease. The *intradermal test* is used to detect infected animals in the herd before an outbreak occurs. The *serological test* is used once an outbreak occurs in order to identify the types of organisms that caused the problem.

FIGURE 9-8.   Dairy heifer with Johne's disease. (Courtesy of the U.S. Department of Agriculture)

Chap. 9   DISEASES OF CATTLE

KETOSIS (ACETONEMIA, HYPOGLYCEMIA, PREGNANCY DISEASE)

The name *ketosis* comes from the fact that affected cattle give off an odor of ketones on the breath and sometimes in the milk. The odor is very offensive and probably best described as something similar to but much stronger than fingernail polish remover. This sweet odor characterizes ketosis, although a definite diagnosis should come from a veterinarian because of other conditions that create the same signs.

The first form of ketosis is referred to as the *wasting form*. Decrease in appetite and milk production over a period of 2 to 4 days, rapid body weight loss, and constipation, but normal temperature, pulse, and respiration are clinical signs of the wasting form. Milk production drops severely, animals may die, and those that recover may never regain former production levels until the next lactation. Although ketosis is more common in dairy cattle, it affects all domestic animals.

A second form of ketosis, the *nervous form*, has clinical signs that include mild staggering, partial blindness, walking in circles, depressed appetite, excessive chewing on nothing, and salivation. The cow may appear to recover, but signs recur intermittently.

This disorder definitely requires the use of an experienced veterinarian. The very complicated metabolic disorder involves the elimination of ketones in the urine and an exhaustion of the glucose stored in the liver. Since cattle have small glucose reserves, the condition can become quite serious.

A disruption in the metabolism of carbohydrates and fatty acids to produce ketones is the cause of the disorder. This is normally associated with a higher milk production demand that can be met through normal digestive and metabolic processes. Although it is most common in dairy cows, all cattle are susceptible to ketosis.

Prevention consists of keeping cattle neither too fat nor too thin before calving because ketosis often occurs just prior to or just after birth. The feeding of sodium proprionate at the rate of 100 g per day for 6 weeks has been effective as a preventive treatment. Keeping an adequate mineral mix available is also recommended.

Only trained veterinarians should attempt treatment of the disorder. Glucose replacement therapy at the rate of 500 to 1000 cc of 50% dextrose is often recommended along with steroid therapy. Administration of steroids will vary according to type used and the manufacturer's direction. Treatment may need to be repeated periodically.

LEPTOSPIROSIS

The kidney is affected in leptospirosis, but in such a variety of ways that the following conditions may occur: abortion (Figure 9-9), mastitis (blood-tinged

FIGURE 9-9.   A cow and her aborted fetus, caused by leptospirosis. (Courtesy of the U.S. Department of Agriculture)

milk coming from a noninflamed udder), high temperature, jaundice (yellowing of the pink membranes, especially around the eye, vulva, penis, and gum), wine-colored urine, and anemia. Very young animals commonly die; older cattle have a death rate of only 5%. Autopsies reveal grey to white lesions on the liver. Recovered animals appear dull and listless. Definite diagnosis is made only through laboratory culture.

The cause is microscopic bacteria. Three major strains—*L. pomona, L. grippotyphosa,* and *L. hardjo*—are the chief infecting agents. It was previously thought that only *L. pomona* was the infecting agent, but it has been learned since that all three types may create the condition. The more expensive vaccines usually cover all three strains. Cheaper vaccines may cover only *L. pomona;* therefore, it is important to read the label. Vaccinations are recommended regardless of location and prevalence of the disease. A one-shot treatment that covers all three strains previously mentioned is widely available. Since deer carry the disease and cannot be vaccinated, it is important to immunize cattle.

Transmission of leptospirosis is mainly through water. It can be spread not only to other animals, but also to humans. Infected deer, cattle, and other animals urinate in streams and contaminated water is carried to low-lying areas. Since the organism passes through the kidney, it is easy to understand how the disease may be spread to other animals through cuts, abrasions, or drinking water. Leptospirosis can be spread through the semen of bulls, both in natural and artificial insemination reproduction.

Tetracycline, streptomycin, and penicillin have helped, although recovered animals often remain carriers. The best treatment is prevention, isolating any suspect immediately from the herd. Then vaccinate all animals that have not been immunized.

LUMPY JAW (ACTINOMYCOSIS, WOODEN TONGUE)

A very hard swelling of the bony tissue about the head, usually the jaw, is characteristic of actinomycosis. It may also affect the tissues of the throat area causing *wooden tongue*. An infection of the bone develops gradually, and by the time the swelling is noticed, it is difficult to treat. Swelling may break through the skin, discharging pus and a sticky, honey-like fluid containing little white granules. This eruption actually forms a pathway for drainage spreading the organism into the surrounding area.

Wooden tongue (actinobacillosis) is characterized by inability to eat, excessive salivation, and a chewing motion of the tongue. The hard, swollen tongue may have ulcers along the edge, and movement of the tongue from side to side by the observer causes obvious pain. The tongue later becomes shrunken and immobile. Lymph nodes may swell, rupture, and exude pus.

Lumpy jaw is caused by the microorganisms *Actinomycosis bovis*. Wooden tongue is caused by *Actinobacillus*. Prevention is the key to control of both these disorders. Eliminate rough, coarse feeds or other objects that could injure the lining of the cow's mouths when they chew. Infected animals should be removed from the herd.

Transmission of the organism is through contamination and injection into the body caused by eating sharp objects such as rough hay or feeds containing awns, trash, or wire. Treatment consists of the use of iodides for both actinomycosis and actinobacillosis (lumpy jaw and wooden tongue). An intravenous 10% solution of sodium iodide is the common treatment. It usually must be repeated in 10 to 14 days. Although lumpy jaw can be cured by this treatment, the disfigurement usually remains. Antibiotics such as tetracyclines may also be helpful as treatment.

MALIGNANT EDEMA

Signs are similar to blackleg except this disease affects cattle of any age and it usually occurs following an injury. High fever, swellings that make a crackling sound when touched, and production of a thin, reddish fluid from inflamed areas are the major signs. The swelling around the area of a wound is caused by gaseous air bubbles under the skin. Body temperature is normally elevated. If the disease progresses throughout the body, there will be depression followed by death.

The cause is the bacterium *Clostridium septicum*. Entrance is gained by this organism through cuts and scratches. Injuries may include difficult births, vaccinations, castrations, and navel infections, surgical procedures, and parasitic infestations, all of which allow the organism entrance. Sanitary conditions are a must because the organism is soilborne and tends to build up in the environment following heavy concentrations of cattle. Treatment consists of massive doses of tetracycline and penicillin both systemically and around the wound. There is a vaccine to effectively prevent the occurrence of malignant edema.

MASTITIS

The udder may become hot and very hard, and may produce lumpy milk sometimes streaked with blood (Figure 9-10). Less severe cases produce only lumpy milk. The cause is either a bacterium that may spread from one animal to another through improper sanitation (especially in dairy milking parlors) or an injury or chilling of the udder. Infection may occur in only one quarter, and if it is not treated, gangrene could develop and prove fatal.

At the first sign of mastitis, treatment consists of thoroughly "stripping clean" the affected quarter or quarters. Systemic antibiotics are helpful, and specially prepared broad-spectrum antibiotics are available in throw-away sterile syringes for injection through the teat canal directly into the infected quarter. By gentle injection and "reverse milking," the antibiotics are forced thoroughly into the infected quarter.

Response to treatment is good, but it works better in dry cows than in lactating cattle. Treatment should begin immediately after signs are seen and continued for at least 5 days or until signs have disappeared. If the condition persists, a sample of the milk should be sent to a laboratory to identify the specific organism involved and then treatment continued with the recommended antibiotic to combat that organism.

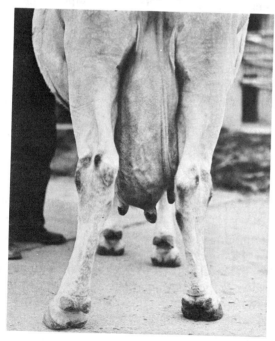

FIGURE 9-10. Mastitis caused the udder of this cow to become hot and very hard, and to produce lumpy milk sometimes streaked with blood. (Courtesy of the U.S. Department of Agriculture)

Occasionally, gangrenous mastitis will set in on an affected quarter. The quarter will turn black or purple and feel cold to the touch because of a loss of blood circulation. If not treated, it could be fatal or the quarter could slough off. In this case, amputation of the quarter is advisable. Usually if only one quarter is lost, the other three will compensate and produce a level of milk similar to that before the disorder occurred. If the cow has raised good calves in the past, her total milk production will not be impaired through the loss of one quarter.

MILK FEVER (PARTURIENT PARESIS)

This problem occurs at or shortly after calving but seldom until the second calf. The cow shows extreme weakness, wild eyes, and a loss of consciousness with the head "curled" toward the body similar to a sleeping dog. Body temperature is generally below normal. The cause is a deficiency of calcium brought on by rapid mammary demands for milk production. Veterinary treatment usually consists of intravenous injection of a calcium salt or injection of air into the udder to suppress milk production, thus lowering calcium requirements. Response of treated animals is excellent, although untreated animals are likely to die.

Recent observations suggest that 20 to 30 million units daily of vitamin D, given in the feed 5 to 7 days before calving, may reduce the incidence of milk fever. If administration is stopped more than four days before calving, the cow may be more susceptible than if not treated at all. Dosing for periods longer than recommended should be avoided because of the possibility of poisoning. A single intravenous or subcutaneous injection of vitamin D, 2 to 8 days before calving, has also proven effective as a preventive measure. However, severe reactions often occur after intravenous injections; thus it is well to be prepared for emergency procedures. Obviously, a veterinarian should make any treatment of this nature.

PINKEYE (INFECTIOUS KERATITIS, INFECTIOUS CONJUNCTIVITIS)

General signs are a reddening of the membranes of the eye, excessive tears, pain when exposed to light. Eyeballs may become covered by a milky film. Temporary or permanent blindness may result. In severe cases, the eyeball may be lost. The course of the infection may run from 4 to 8 weeks or longer.

Pinkeye rarely causes death in cattle but is extremely serious because of weight loss and decreased milk production. This disease appears suddenly, spreads rapidly, and is almost always associated with large numbers of flies.

The bacterium *Morexella bovis* was previously thought to be the sole cause of pinkeye, but recently other factors have been discovered. The most frequently involved causes are the bacterium *Morexella bovis* plus the irritation brought on by bright sunlight, dust, wind, flies, or weed seed; the IBR virus (involved mostly in the fall and winter rather than the customary summer incidents); ultraviolet light; an overdose of phenothiazine, a worming compound that creates sensitivity to light; marginal amounts of vitamin A.

Pinkeye is more common in Hereford cattle because the reflection of light from the white unpigmented face produces irritation from ultraviolet and other sources of light and irritation increasing susceptibility to the disease. However, all cattle are susceptible to pinkeye regardless of their eye pigmentation. Obviously, those breeds of cattle with white faces, white around the eyes, or light-colored eye pigmentation would stand a higher chance of this condition. Prevention is directed mainly toward control of flies, insects, and dust, and clipping of pastures to reduce irritation of weed seed to the eyes. Providing shaded areas to allow cattle to seek relief from ultraviolet light is also helpful.

Vaccines are not very effective. Supportive treatment in the form of vitamin A injections at the rate of 3 to 6 cc injected twice a year for vaccination on a 6- to 12-month schedule for IBR is recommended to possibly reduce inflammation from the form of pinkeye caused by the IBR virus. Bright sunlight, dry dusty conditions, and feeding in tall grass are common methods of transmission. Flies spread the disease from one cow to another.

Treatment is difficult but involves the following recommendations:

1. *Ophthalmic sprays.* Usually, an anesthetic plus a blue dye to act as a filter for some of the sun's rays is used. The product usually comes in a self-contained vessel for spraying directly into the eye. Treatment is most effective if applied daily.

2. *Ophthalmic ointment and drops.* An eyewash consisting of 5% solution of boric acid, followed by a 2% solution of mercurochrome ointment has shown good results.

3. *Cauterization.* Silver nitrate used in 1½% aqueous solution, 5 to 10 drops in each eye, two treatments given 4 days apart, has proven effective. More advanced cases may require a stronger solution.

4. *Suturing of eyelids.* Veterinarians sometimes suture the eyelids together to keep the eyes closed, shutting out irritants and sunlight, and bathing the eye in tears, which have antibacterial action.

5. *Eye patches.* A special adhesive and patch is placed over the eye and remains for 1 to 3 weeks before falling off. Thus sunlight and other irritants are kept out allowing for faster healing.

Although only one eye may be affected, both eyes should be treated because of the contagious nature of the organisms that produce pinkeye. Of course, a patch or suturing should be used only on the infected eye.

PNEUMONIA

Signs include wide leg stance, wheezing sounds from chest and lungs, nasal discharge, extension of tongue, and labored breathing. The cause is usually a virus condition brought on by exposure to cold or the water vapors inhaled into the lungs during improper drenching, and inhalation of chemicals or dust. Drenching with the head held high rather than the normal, level condition should be especially guarded against.

Treatment with antibiotics is a standard practice. Broad spectrum antibiotics, such as penicillin and tetracycline, are often recommended, along with isolation and prevention of undue stress. Steroid drugs may be recommended if the problem was precipitated by chemical or dust inhalation. Death rate is usually 10 to 20% of affected animals if treated, as high as 75% if untreated. A veterinarian should be in charge of specific treatment recommendations.

RED WATER DISEASE (BACILLARY HEMOGLOBINURIA)

Red water is an acute, highly fatal disease. It usually occurs in the spring of the year and is characterized by high fever and a breakdown of hemoglobin in the blood, which travels through the kidneys to cause a port-wine-colored, blood-tinged urine (hemoglobinuria). The blood-tinged urine is the key sign to watch for in this disease. In addition, jaundice (a yellow coloring of the pink membranes) is usually seen around the eye and vulva, or the prepuce in males.

This disease is rapidly fatal (usually within 12 hours from onset of signs) with nearly 100% mortality. Signs of the disease are labored breathing and grunting, an arched back indicating abdominal pains, a weak and rapid pulse, temperature elevation, and a swollen brisket. The manure of an affected animal is very dark, and the urine, as previously mentioned, wine-colored. Because the disease is so rapidly fatal, signs are usually not observed until one or more have died from the disease and closer observation takes place. The cause is a bacterium, *Clostridium hemolyticum.*

The disease is most commonly spread by flooding from an infected area to a clean one. It is also thought to be associated with liver flukes, which are known to be spread by snails, and the disease may very likely be spread at the same time by this same carrier. Treatment consists of very high doses of tetracycline with very high levels of dextrose and electrolyte solutions given intravenously. Even then, the outcome of the treatment is doubtful.

Prevention is the key to eliminating losses from this disease. A red water vaccine is available for herds in areas of high risk. Vaccination is recommended every 6 months, or if only one vaccination per year is desired, it is recommended to do so in the spring when the disease is most prevalent. Adequate drainage of pasture to control flooding and snails, thought to be "spreaders," is another method of control.

SHIPPING FEVER (HEMORRHAGIC SEPTICEMIA)

The first signs of this disease occur usually after an animal has undergone the stress of castration, vaccination, dehorning, weaning, working, perhaps after it has been chilled or wet, or especially if it has been moved a considerable distance to a new location. A sudden onset of high temperatures (104 to 106 °F), depression, going off feed, difficult breathing, coughing, runny nose and eyes, unthrifty appearance, and diarrhea characterize the disease (Figure 9-11). Affected animals may die within 3 weeks or may recover, but seldom do well in a feedlot situation.

Although stress is the precipitating factor, the disease is actually caused by the *Pasteurella multocida* organism. This is a complex disorder that may be brought on by a multitude of other organisms. Rough handling and shipping long distances are usually associated with the susceptibility of the animal to bacterial infection.

Treatment consists of high levels of antibiotics such as penicillin, streptomycin, tetracycline, and tylocin, combined with steroid therapy. Vitamin injections may also be used as a supportive therapy. Isolation of the sick animals in a warm dry area and treatment the same as if they had pneumonia is often recommended. The key to reducing losses from shipping fever is a preconditioning program. Since it is known that stress is the major contributing factor to this disease, this part of the complex disorder is reduced by not doing all

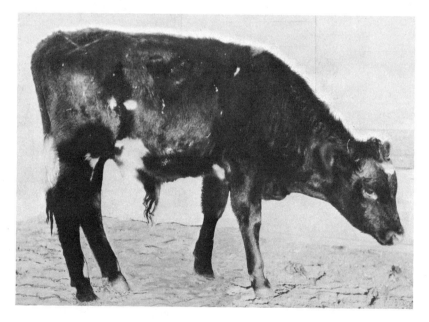

FIGURE 9-11. A calf infected with shipping fever (hemorrhagic septicemia). (Courtesy of the U.S. Department of Agriculture)

the jobs of weaning, dehorning, castration, and so on, at one time. Cattle are worked moderately and gently through a series of jobs over a period of time. The shock to the system is apparently reduced in this way, allowing the body's natural systems to resist the organisms gaining a foothold.

TRICHOMONIASIS

Abortion in heifers or cows at very early stages (2 to 4 months) of gestation is the key sign (Figure 9-12), and an infected uterus often follows abortion to keep cows from coming back into heat. This is a true venereal disease of cattle caused by a protozoal microorganism, *Trichomonas fetus*. The organism is passed from cow to cow by bulls that are infected with it. An inflamed prepuce on the bull is exhibited in fully developed stages of the disease. Although the disease may be transmitted by artificial insemination, commercial semen can be treated with antibiotics and poses no threat of infection.

Treatment of trichomoniasis is effective through administration of broad-spectrum antibiotics to both males and females. Culling, isolation, and periods of sexual inactivity are recommended practices. This is normally a rare disease, but if an outbreak is confirmed, close observation of the herb and careful checking of replacements, especially bulls, are in order. The bull spreads the disease.

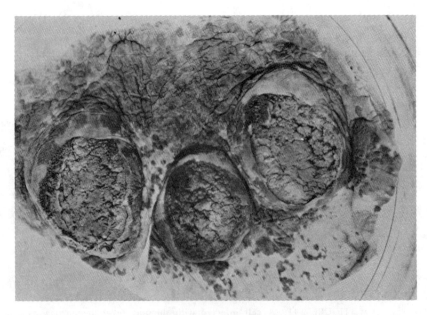

FIGURE 9-12. Part of the fetal membranes, showing cotyledons or "buttons" extensively diseased as a result of infection with an abortion organism. (Courtesy of the U.S. Department of Agriculture)

Chap. 9 DISEASES OF CATTLE

VIBRIOSIS

Although vibriosis is difficult to detect, the first signs of the problem may be a breakdown of a definite calving season to one of unmanageable order. Calves may be dropped every month of the year. This is caused by undetected abortions at gestations of less than 5 months. Aborted fetuses may go undetected because of their small size or they may be eaten by scavengers. Cows or heifers may abort once or twice, become immune, then breed normally, but continue spreading the disease to others. Calving may be delayed 10 to 12 months, causing tremendous loss and throwing the planned calving season out of sequence. The cause is a microscopic organism, *Vibro fetus,* or *Campylobacter fetus.* Vibriosis, like trichomoniasis, is a veneral disease spread mechanically by the bull from female to female through the normal act of breeding.

Treatment is questionable at present. It is recommended, in areas where vibriosis is a known problem, that cows and heifers be vaccinated annually with a good quality vaccine about two months before turning in the bulls. Controversy exists in the United States over the effectiveness of vaccination of the bulls. Some authorities say it has no effect, because the bulls spread the disease mechanically without themselves being visibly affected by it.

X DISEASE (HYPERKERATOSIS)

This is a skin condition resulting in thickened hide, loss of hair, watery eyes, and diarrhea (Figure 9-13). Death losses can exceed 75%. The cause is con-

FIGURE 9-13. An animal showing an advanced case of X disease. (Courtesy of the U.S. Department of Agriculture)

sumption of oil- or grease-type products for which cattle have a peculiar attraction. No treatment is recognized. Prevention by keeping cattle away from sources appears the best solution.

STUDY QUESTIONS

1. The normal temperature of cattle is _____ to _____ .
2. Anaplasmosis is caused by a minute parasite that affects the _____ .
3. An unexpected death, a bloated carcass with bloody discharges from body openings, and "the animal with all four feet up" is a sign of _____ .
4. Blackleg has been prevented by _____ at 3 to 4 months of age.
5. Gaseous bloat is often relieved by _____ , passing a tube down the esophagus, or puncture of the rumen.
6. Abortion in the last third of pregnancy is a symptom of _____ .
7. Foot-and-mouth disease is best controlled by _____ .
8. Deformed hooves, lameness, and red and swollen skin between the toes are symptoms of _____ .
9. Johne's disease is caused by _____ .
10. A sweet, offensive odor on the breath of a down milk cow is a symptom of _____ .
11. Lumpy jaw is caused by an organism that invades the bones of the _____ and _____ .
12. Swollen udders and lumpy milk are a sign of _____ .
13. Milk fever is a problem occurring only shortly after _____ .
14. Pinkeye only affects breeds with white pigmented hair around the eye. (True, false) _____ .
15. Labored breathing, extension of tongue, and nasal discharge after cold weather probably indicate _____ .
16. A disease that is often contracted from swampy, poorly drained areas and that is carried by snails is _____ .
17. _____ is a complex disease that occurs after stress conditions.
18. Abortion as a result of trichomoniasis will occur during the first _____ to _____ months of pregnancy and vibrionic abortion occurs during the first _____ to _____ months.
19. Consumption of _____ products can cause X disease.

DISCUSSION QUESTIONS

1. Discuss the signs in both cattle and humans of cattle diseases that are contagious to bovines and humans.

2. An apparently healthy herd of cattle has just been turned in to a spring pasture. One cow is found dead a few hours later. She is greatly swollen and discharging clear fluids from the body openings. What would you suspect as possible causes and what precautions would you take?

3. How can founder be distinguished from foot rot and grass staggers?

4. Which diseases are likely to cause abortion in cattle, and in what stage of gestation?

5. Discuss the diseases and their signs that are associated with any form of diarrhea.

6. Describe the diseases or disorders usually associated with milk production.

7. What are blackleg-type diseases, and what causes them? What practical use can be made of this information?

Chapter Ten

THE DAIRY INDUSTRY

The basis of the diary industry is milk—one of nature's most perfect foods. Milk is the primary source of food for all newborn mammals and can be a very important part of the human diet regardless of age. Its composition of easily digestible high-quality protein, minerals, and vitamins makes it an essential diet resource. Milk is also a flexible food source which can be easily modified in fat content to meet today's consumer demands.

THE HISTORY OF DAIRYING

Milk has always been associated with goodness, youth, and life. Terms such as "the land of milk and honey" are included in the Bible along with 50 other references to milk and milk products, which are described as very desirable foods. One of the early purposes of domesticating cattle and other livestock was for milk as much or more so than meat. Records of cows being milked date to 9000 B.C. Cow's milk was used for food, sacrifices, offerings, cosmetics, and medicine. India was producing butter for food and holy offerings as early as 2000 B.C. Egyptians used milk, cheese, and butter as early as 3000 B.C.

The first cows were brought to the New World on Columbus' second voyage. The early dairy industry in America, however, mainly relied on one or two cows cared for by family labor. The perishable nature of milk and problems in transporting fluids limited early growth. By the mid-nineteenth century, the modern dairy industry had begun, with excess milk being "pooled" for the manufacture of cheese. The first rail shipment of milk in 1841 from Orange County, New York, to New York City (a distance of 80 miles) was a major achievement. Major inventions and developments which helped shape the early dairy industry include:

1. The experiments of Louis Pasteur on principles of pasteurization (heating milk to kill harmful bacteria).
2. The invention of condensed milk by Gail Borden (1856), which led to concentration of milk solids in less volume, widened markets, and allowed the use of surplus milk.
3. The invention of the cream separator (1878), which separated milk and cream by centrifugal force rather than by letting cream rise to the top of the milk.
4. The Babcock Cream Test (1892).
5. Mechanical refrigeration allowing for transport of milk.
6. Homogenization (breaking up fat molecules for distribution throughout milk, which keeps fat from coming to the top).
7. Milking machines.
8. A host of other products and methods for processing milk, such as mechanical churns, advancements in cheese manufacturing, and improved ice cream and yogurt methods.

THE DAIRY INDUSTRY TODAY

Today milk is produced and processed in all 50 states. However, the major areas of milk production are located near dense urban populations. Major milk producing states are Wisconsin, California, New York, Minnesota, Pennsylvania, Michigan, Ohio, and Iowa. Although milk is used as a basis for many products, about 44% of the milk produced in the United States is still consumed as fluid or fresh milk and cream. Milk by-products make up the remaining 56%, with cheese 24%, butter 18%, ice cream products 10%, and evaporated and condensed milk 2%. The remaining 2% is still used on the farm where it is produced. The percentages vary slightly from year to year depending on production of milk and consumption trends. Table 10-1 indicates some typical dairy products available to today's consumer.

Total milk production has kept up with population demands in the United States. This has been made possible by increased milk production per cow and by a decrease in per capita consumption of milk and milk products (from 730 lb per person in 1950 to about 540 lb today). Since 1950, milk production per cow has doubled, with 10,000 to 12,000 lb of milk per cow per year

TABLE 10-1. MODERN DAIRY PRODUCTS

Fluid milk products	Solid products
Homogenized milk (3.25% fat or more)	Dried skim milk
Lowfat milk (1 or 2% fat)	Butter
Skim milk (0.5% or less fat)	Cheese
Half and half (10.5% fat)	Cheddar (Colby, Monterey
Whipping cream (32% fat)	Jack)
	Swiss
Flavored milk products	Italian types
Chocolate milk (3.25% fat)	Brick
Chocolate drink (less than 3.25% fat)	Limburger
Eggnog (6% fat)	Blue
Fermented milk products	Cottage
Buttermilk	Frozen Desserts
Yogurt	Ice cream (10% fat or more)
Sour cream	Sherberts
Sour half and half	Ice milk (2–7% fat)
Sour cream dips	

now being common. Many top cows today produce more than 30,000 lb of milk per year. The size of the dairy farm has also greatly increased, with the trend toward larger farms and more cows. The number of dairy cows in the United States reached its maximum of about 25 million in 1945 and has since steadily declined to less than one-half that figure today.

DAIRY BREEDS

As mentioned previously, many beef cattle breeds were developed for a dual (milk and meat) or even triple purpose (milk, meat, and draft). Some still carry such distinctions, while others have been selected for either dairy or beef characteristics only. In the United States today there are five commonly recognized dairy breeds: Ayrshire, Brown Swiss, Guernsey, Holstein, and Jersey. Table 10-2 condenses some qualities for which each breed is noted.

Ayrshire

The Ayrshire breed was developed in the county of Ayr in southwestern Scotland. This area is cold and damp with relatively little forage available. Hence animals were selected for hardiness and excellent grazing ability. The color pattern of the Ayrshire breed varies from red and white to mahogany and white (Figure 10-1). The breed is more nervous than other dairy breeds. Early breeders were noted for careful selection in type, and the breed is still noted for its style, symmetry, strong attachment of udder, and smooth, clean "dairy-type" bodies. The Ayrshire has only a fair rating for beef and veal.

TABLE 10-2. CHARACTERISTICS OF MAJOR DAIRY BREEDS

	Ayrshire	Brown Swiss	Guernsey	Holstein	Jersey
Cow size (lb)	1200	1400	1100	1500	1000
Color	Red or mahogany and white	Brown	Fawn with clear white markings	Black and white	Fawn, with or without white
Temperament	Nervous	Docile	Docile	Docile	Somewhat nervous
Grazing ability	Excellent	Excellent	Fair	Fair	Good
Maturity	Medium	Late	Early	Late	Early
Value for beef	Good	Excellent	Poor	Excellent	Poor
Milkfat (%)	4.0	4.0	4.7	3.7	5.0
Milk nonfat solids (%)	9.0	9.2	9.5	8.5	9.5
Average milk production (approx. lb/year)	11,000	11,000–12,000	10,000	13,000–14,000	9000
Origin	Scotland	Switzerland	Isle of Guernsey	Holland	Isle of Jersey
Size at birth (lb)	75	90	75	95	60

Brown Swiss

The Brown Swiss (Figure 10-2) was developed on the mountain slopes of Switzerland. They were grazed from the foot of the mountains in the spring to the highest slopes during summer. Such terrain produced hardy animals with excellent grazing ability. Their large size and white body fat makes them also desirable for beef and veal.

The Brown Swiss color varies from light brown to dark brown. The breed is noted for being very docile with a slight tendency to be stubborn at times. Brown Swiss were developed for cheese production and meat, and they produce large volumes of milk with a relatively high percentage of milkfat and solids.

Guernsey

The Guernsey (Figure 10-3) was developed on the island of Guernsey, one of the channel islands between France and England. The island is noted for luxuriant pastures; hence, grazing ability was not a trait that was selected in early breeding.

FIGURE 10-1.   The red to mahogany and white Ayrshire is noted for style as well as milk. Pictured is Leete Farms Betty's Ida, which holds the Ayrshire World Production Record of 37,170 lb of milk and 4.3% milkfat for a 305-day lactation. (Courtesy of the Ayrshire Breeders Association)

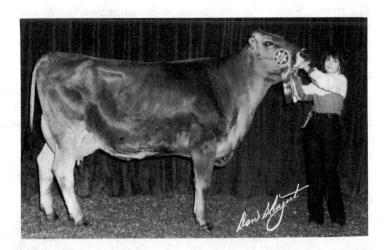

FIGURE 10-2.   The Brown Swiss is noted for milk production as well as veal production. (Courtesy of Don Shugart Photography, Grapevine, Texas)

FIGURE 10-3. The fawn and white Guernsey breed is noted for milk with a deep yellow color that is high in milkfat and solids content.

Guernsey cattle are fawn colored with clear white markings. They are very docile, but their yellow body fat and small size make them undesirable for veal production. They are noted for producing milk with a deep yellow color indicating high carotene (vitamin A precursor) content. Guernsey milk is high in milkfat and total solids content.

Jersey

The Jersey was developed on the island of Jersey located only 22 miles from the island of Guernsey. This island also has luxuriant pastures, making it unnecessary to stress selection for grazing ability. Butter was the major product produced on the island; hence the Jersey was bred for large amounts of milkfat, a trait for which it is still noted today. During development of the breed, only animals of excellent type were allowed on the island, and Jerseys are still known for uniformity in type (Figure 10-4).

Milk from the fawn-colored cattle is yellow due to carotene content and is very high in percentage of milkfat and milk solids. The body fat of the Jersey is yellow, and this, combined with its small size, makes the Jersey undesirable for veal and beef.

Holstein–Friesian

The Holstein–Friesian (Figure 10-5) is the most prevalent breed in the United States, accounting for 80 to 90% of the dairy cattle. Its origin was in the Netherlands in the province of North Holland and West Friesland, an area noted

FIGURE 10-4. The solid fawn Jersey is noted for milk with a high percentage of milkfat. (Courtesy of Don Shugart Photography, Grapevine, Texas)

for excellent pastures. The breed was not forced to develop good grazing abilities during its early stages. Milk was used for cheese, so selection for animals with large volumes of milk was important.

The black and white cattle (red and white Holsteins also exist) are still noted for production of large volumes of milk with the lowest milkfat—a trait that is desirable in today's milk market. Its size, fast growth, and good carcasses make it a very desirable breed for veal and beef production.

Other Dairy Breeds

Other breeds of cattle are also used for milk production in the United States. These are usually dual-purpose breeds where selection was based more on milk

FIGURE 10-5. The black and white Holstein–Friesian is the prevalent dairy breed in the United States. (Courtesy of Don Shugart Photography, Grapevine, Texas)

Chap. 10 THE DAIRY INDUSTRY

production than beef. Good examples are the milking Shorthorn breed (a strain of beef Shorthorn bred for milk production) and the Red Poll. Several other dual- and triple-purpose breeds can be used for milk production. Selection, based on milk rather than beef, may make them candidates for addition to dairy herds.

STUDY QUESTIONS

1. Four of the leading states in milk production are _____ , _____ , _____ , and _____ .
2. Milk consumption per person has _____ (increased, decreased) during the last 20 years.
3. There are _____ main dairy breeds in the United States today.
4. The Ayrshire breed is colored _____ and white and is noted for large volume of milk.
5. The dairy breed which is used mostly for veal and milk is the _____ .
6. The golden _____ breed still produces milk high in milkfat and characteristic yellow color.
7. The Jersey breed is noted for milk high in _____ .
8. Holsteins are _____ and white and produce large quantities of milk with a lower milkfat percentage.

DISCUSSION QUESTIONS

1. Name 15 products derived from milk other than fluid milk.
2. How many pounds of milk annually might be expected from a herd of 100 cows producing top yields?
3. Discuss the merits of each of the major dairy breeds in the United States including level of production and percent of butter fat.
4. Many dairy herds in the United States consist of purebred cows but of more than one breed. Why?

DAIRY CATTLE ASSOCIATIONS

AYRSHIRE
Ayrshire Breeders
    Association
Brandon, Vermont 05733

BRAHMANSTEIN
Brahmanstein Breeders
    Association
P.O. Box 798
Canton, Texas 75103

BROWN SWISS

Brown Swiss Cattle Breeders
Association
Box 1038
Beloit, Wisconsin 53511

DUTCH BELTED

Dutch Belted Cattle Association of
America
Box 358
Venus, Florida 33960

GUERNSEY

American Guernsey Cattle Club
P.O. Box 27410
Columbus, Ohio 43227

HOLSTEIN

Holstein–Friesian Association of
America
P.O. Box 808
Brattleboro, Vermont 05301

JERSEY

American Jersey Cattle Club
Box 27310
Columbus, Ohio 43227

MILKING SHORTHORN

American Milking Shorthorn
Society
1722-JJ South Glenstone
Springfield, Missouri 65804

Chapter Eleven

FEEDING, REPRODUCTION AND MANAGEMENT OF DAIRY CATTLE

Since dairy cattle are physiologically the same as beef cattle, the basic principles of feeding and reproduction previously learned for beef cattle apply. Gestation time, estrous cycles, principles of reproduction, ruminant parts and function, and the general usage and requirement of nutrients are the same for both species. In this chapter, only those feeding, reproduction, and management practices unique to dairying are discussed.

CARE OF NEWBORN DAIRY CALVES

Although some dairy operations obtain replacements for sales most dairymen still rely on raising their own herd replacements. Care of the newborn calf from birth to weaning, therefore, becomes a necessary part of dairy herd management. The dam should be removed from other cows and placed in an area where she can be carefully watched in case she needs assistance in calving, but do not disturb her unnecessarily. Soon after birth, remove any mucus from the calf's nostrils and mouth if necessary. In cold weather, dry the calf or let the cow lick the calf dry so that the calf does not become chilled. Add tincture

TABLE 11-1. DILUTION RATE FOR VARIOUS LIQUID DIETS FOR CALVES FED ONCE OR TWICE DAILY

Type of Milk (Ingredient)	Percent Dry Matter	Once-Daily Feeding[a] Ingredient + Water = lb Dry Matter (lb/feeding once daily)	Twice-Daily Feeding[a] Ingredient + Water = lb Dry Matter (lb/feeding twice daily)
First milk colostrum	28	3.5 + 3.5 = 1.0	2.0 + 2.0 = 1.1
Pooled excess colostrum	16	6.0 + 0 = 1.0	3.0 + 1.5 = 1.0
Whole milk, Holstein	12	7.0 + 0 = 0.8	4.0 + 0.0 = 1.0
Milk replacer	88	0.8 + 5.0 = 0.7	0.5 + 3.5 = 0.9

Source: Courtesy of the University of Minnesota, reprinted from *Feeding the Dairy Herd,* Extension Bulletin 218.

[a]Use 75 to 80% of these amounts for Jersey or Guernsey calves.

of iodine to the calf's navel to prevent entry of bacteria into the body through the navel cord.

Since the object of the dairy is milk, little is wasted on nursing calves from birth to weaning. Newborn calves are usually allowed to suckle their dams only once. The cow is then milked and managed with the rest of the producing herd. It is important for dairy calves, and for all newborn animals, to receive some colostrum (first milk) within an hour after birth. The colostrum provides antibody protection against diseases. Dairy calves must receive about 4 lb of colostrum as soon after birth as possible. Some dairymen prefer to remove the calf immediately at birth and hand feed colostrum from a milk pail or nipple to ensure consumption of the proper amount. The calf continues to receive colostrum for another 2 or 3 days, usually hand fed from a milk pail or nipple. Extra colostrum cannot be sold as milk, so it is usually diluted with water (2 parts colostrum to 1 part water) and fed to other young calves. It can also be frozen or stored in the fermented state and used for other calves or sick calves at a later date. See Table 11-1 for suggested dilution rates and feeding levels.

FEEDING DAIRY CALVES TO WEANING

Although there are numerous calf-rearing systems, the recommended feeding practices are generally:

1. Keep calves in dry, draft-free housing.
2. Make sure calves receive colostrum for at least 1 day and preferably 2 or 3 days.
3. Change to whole milk or milk replacer at 2 or 3 days of age.

4. Do not overfeed; overfeeding is a serious cause of scours and loss of calves.

5. Feed milk or mixed milk replacer daily at a general rate of 8% of body weight (8 lb for a 100-lb calf).

6. Feed regularly, usually twice daily. Once-a-day feeding has been successful but requires more attention to health problems, especially scours, at an early age.

7. Use clean feeding utensils and make sure sanitary conditions exist around pens at all times.

8. Milk or milk replacer should be fed at the same temperature each feeding, usually 95 to 100 °F (35 to 38 °C).

9. Calves must receive milk or milk replacer until they are at least 3 to 8 weeks of age.

10. Age at weaning is dependent on the time required for calves to develop functional rumen and eat from 1.5 to 2 lb of calf starter per day (usually 6 to 8 weeks).

11. Feed calf starter and high-quality hay starting at 7 days of age. Feed free choice, allowing fresh feed to be available at all times. Fresh, clean water must be available. See Table 11-2 for example starter rations.

TABLE 11-2. EXAMPLE CALF STARTER RATIONS

Ingredients	Ration[a]					
	A	B	C	D	E	F
Corn, coarse grind (lb)	50	39	54	44	34	24
Oats, rolled or crushed (lb)	35		12	22	34	24
Barley, rolled or coarse grind (lb)		39				
Beet pulp, molasses (lb)						20
Corncobs, ground (lb)					15	
Wheat bran (lb)		10	11			
Soybean meal (lb)	13	10	8	26	15	25
Linseed meal (lb)			8			
Molasses, liquid (lb)			5	5		5
Dicalcium phosphate (lb)	1	1	1	1	1	1
Trace mineral salt (lb)	1	1	1	1	1	1
Vitamin A (IU)	200,000	200,000	200,000	200,000	200,000	200,000
Vitamin D (IU)	50,000	50,000	50,000	50,000	50,000	50,000
Total (lb)	100	100	100	100	100	100
Protein (% of DM)	16	16	16	20	16	20
Fiber (% of DM)	6	5	5	5	11	9

Source: Courtesy of the University of Minnesota, reprinted from Feeding the Dairy Herd, Extension Bulletin 218.

[a]Rations A, B, and C recommended for calves weaned after 4 weeks of age and receiving forages.

Ration D recommended for calves weaned before 4 weeks and receiving forage.

Ration E recommended for calves weaned after 4 weeks and not consuming forage.

Ration F recommended for calves weaned before 4 weeks and not receiving forage.

CALF-REARING SYSTEMS

Calf-rearing systems to be used after feeding colostrum are as follows:

1. *Nurse cow system.* Two or more calves are allowed to nurse one cow. This is the most expensive system due to the loss of income from one cow. It is usually recommended for veal production only.
2. *Whole milk system.* Milk is hand fed to calves at a rate of 6 to 8 lb per day. Calves are usually weaned at about 60 days and will eat 300 to 500 lb of whole milk.
3. *Milk replacer system.* This method can cut the cost of the whole milk system in half. Calves are hand fed milk replacer instead of whole milk at the rate of 6 to 8 lb per day. They are weaned at 28 days (early weaning) to 60 days, depending on how fast calves eat solid concentrates and forage.
4. *Combination of milk replacer and whole milk systems.* This reduces the amount of whole milk used by gradually switching from whole milk to milk replacer at 2 weeks of age.

FEEDING DAIRY REPLACEMENTS FROM WEANING TO PUBERTY

The main management purpose of this period is to produce large, fast-growing heifers at a minimum cost, capable of beginning lactation at an early age. Concentrates are fed at a rate of 3 to 4 lb per day up to 1 year of age, with hay and/or high-quality pasture fed free choice. More grain is fed if roughage quality is fair or poor. Animals are kept in good condition for proper growth but are never fat or overly conditioned. Tables 11-3 and 11-4 give specific ration suggestions.

CARE AT BREEDING AND DURING GESTATION

Well-cared-for heifers should be of sufficient size for breeding at 13 to 15 months of age so they can enter the milking herd at about 2 years of age. Most dairymen breed according to weight or size, not age, with the following minimum weights given as a guide for breeding:

Holstein and Brown Swiss	750 lb
Ayrshire	600 lb
Guernsey	550 lb
Jersey	500 lb

TABLE 11-3. GROWER RATIONS FOR 400-LB CALVES

Ration 1

6 lb alfalfa-grass hay, free choice (16–18% CP)
4 lb grain mix (9.8% crude protein):
    1500 lb coarsely ground shelled corn
     455 lb rolled or ground oats
      20 lb trace mineral salt
      20 lb monosodium phosphate
       5 lb vitamin premix

Ration 2

5 lb alfalfa-grass hay,free-choice (12–16% CP)
5 lb grain mix (12.8% crude protein):
     900 lb rolled barley
   1000 lb rolled oats
      55 lb dry molasses
      20 lb trace mineral salt
      20 lb dicalcium phosphate
       5 lb vitamin premix

Ration 3

5 lb grass hay, free-choice (10–14% CP)
5 lb grain mix (10.9% crude protein):
    1800 lb corn and cob meal
     100 lb soybean meal
      55 lb dry molasses
      20 lb trace mineral salt
      20 lb dicalcium phosphate
       5 lb vitamin premix

Ration 4

6 lb corn silage (8–9% CP)
3 lb grass hay (12–14% CP)
4 lb grain mix (17% crude protein):
   1000 lb coarsely ground shelled corn
     655 lb rolled or ground oats
     300 lb soybean meal
      20 lb trace mineral salt
       5 lb limestone
      15 lb dicalcium phosphate
       5 lb vitamin premix

Source: Courtesy of the University of Minnesota, reprinted from *Feeding the Dairy Herd,* Extension Bulletin 218.

Recent research has shown that breeding at 13 months of age regardless of weight is a simple, effective method and is being used by large dairies. In breeding heifers, attention must also be given to planned calving time of replacement heifers to fill low milk production gaps. Breeding time is correlated with desired calving time and entry of replacement heifers into the milking herd.

The bred heifers are kept in good condition on pasture and/or hay with concentrates supplied only if needed. Heifers are expected to continue growth until calving and may require additional nutrients to obtain desired condition at calving. If supplemental feed is needed, use any of the suggested rations in Table 11-4 as guidelines.

FEEDING THE DRY COW AND SPRINGING HEIFERS

Heifers within 2 to 3 months of calving (springers) are often kept with the dry cows (cows not being milked). Cows are usually dried up (milking stopped) 50 to 60 days before the expected calving date to allow the mammary system and cow to recover from the stress of lactation before starting another milking cycle. Dry cows are separated from the milking herd and grazed on good pasture or fed, free choice, its equivalent in hay and/or silage. Concentrates are

TABLE 11-4. RATIONS FOR 700-LB HEIFERS THAT ARE GAINING 1.5 LB PER DAY

Ration 1

42 lb corn silage (33% DM)
  1 lb grain mix:
    160 lb corn and cob meal
    1705 lb 44% supplement
    98 lb dicalcium phosphate
    12 lb limestone
    20 lb trace mineral salt
    5 lb vitamin premix

Ration 2

50 lb sweet corn cannery silage (20% DM)
  5 lb grain mix
    1274 lb corn
    425 lb oats
    247 lb 44% protein supplement
    10 lb dicalcium phosphate
    19 lb limestone
    20 lb trace mineral salt
    5 lb vitamin premix

Ration 3

28 lb oat silage
  2 lb grain mix:
    1960 lb corn and cob
    5 lb limestone
    10 lb dicalcium phosphate
    20 lb trace mineral salt
    5 lb vitamin premix

Ration 4

15 lb grass hay
  5 lb grain mix:
    1975 lb corn and cob meal[a]
    20 lb trace mineral salt
    5 lb vitamin premix

Ration 5

 7 lb alfalfa hay
20 lb corn silage
 2 lb grain mix:
    1940 lb corn and cob
    35 lb monosodium phosphate
    20 lb trace mineral salt
    5 lb vitamin premix

Ration 6

 7 lb grass hay
20 lb corn silage
 3 lb grain mix:
    1314 lb shelled corn
    438 lb oats
    211 lb 44% supplement
    8 lb dicalcium phosphate
    4 lb limestone
    20 lb trace mineral salt
    5 lb vitamin premix

Ration 7

15 lb alfalfa hay
 3 lb grain mix[b]:
    955 lb barley
    1000 lb oats
    20 lb trace mineral salt
    20 lb monosodium phosphate
    5 lb vitamin premix

Ration 8

20 lb corn stover (stalklage)
 3 lb grain mix:
    1100 lb corn and cob meal
    865 lb 44% supplement
    20 lb trace mineral salt
    10 lb dicalcium phosphate[3]
    5 lb vitamin premix

Source: Courtesy of the University of Minnesota, reprinted from *Feeding the Dairy Herd,* Extension Bulletin 218.

[a]Could substitute barley–oats (50–50 mixture).
[b]Could substitute corn and cob meal.

generally fed only if required to maintain condition during this last 2 months of gestation. Mineral mix and salt are also provided free choice.

Dairy cows are usually in very good condition during the dry period. High-producing cows cannot eat enough feed to fulfill nutritive requirements in early lactation and rely on stored body fat for milk requirements. This loss of weight must be restored prior to the next calving and lactation. This is usually accomplished by feeding extra in late lactation just prior to the dry period. The resulting weight gain must be carried through the dry period.

At 2 to 3 weeks prior to expected calving, cows and/or heifers are placed in a maternity barn or small trap depending on the climate of the area. Good pasture or high-quality grass hay is provided free choice along with a gradual increase in concentrate ration until cows are eating about 1½ lb of concentrates per 100 lb of body weight. This feeding program is maintained until the animal enters the milking herd. At calving, the proper care as described for beef cattle in Chapter 3 is required.

MANAGING THE MILKING HERD

Ideally, the normal lactation is 305 days with a 60-day dry period. The typical lactation curve is shown in Figure 11-1. In practice, the length of a cow's lactation may vary from 270 days or less to over 400 days. Shorter periods normally result when the cow is bred back soon after calving or is dried up due to illness. Longer periods result primarily from problems in getting the cow bred. The average calving interval is about 400 days.

Milk production is fairly high immediately after calving. This amount increases for approximately 4 to 6 weeks until the cow reaches maximum production. From this point, there is a gradual decline until the end of lactation. High-producing cows with high 305-day lactation records have good persistence or the ability to maintain relatively high levels of production throughout the entire lactation period. The average decline in production after peak lactation is about 6% per month. Lactation records of 15,000 to 25,000 lb of milk are not uncommon today.

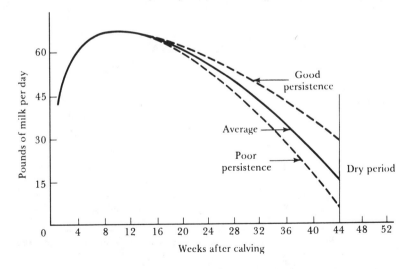

FIGURE 11-1. Typical lactation curve.

Production records also vary according to age of the cow. Cows that freshen at 2 years of age can be expected to increase production by 25% to peak production years. Most cows reach maximum production at 6 to 8 years of age (fourth or sixth lactation). Production per year decreases thereafter.

Production records (DHI, Dairy Herd Improvement, and DHIR, Dairy Herd Improvement Registry) usually take into account those factors which vary lactations and report a lactation corrected to 6 years of age, 305 milking days, milked two times a day. Other corrections are made for month calved and location of cow. Additional correction factors are used to equalize records for cows milked three times a day, instead of two, for cows milked for more or less than 305 days, for cows at an age other than 6, or for milk produced with a specific milkfat. Details can be obtained from breed associations and Agriculture Extension Services. Thus, when cows' records are reported, they are usually corrected to reflect the norm for fair comparison.

FEEDING THE MILKING HERD

Since dairy cows are ruminants, dairy cattle rations normally consist of high-quality legume and nonlegume roughages (pasture, silage, and/or hay), with high-quality, palatable concentrates to supplement the roughage for maximum production. Although all nutrients are important, the main concern of the dairyman is energy. It is difficult, if not impossible, for a high-producing dairy cow to consume enough of any ration to meet the tremendous energy required to produce large quantities of milk. Table 11-5 serves as a guideline to meet these energy requirements. Table 11-6 covers suggested protein levels throughout the changing lactation period.

TABLE 11-5. AMOUNT OF GRAIN TO FEED BY PERIODS (1400-LB COW, 4% MILK)

	Production Ability of the Cow (lb)[a]			
Average daily first period:	50	60	80	90–100
Lactation total:	10,000	12,000	15,000	18,000
	Grain-to-Milk Ratio			
Period of lactation				
1. (First 10 weeks)	1:4	1:3	1:3	1:3
2. (Second 10 weeks)	1:4	1:3	1:2.5	1:2.5
3. (Last 24 weeks)	1:4	1:4	1:3	1:3
4. (Dry, 6–8 weeks)	3–4 lb daily	4 lb daily	4 lb daily	6 lb daily
Total grain (approximate)	3000 lb	4000 lb	5000 lb	6000 lb

Source: Courtesy of the University of Minnesota, reprinted from *Feeding the Dairy Herd,* Extension Bulletin 218.

[a]Ratios based on 100% dry matter basis, grain containing 80% TDN, and forage 60% TDN.

TABLE 11-6. PROTEIN LEVELS OF TOTAL RATIONS DURING FOUR PERIODS

	Production Ability of the Cow (lb)		
Average first period:	60	80	90
Lactation total:	12,000	15,000	18,000
	Percent of Ration Dry Matter		
Period of lactation			
1. (First 10 weeks)	16	19	20
2. (Second 10 weeks)	13	15	16
3. (Last 24 week)	12	12	12
4. (Dry, 6–8 weeks)	9	9	9

Source: Courtesy of the University of Minnesota, reprinted from *Feeding the Dairy Herd,* Extension Bulletin 218.

Since the highest level of milk production occurs in the first 6 to 12 weeks of lactation, this is the critical period in feeding. Fresh cows entering the herd should receive all the concentrates they can handle safely (2 to 2½ lb per 100 lb of body weight). This is in addition to the high-quality roughage that the cow will eat free choice. Combining forage and concentrate, the cow will be able to eat about 3 to 4 lb of dry ration per 100 lb of body weight. To supply as much energy as possible, concentrates are used at maximum levels, usually 60% of the ration. When concentrates exceed 60%, a sharp depression in the percentage of milk fat occurs. Thus the total ration is usually 60% concentrate and 40% roughages in early lactation.

As mentioned earlier, a good dairy cow is expected to lose body weight during peak production. This weight is replaced later in lactation. As dairy milk production drops, nutritive requirements decline, and cows are fed accordingly. Cows are fed less concentrate and more roughage in later lactation. Once weight loss ceases, feed allowance is reduced to slightly above requirements for milk and body maintenance.

Dairy Concentrates

The major function of a concentrate feed, such as corn, is to supply the additional energy required for maximum milk production that cannot be provided by forages. Second, protein concentrates, such as soybean meal, adjust the protein level of the feed. Therefore, the percent of protein needed will vary directly with the protein content of the roughage. The concentrate mixture of energy and protein feeds usually varies between 12 and 18% crude protein, with 14 to 16% most commonly utilized on a dry matter basis. Concentrates are fed either in the milking barn or drylot prior to milking. For those not interested in balancing rations, a good rule of thumb for concentrate feeding is 1 lb of concentrate for each 4 lb of milk produced from large breeds (Holstein, Ayshire, and Brown Swiss) and 1 lb of concentrate for each 3 lb of milk produced from smaller breeds (Jersey and Guernsey).

Dairy Feeding Systems

Recently, feeding systems other than those previously mentioned have proven successful. These consist of lead feeding, challenge feeding, and the complete ration system.

Lead Feeding. Lead feeding refers to feeding high levels of concentates starting during the dry period. Concentrate rations are fed at a rate of 4 to 6 lb per day starting at 2 weeks before calving. This is increased by 1 lb per day so that cows are eating approximately 20 pounds at calving. This helps eliminate problems with rapid ration changes at or right after calving and helps increase the freshening cow's appetite for concentrates.

Challenge Feeding. In this system, the cow is offered all the feed she will eat and culled if she fattens instead of producing milk. This system is started about 2 weeks before calving, with the cow receiving concentrate feed at about 1% of body weight. After calving, the cow is fed roughage at 1 to 2% of body weight and given concentrates free choice. Such feeding for the first three or four months of lactation will allow the cow to perform at her genetic maximum for milk production. At 4 months after calving, she is fed according to that milk production requirement until the dry period.

Complete Rations. A complete ration is prepared by mixing the concentrates and forages into one single feed. The ration is balanced for energy and protein and fed free choice to the milking herd. The milking herd is usually separated by milk production levels, and a complete ration is formulated for each group. Thus a separate complete ration is fed to dry cows and cows producing 40 lb of milk per day or less, 41 to 60 lb of milk, and above 60 lb of milk per day. With this system, no additional grain is fed in the milking barn. Complete ration examples are given in Table 11-7.

BREEDING AND SIRE SELECTION IN DAIRY CATTLE

As for beef cattle, about 50 to 60 days are required before a dairy cow is re-bred. Usually, the reproductive tract is manually checked for abnormalities by a veterinarian at 45 days after calving. If the tract is normal, the cow is bred at the first heat period 60 days after calving.

The dairy industry has always led the way for AI (artificial insemination), previously described for beef cattle. Competition between breeding services has resulted in very thorough testing of dairy sires and their effectiveness. As explained before, dairy cattle performance is measured by corrected lactation records. Sire proofs are calculated from this information. The USDA calculates and publishes data on all the cows in such testing programs and their relationship to bulls. A sire with many high-producing daughters is sought after for AI breeding.

TABLE 11-7. SAMPLE RATIONS FOR COWS IN VARIOUS FEEDING PERIODS[a] WITH VARIOUS FORAGE TYPES AND COMBINATIONS

	Period			
Type of Forage	1	2	3	4
Legume Forage				
Alfalfa hay (lb)	22	29	29	22
Grain mix (lb)	41	33	19	6
Oats (lb)	580	630	660	630
Shelled corn (lb)	1180	1280	1300	1300
44% Supplement (lb)	200	60	0	0
Dicalcium (lb)	20	10	20	50
Trace mineral salt and vitamins (lb)	20	20	20	20
Corn Silage, Limited Hay				
Alfalfa hay (lb)	6	6	6	6
Corn silage (lb)	45	63	63	45
Grain mix (lb)	36	27	14	2
Oats (lb)	465	460	460	650
Shelled corn (lb)	1000	935	900	1300
44% Supplement (lb)	480	550	585	0
Dicalcium (lb)	20	25	30	30
Limestone (lb)	15	10	5	0
Trace mineral salt and vitamins (lb)	20	20	20	20
Legume (½), Corn Silage (½)				
Alfalfa hay (lb)	11	15	15	11
Corn silage (lb)	30	40	40	30
Grain mix (lb)	38	29	16	4
Oats (lb)	520	540	580	660
Shelled corn (lb)	1075	1095	1200	1300
44% Supplement (lb)	360	325	185	0
Dicalcium (lb)	15	20	0	0
Monosodium phosphate (lb)	0	0	15	20
Limestone (lb)	10	0	0	0
Trace mineral salt and vitamins (lb)	20	20	20	20
Grass				
Grass hay (lb)	22	29	29	22
Grain mix (lb)	42	34	21	7
Oats (lb)	550	575	635	655
Shelled corn (lb)	1113	1170	1265	1300
44% Supplement (lb)	290	210	60	0
Dicalcium (lb)	10	10	5	0
Limestone (lb)	17	15	15	25
Trace mineral salt and vitamins (lb)	20	20	20	20

Production: period 1 = 90 lb, period 2 = 80 lb, period 3 = 50 lb; fat test = 3.8% all periods and 1300-lb cow.

Dry matter levels: hay = 90%; corn silage = 33%; and grain = 88%.
Forage content (100% DM): alfalfa = 16% CP and 33% CF; corn silage = 8% CP and 26% CF; and grass = 12% CP and 37% CF.

Source: Courtesy of the University of Minnesota, reprinted from *Feeding the Dairy Herd*, Extension Bulletin 218.

[a]The amount of feed indicated meets the cow's needs. Cows may not be able to consume the indicated amounts in periods 1 and 2.

Based on a comparison of the bull's daughters with their contemporaries (herdmates that are in similar lactation and calve in the same season), a predicted difference (PD) is calculated. This estimates a sire's ability to transmit production to future daughters. The USDA publishes PD values for dollars, milk, milkfat percentage, and fat. Some breed associations also calculate a PD for type. Thus a PD of +1200 milk means that mature daughters of this bull will produce 1200 lb of milk more per lactation than mature daughters of a bull with a PD of +0 milk. Some bulls will have PDs of greater than +2000 milk. Another term used is *percent repeatability of predicted difference*, which gives an indication of the reliability of the PD values. This number may range from 1 to 99% and increases as the number of daughters and the number of herds in which the daughters are located increase. The higher the percent repeatability, the more confidence that can be placed in selecting an outstanding sire.

Selected sires should have a high PD (+$100 or more; +1200 milk or more) and a repeatability of at least 60%. Among these sires, the final selection is based on other traits, such as milking ease, udder conformation, and so on. USDA sire summaries are available from state Agriculture Extension Services or AI breeding services. Figure 11-2 shows an example of such information.

FIGURE 11-12. An excellent example of sire selection using predicted differences and repeatability, as well as dairy characteristics. Pictured is 9H107 Arlinda Jet Stream, bred by Wallace N. Lindskoog of Arlinda Farms, Turlock, California. July 1978 USDA proof is PD +1930 milk, −0.21% milkfat +37% fat, +149.519 daughters, 326 herds, 95% reporting. (Courtesy of Sire Power, Inc., Tunkhannock, Pennsylvania)

STUDY QUESTIONS

1. Dairy cattle have a _____ digestive system and are fed similar to beef cattle.
2. Young dairy calves are removed from dams shortly after birth and fed either _____ milk or _____ .
3. Hand-fed calves are weaned from milk at _____ to _____ weeks of age.
4. Replacement heifers should be bred according to _____ not age.
5. Milking cows are usually milked _____ days and "dried up" _____ days before the next lactation.
6. It is undesirable for a good dairy cow to lose weight in early lactation. (True, false) _____ .
7. Cows are usually rebred at _____ days after calving.
8. A sire is usually selected from AI services by use of USDA _____ and _____ figures.

DISCUSSION QUESTIONS

1. Suppose you have eight newborn calves all born on the same day. How much frozen colostrum would need to be thawed? What rate of consumption could be expected? Could excess colostrum be continued to be fed? When and how should it be diluted?
2. Name the ingredients and calculate the pounds needed of each to make 1 ton of ration for dairy calves weaned before 4 weeks and not receiving forage.
3. Calculate the total feed (grain, concentrate protein supplement, and roughage) needed for one lactation for a 1200-lb cow giving 15,000 lb of milk.
4. How might a sire be selected that would have some credibility for increasing milk production in your herd?

Chapter Twelve

JUDGING AND SELECTING DAIRY CATTLE

Judging or selecting dairy cattle involves observation to correlate proper dairy type with the milk production function. A cow with the correct dairy characteristics will perform the function of efficient milk and milk fat production over a long lifetime. This chapter will familiarize the student with the description of the appearance of an ideal dairy cow and the relationship between conformation and milk production. Whether judging (comparing several animals) or selecting (comparing one animal to an ideal type), observation must be based on knowledge of animal parts and function. The student must be familiar with the parts of the dairy cow and recognize them by name. Figure 12-1 gives the proper name and location of such parts.

PROPER FUNCTIONAL TYPE

The Dairy Cow Unified Score Card is used by all dairy breed associations for a description of proper dairy type (Figure 12-2). The score card is broken down into four main divisions: general appearance (30 points), dairy character (20 points), body capacity (20 points), and mammary system (30 points). A detailed study of the Dairy Cow Unified Score Card indicates the appearance and point values of desired dairy type. Although the function of each division

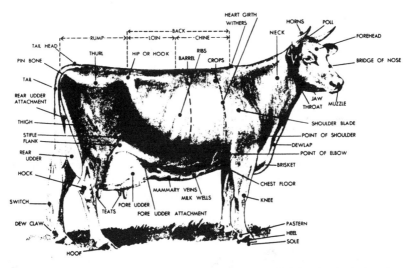

FIGURE 12-1. Parts of a dairy cow. (Courtesy of the Purebred Dairy Cattle Association)

FIGURE 12-2. The Dairy Cow Unified Score Card is used to judge any breed of dairy cattle either in the show ring or on individual farms. (Courtesy of the Purebred Dairy Cattle Association)

DAIRY COW UNIFIED SCORE CARD

Copyrighted by The Purebred Dairy Cattle Association, 1943. Revised, and Copyrighted 1957
Approved — The American Dairy Science Association, 1957

Breed characteristics should be considered in the application of this score card	Perfect Score
Order of observation **1. GENERAL APPEARANCE**	**30**

(Attractive individuality with, feminity, vigor, stretch, scale, harmonious blending of all parts, and impressive style and carriage. All parts of a cow should be considered in evaluating a cow's general appearance)
    BREED CHARACTERISTICS — (see reverse side)
    HEAD — clean cut, proportionate to body; broad muzzle with large, open nostrils; strong jaws; large, bright eyes; forehead, broad and moderately dished; bridge of nose straight; ears medium size and alertly carried
	10

SHOULDER BLADES — set smoothly and tightly against the body
    BACK — straight and strong; loin, broad and nearly level
    RUMP — long, wide and nearly level from **HOOK BONES** to **PIN BONES**; clean cut and free from patchiness; **THURLS**, high and wide apart; **TAIL HEAD**, set level with backline and free from coarseness; **TAIL**, slender
	10

LEGS AND FEET — bone flat and strong, pasterns short and strong, hocks cleanly moulded. **FEET**, short, compact and well rounded with deep heel and level sole. **FORE LEGS**, medium in length, straight, wide apart, and squarely placed. **HIND LEGS**, nearly perpendicular from hock to pastern, from the side view, and straight from the rear view	10
2. DAIRY CHARACTER	**20**
(Evidence of milking ability, angularity, and general openness, without weakness; freedom from coarseness, giving due regard to period of lactation) **NECK** — long, lean, and blending smoothly into shoulders; clean cut throat, dewlap, and brisket. **WITHERS**, sharp. **RIBS**, wide apart, rib bones wide, flat, and long. **FLANKS**, deep and refined. **THIGHS**, incurving to flat, and wide apart from the rear view, providing ample room for the udder and its rear attachment. **SKIN**, loose, and pliable	20
3. BODY CAPACITY	**20**
(Relatively large in proportion to size of animal, providing ample capacity, strength, and vigor) **BARREL** — strongly supported, long and deep; ribs highly and widely sprung; depth and width of barrel tending to increase toward rear	10
HEART GIRTH — large and deep, with well sprung fore ribs blending into the shoulders; full crops; full at elbows; wide chest floor	10
4. MAMMARY SYSTEM	**30**
(A strongly attached, well balanced, capacious udder of fine texture indicating heavy production and a long period of usefulness) **UDDER** — symmetrical, moderately long, wide and deep, strongly attached, showing moderate cleavage between halves, no quartering on sides; soft, pliable, and well collapsed after milking; quarters evenly balanced	10
FORE UDDER — moderate length, uniform width from front to rear and strongly attached	6
REAR UDDER — high, wide, slightly rounded, fairly uniform width from top to floor, and strongly attached	7
TEATS — uniform size, of medium length and diameter, cylindrical, squarely placed under each quarter, plumb, and well spaced from side and rear views	5
MAMMARY VEINS — large, long, tortuous, branching	2
"Because of the natural undeveloped mammary system in heifer calves and yearlings, less emphasis is placed on mammary system and more on general appearance, dairy character, and body capacity. A slight to serious discrimination applies to overdeveloped, fatty udders in heifer calves and yearlings."	
Subscores are not used in breed type classification. **TOTAL**	**100**

for milk production is briefly given, further explanation is useful in connecting dairy type and milking ability.

General appearance denotes breed characteristics and femininity. A cow in milk should give the general appearance of milkiness and feminine refinement (Figure 12-3). She should have symmetry, balance of the body and mammary system, barrel capacity, and a straight, long back and topline to represent strength for milking longevity.

"Dairy character" denotes an angular body with a lack of beefiness (Figure 12-4). This angularity indicates the cow will convert feed to milk and not fat. A dairy cow should carry enough flesh to be in presentable condition but

FIGURE 12-3. A dairy cow head should be clean cut and have a broad muzzle with large, open nostrils and large eyes; the forehead should be broad and moderately dished, showing femininity. (Courtesy of the Holstein-Friesian Association of America)

Chap. 12   JUDGING AND SELECTING DAIRY CATTLE

FIGURE 12-4. The dairy character and body capacity essential for milk production are illustrated in this picture. An angular body with openness, a long, wide barrel, and large heart girth are pictured. (Courtesy of the Holstein-Friesian Association of America)

not be too thin (representing unthriftiness) or too fat (representing more meat than milk production). Body smoothness and openness is evidence of milking ability.

The body capacity (Figure 12-4) should be exemplified by a long, wide, and deep barrel, strongly supported by well-sprung ribs, along with a large heart girth. Body capacity is essential for heavy milk production. It indicates heart and lung capacity, as well as capacity to handle the feed needed for milking.

The mammary system must be large, strongly attached, and well carried for a long milking career (Figure 12-5). Large, soft, pliable udders indicate a sufficient amount of active milk glands and storage capacity. Blood veins should be prominent due to the large amount of blood required to produce milk.

JUDGING DAIRY CATTLE

Dairy cattle are judged in show rings, where one cow is compared to other cattle, or on farms, where the cow is compared to the ideal cow indicated by

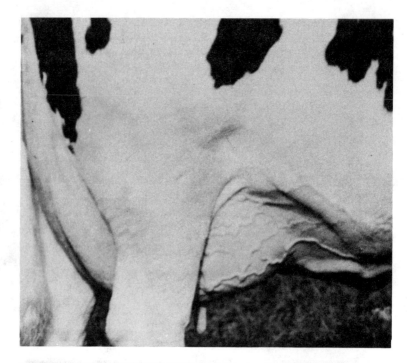

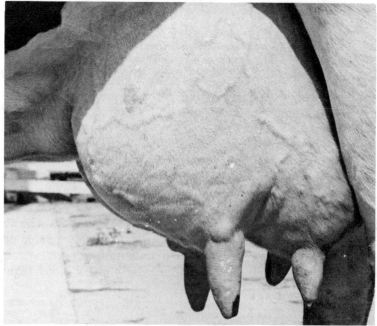

FIGURE 12-5. (a) An excellent udder. Note the strength of attachments. Teats are squarely placed and veining is very good. (b) An undesirable udder. This fore udder is short, bulgy, and loose, and the teats are undesirable in size and shape. (Courtesy of the Holstein-Friesian Association of America)

the Unified Score Card. This second type of judging is a form of selection called classification and is usually done by a person representing a breed association. A numerical score of points from the Unified Score Card is given to the cow, and the cow is classified according to the breed standards. Type classification for breeds is excellent (90 and over points), very good (85 to 90 points), good plus (80 to 84 points), good (75 to 79 points), fair (65 to 74 points), and poor (below 65 points). This may vary among the different breeds and is given as an example.

Dairy cattle, once classified, may carry their highest classification as a part of their production record. In most breed associations, cows' records reflect highest classification only. Thus, reclassification only for animals who may go into higher classification scores is normal. Milk production records (see Chapter 11) and cow classification scores are used in selecting cows and cow daughters as replacements. Figure 12-6 illustrates a cow with both production records and classification scores.

Overall, the correlation between milk production and type classification of a cow is not high. However, if a cow has good type, she will probably have greater longevity since her chances of being culled for a poor udder, bad legs, or some other trait are minimized.

FIGURE 12-6. Leete Farms Betty's Ida of the Ayrshire breed was classified as excellent and had a 305-day production record of 37,170 lb of milk containing 4.3% fat. (Courtesy of the Ayrshire Breeders Association)

STUDY QUESTIONS

1. The _____ is used by all dairy breed associations as the standard for judging and classifying cattle.
2. The main points of a dairy cow are general appearance, dairy character, _____ , and _____ .
3. Unlike beef cattle, a dairy cow's body should be _____ shaped.
4. Body capacity is seen by the _____ and heart girth.
5. On-the-farm judging of individual cows as compared to use of the score card is termed type _____ .
6. The type classification of _____ is the highest given to a dairy breed.

DISCUSSION QUESTIONS

1. Explain what one looks for in dairy cattle in (a) dairy character, (b) body capacity, (c) udder development.
2. Using the Unified Score Card System, give a fictitious, detailed example of judging a 1200-lb Holstein heifer.
3. From memory, name as many of the 47 parts of a dairy cow as you can. Give yourself a Fair rating if you can name 10; Good for 11 to 30; Very good for 31 to 40; Excellent for 41 or more.

Chapter Thirteen

MILK SECRETION AND MILKING MACHINES

For the student to understand the milking operation, it is necessary to have knowledge of the internal anatomy of the udder and of the physiology involved in milk secretion and milk letdown.

INTERNAL ANATOMY OF THE UDDER — THE PATHWAY OF SYNTHESIZED MILK

The udder of a cow is divided into four separate quarters. The fore quarters are usually about 20% smaller than the rear quarters, and each quarter is independent of the other three (Figure 13-1). To review the basic internal anatomy of one of the quarters of the udder, we will follow the path of milk from the point of synthesis to the end of the teat.

Milk is secreted by individual secretory "grape-like" units called *alveoli*. These small units range from 0.1 to 0.3 mm in diameter and consist of an inner lining of epithelial cells that surround a hollow cavity, the *lumen*. The epithelial cells actually secrete the milk by taking raw materials from the blood supply and synthesizing them into milk. The synthesized milk is secreted into the lumen of the alveolus, which when full contains about ⅕ of a drop of milk. A group of such alveoli in a "grape-like" cluster is termed a *lobule*. The lumen

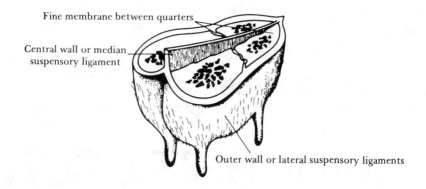

Fine membrane between quarters

Central wall or median suspensory ligament

Outer wall or lateral suspensory ligaments

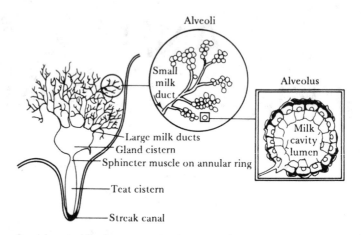

Alveoli

Small milk duct

Alveolus

Large milk ducts
Gland cistern
Sphincter muscle on annular ring

Milk cavity lumen

Teat cistern

Streak canal

FIGURE 13-1.   The udder is divided into four separate quarters, each independent in its milk-producing function.

of each alveolus connects directly with the stem of the cluster of *tertiary* ducts of the lobule. The tertiary ducts drain into the *secondary* ducts, which drain into the large ducts or *primary* ducts. A slight constriction at the junction of one duct to another prevents complete drainage of milk. The primary ducts carry the milk to the *gland cistern.* The gland cistern is the collection point of all ducts and holds about 1 kilogram (2 lb) of milk. The gland cistern drains through the *annular ring* of the upper teat into the *teat cistern* or cavity inside the teat. Milk is prevented from leaking from the teat cavity by the action of a *sphincter muscle,* which surrounds and closes the *streak canal.* The streak canal is the opening from the teat cistern to the outside of the teat.

PHYSIOLOGY OF MILK SECRETION

Since little or no milk is secreted or synthesized during the milking process, all of the milk is present in the udder at the time of milking. Milk is formed or secreted by the cow in the interim between milkings. In this interim, milk is synthesized in each functioning epithelial cell of the alveolus and is expelled into the lumen of the alveolus. Since all milk constituents and precursors are transported to the alveolus by the bloodstream, a great amount of blood must pass through the udder in the synthesis of milk. It has been estimated that from 300 to 500 lb of blood pass through the udder to synthesize 1 lb of milk.

Milk synthesis is most rapid immediately after milking, with the first synthesized milk filling the normal storage places in the udder. No increase in size of the udder or significant increase in mammary pressure occurs in the first hour after milking. The natural storage spaces of the udder hold about 40% of the milk present at milking time with the other 60% being accommodated through stretching of the udder. The udder increases about ⅓ in size during the interim between milkings. With the filling of the natural storage spaces of the udder and the initiation of stretching, mammary pressure increases. With the increased pressure, the secretion rate is slowed so that after 6 hours, it is slightly less than it was in the previous hour. The secretion rate continues to slow with increased pressure until an equilibrium is reached, and, if milk is not removed and mammary pressure exceeds 40 mmHg (mercury), reabsorption of milk occurs.

So at the time of milking, the milk has been previously synthesized and stored in the udder. Forty percent of the milk is stored in the large duct system and cisterns and 60% of the milk is stored in the small duct systems and the alveoli.

Milk Letdown

Only a small quantity of the milk present in the udder is immediately available by natural drainage to the milker. This is the milk located in the cisterns of the udder. Most of the milk is stored in the small ducts and alveoli where natural drainage is prevented. Some mechanism is necessary, therefore, to force the milk into the large ducts and cisterns. This expulsion of milk from the alveoli and small ducts is termed *milk letdown* (Figure 13-2). Without milk letdown, only about 2 lb of milk per quarter could be obtained in milking.

Milk letdown is a nervous reflex produced by various stimuli. These include sucking of the teat by a calf, manipulation of the teats in washing or milking, auditory and visual stimuli, and other pleasant sensory stimuli regularly associated with milking. Such stimuli cause the release of the hormone *oxytocin* from the posterior lobe of the pituitary gland into the bloodstream. Oxytocin is produced by the hypothalamus and stored in the posterior pituitary. Oxytocin reaches the udder within a few seconds and causes contraction of the tissue of the alveoli and small ducts, forcing milk into the larger duct system. Following milk letdown, mammary pressure increases more than 25% due to this expulsion mechanism. Since milk letdown lasts only 6 to 8 minutes, milking must be accomplished within this letdown period for maximum production.

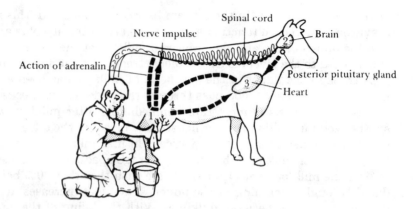

FIGURE 13-2. Milk letdown is caused by various stimuli causing the cow to force milk from the storage places of the udder.

Inhibition of Milk Letdown

When cows are frightened, angered, in pain, or ill-treated, they will not let down their milk. This inhibition of milk letdown is caused by the release of *epinephrine* (adrenalin) from the adrenal gland. Epinephrine inhibits the action of oxytocin and lasts for 20 to 30 minutes.

Residual Milk

At a normal, thorough milking, not all of the milk is removed. Additional milk, termed *residual milk,* may be obtained after normal milking by an injection of oxytocin. The amount of residual milk is variable but is usually about 20% of the total milk produced.

THE MILKING PROCESS

Following milk letdown, milk is under pressure in the large ducts and cisterns but is prevented from draining through the streak canal by the sphincter muscle, which does not relax in milk letdown. In the milking process, some method must be used to force open the streak canal and allow milk to flow from the teat.

In hand milking, the opening between the gland cistern and the teat cistern is closed by squeezing the teat between the index finger and the thumb. Milk trapped in the teat cistern is then forced downward and through the streak canal by compressing the teat against the palm with the fingers.

Machine Milking

Unlike hand milking where the milk is forced through the streak canal, machines use a vacuum or negative pressure to remove milk and to massage the teat. This is accomplished by the use of two vacuum systems: a continuous vacuum and an alternating vacuum.

These vacuum systems are utilized in a double-chambered teat cup assembly consisting of a teat cup shell forming the outer wall and a flexible inflation forming the inner wall. An airtight chamber is formed between the inflation and the teat cup shell. The inner chamber, or inside the inflation, supplies a constant negative pressure of 13 inches Hg to the teat end. (Normally, the vacuum gauge at the pump may read 15 inches Hg. However, this level decreases as the distance to the teat cup increases.) This constant negative pressure causes a pressure differential between the inside of the udder and the inside of the inflation or liner. The higher pressure in the teat forces the teat orifice open, and milk is allowed to flow through the streak canal. Thus milk is not squeezed or forced through the streak canal but is pulled or sucked from the teat by a constant negative pressure or vacuum.

Unless relieved, the constant vacuum will cause internal hemorrhaging and irritation to the teat tissue. An alternating vacuum and atmospheric air operates in the chamber between the inflation and the shell to alleviate this problem. When there is a negative pressure in the outer chamber, it counteracts the continuous vacuum inside the inflation, and the inflation remains open. This is termed the *milking phase*. The vacuum inside the inflation draws the milk from the teat as described previously (Figure 13-3).

When atmospheric air is allowed to enter the chamber, the vacuum inside the inflation causes it to collapse. This is termed the *resting phase*. The closed inflation relieves the teat of the constant vacuum, thus preventing congestion of blood and massaging of the teat to maintain stimulation and proper milk letdown.

This vacuum and atmospheric air is alternated to the teat assembly by a pulsator, which changes the vacuum and atmospheric air through use of

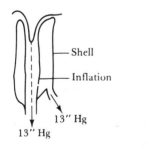

Milking phase

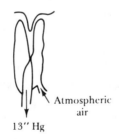

Resting phase

FIGURE 13-3. The milking and resting stages of a milking machine are created by alternating vacuum and air between the shell and inflation.

electrical current and a magnetic valve. The *pulsation rate* is the number of cycles of alternating vacuum and atmospheric air which occur per minute. This pulsation rate is 45 to 60 cycles per minute on most machines. The milking or pulsation ratio is the proportion of time spent under vacuum and atmospheric air and is usually approximately 60:40. This means that the inflation is open 60% (vacuum is on and in milking phase) and closed 40% (atmospheric air is present and in resting phase).

Milking Machine Parts and Terms

Following is a listing of milking machine parts and their functions:

1. *Milking machine.* The mechanical milking system and all auxiliary equipment.
2. *Vacuum pump.* Either a piston or rotary pump which produces the vacuum used in milking. Since the vacuum used is only a partial vacuum of ½ atmospheric pressure (13 in. Hg), the function of the pump is to remove part of the air as it comes into the air inlets.
3. *Airflow meter.* The metering device that measures cubic feet per minute of air at a given vacuum level.
4. *Vacuum level.* The degree of vacuum in a milking system during operation, expressed as inches of mercury differential measured from atmospheric pressure and indicated by a conventional vacuum gauge.
5. *Vacuum tank.* A vessel or chamber in the vacuum system between the pump and the point of air admission that reduces and stabilizes pressure differentials.
6. *Vacuum controller or regulator.* An automatic air valve in the vacuum system that prevents the vacuum from exceeding a preset level by admitting atmospheric air as needed.
7. *Vacuum reserve.* The additional air-moving capacity of the vacuum pump after the requirements of the milking units, bleeder holes, operating accessories, and air leaks have been met—equal to the volume of air entering through the controller.
8. *Pulsator.* The mechanism that permits alternate vacuum/atmospheric pressure to exist between the rubber teat cup liner and the metal shell. This unit creates the massaging action.
9. *Claw.* The sanitary manifold that spaces and connects the four teat cup assemblies and the milk hose.
10. *Teat cup assembly.* Made up of the shell, inflation, and air hose.
11. *Liner (inflation).* The rubber part of the milking machine in actual contact with the cow's teat.
12. *Shell.* The cylindrical metal part of the teat cup.

13. *Milk cup.* A milk reservoir adjoining the claw between the milk tubes and the milk hose.

14. *Milk hose.* Connects buckets or pipeline to the claw.

15. *Receiving vessel or jar.* Receives milk from pipeline. Source of vacuum from the vacuum pump.

16. *Releaser.* Releases milk from under vacuum and discharges it to atmospheric pressure.

17. *Milk pump.* A high-speed pump which pumps milk from the releaser to the milk tank.

18. *Milk tank.* A refrigerated storage tank for "on the farm" storage of milk.

Figure 13-4 illustrates a typical milking parlor.

Suggested Milking Procedure

With the physiology of milk letdown and the mechanics of milking machines in mind, the following milking procedures are recommended.

1. Wash the udder and teats thoroughly with an individual towel for cleanliness and proper stimulation.

2. Use a strip cup to check for abnormal milk (possibility of mastitis).

FIGURE 13-4. A modern herringbone milking parlor is designed for convenience of both the cow and the milker.

3. Attach the milking machine promptly within 30 to 60 seconds after washing. This allows maximum use of the milk letdown reflex.
4. Return to the cow in 2½ to 3 minutes.
5. Machine strip for only a few seconds on the teat cups and gently massage the udder after the cow has almost completely milked out.
6. Remove the teat cups as soon as each quarter is milked out. The milkers should be on the cow for about 5 minutes, depending on your production level and milking system.
7. Dip each teat in a sanitizer solution.
8. Sanitize the teat cups between cows.
9. Milk at regular hours, usually at 12-and 12-hour intervals or 10- and 14-hour intervals.

MASTITIS

Mastitis is an inflammation of the udder. It creates losses to the dairy industry stemming from decreased milk production and cost of medication, withholding milk from treated cows, and replacing animals culled from the herd because of infection and low production. Mastitis occurs in either clinical or subclinical forms. Clinical forms represent about 25% of the cases and are characterized by fever, depression, weakness, loss of appetite, reduced milk production, abnormal milk, and possibly one or more swollen, hardened, or sensitive quarters of the udder. Abnormalities, such as clouding, flakes, blood, or lumps, occur in the milk. Subclinical mastitis represents 75% of the cases, but no signs of it are apparent without special testing. There is, however, a loss in milk production. Subclinical cases can develop into clinical ones unless found and treated.

Mastitis is caused by microorganisms, especially streptococcus and staphylococcus bacteria. Strep organisms that cause mastitis depend on transmission from cow to cow. They are found in the udder of a cow with clinical or subclinical mastitis and are transmitted to other cows if management practices are not followed to prevent spread of the bacteria. Antibiotic treatment is usually very effective in combating this type of mastitis. Staph organisms are found throughout the environment and usually enter the mammary gland through the teat canal (the opening at the end of the teat). Once in the mammary system, they invade healthy secreting tissues and cause the subclinical form, a wasting away of the milk-producing tissue of the cow. Treatment with antibiotics at the time of drying off (quit milking) is usually needed to combat this type of mastitis.

Mastitis prevention must include a mastitis testing and control program, milking machine maintenance, and care taken during the milking procedure by the person doing the milking. Subclinical mastitis must be identified and controlled. This is done by a routine CMT (California Mastitis Test) done at least once a month. The CMT test is simple, economical, and detects mastitis while still in the subclinical stage. It can tell the dairyman about the status of the mastitis problem and indicate what steps need to be taken in preventing further spread. The CMT test causes white blood cells produced due to the infection to form a jell-like substance. The amount of jell and consistency is in direct proportion to the number of white cells in the milk. The reaction is scored as negative (no infection), trace, one, two, or three (definite mastitis present).

Control of mastitis centers around keeping the disease from spreading and treating present infections. Anything that causes injury to the udder, such as improper milking procedure, improper function of milking machines, dragging the udder through mud, and so on, or any contact the udder has with unsanitary conditions will increase mastitis in the herd. Thus caution during the milking procedure, maintenance of the milking machine, and especially care on the part of the milker are important in mastitis control.

STUDY QUESTIONS

1. The correct term for a "cow's bag" or mammary system is an _____ .

2. The mammary system is divided into _____ independent quarters.

3. Milk is actually secreted in grape-like structures called the _____ .

4. Secreted milk is stored in the udder in the _____ of the alveoli, duct system, and the gland and _____ cisterns.

5. Milk is secreted from raw materials brought to the cells of the alveoli by the _____ system.

6. The process of letting loose stored milk by the cow is termed _____ .

7. If a cow is frightened, angered, or ill treated _____ will not occur and very little milk can be harvested.

8. In milking, milk is removed from the udder by use of a _____ inside a milking machine inflation.

9. A regular, uniform milking procedure must be followed for maximum production. (True, false) _____ .

10. The disease _____ causes loss of milk in most dairies and must be constantly combated by good management.

DISCUSSION QUESTIONS

1. Trace the secretion of milk, starting with the alveoli. Name and describe the function of each structure from the alveoli through the streak canal.
2. How is milk letdown and production influenced by the hypothalamus gland?
3. Compared to the old-fashioned form of hand milking, what precautions must be taken to prevent udder injuries when using milking machines?
4. What causes mastitis? How can it be detected early? Give specific recommendations for treatment.

Chapter Fourteen

SHEEP— GENERAL VIEW

HISTORY

Sheep were domesticated long before the dawn of recorded history. Wool fibers have been found in remains of primitive villages in Switzerland that date back an estimated 20,000 years. Egyptian sculpture dating 4000–5000 B.C. portrays the importance of this species to man. Much mention is made in the Bible of flocks, shepherds, sacrificial lambs, and garments made of wool. The Roman Empire prized sheep, anointed them with special oils, and combed their fleece to produce fine quality fibers that were woven into fabric for the togas of the elite.

Little is known about the original selection and domestication of sheep, but they are thought to have descended from wild types like the Moufflon, a short-tailed sheep. Wild varieties in Europe and Asia probably served as foundation stock to produce wool, meat, skins, and milk. It appears that selection practices not only removed most of the wild instincts, leaving the species completely dependent on man for management and protection, but the tail also lengthened. Nearly all domestic breeds today have long tails before docking.

As weaving and felting began to develop as an important arm in the advancement of civilization, more definite types and breeds of sheep began to emerge to produce quality fibers at the expense of other traits. The Merino breed of Spain developed into one of the first recognizable, fine wool breeds. It was so prized that the king of Spain made it a crime punishable by death to send any out of the country without his permission.

The English also developed many breeds very early that would adapt to their varying climate. Thus between the years A.D. 1000 and 1500 the two great powers in wool production were Spain and England, and most of the breeds in the United States today can be traced back to breeds or stock exported from those countries.

Domestic sheep were foreign to the New World and were first introduced by Columbus on his second voyage in 1493 to the West Indies. Cortez brought sheep into Mexico in 1519, and Spanish missionaries contributed to their popularity through the teaching of weaving arts to the Indians.

Napoleon invaded Spain, overthrew the government, and scattered the prized Merinos to other countries. Some, along with many English breeds, found their way to the United States where numbers increased and flocks spread westward the way the country was growing. By 1890, sheep were to be found in virtually every state and territory in the United States.

Because of the vast size of the United States with its great variety of terrain, elevations, and climate, it is not surprising to find so many of the 200 or more breeds estimated to be scattered throughout the world right here at home. All sheep, however, have several characteristics in common. As members of the animal kingdom, they are of the phylum *Chordata* (backbone), class *Mammalia* (suckle their young), order *Artiodactyla* (hooved, even-toed), family *Bovidae* (ruminants), genus *Ovis* (domestic and wild sheep), and species *Ovis aries.*

Within this species many different breeds exist, developed of necessity. No matter if purebred or crossbred, each breed has its place. Successful breeders choose their stock carefully. Some breeds and types available for utilization follow. It is important to note that economic considerations such as the price of wool or lamb may cause a shift from one type or breed to another.

BREEDS OF SHEEP

The most common classification of breeds is by type of wool produced. Other factors such as meat type, color, horned or polled, and adaptability characteristics are considered within each type; the broad classifications are fine wool, medium wool, long wool, crossbred wool, carpet wool, and fur.

FINE WOOL BREEDS

All the fine wool breeds in the United States can be traced back to the Spanish Merino, a breed selected primarily for high-quality wool production.

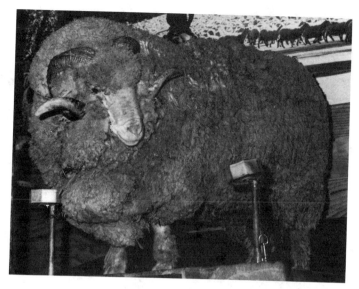

FIGURE 14-1. The fine wool breed most famous in the world is the Merino. (Courtesy of Raewyn Saville, Rotorua, New Zealand)

American Merino

About 1800, importations of Merinos from Spain to Vermont played a major role in spreading the wrinkled, blooded artistocrats through native stock thereby upgrading it. Vermont played a major role in developing the American Merino.

White is the characteristic color; the Merino is usually horned (Figure 14-1), hardy, gregarious (flocks together), and long lived. The type A Merino is very wrinkled, the type B is slightly wrinkled. The Merino ewe will breed year round.

Debouillet

Roswell and Tatum, New Mexico, originated this breed by crossing Rambouillet and Delaine Merino on the Ames Dee Jones ranch. An association of breeders was formed in 1954. The Debouillet (Figure 14-2) is white, horned, hardy, rugged, and adapted to sparse grazing conditions.

Delaine Merino

A strain of Spanish Merino (Figure 14-3) that had more mutton characteristics than the American Merino was imported about 1800 giving rise to the Delaine Merino. The characteristic color is white. This breed is horned, strongly gregarious, and the Delaine Merino ewe will breed year round. The breed is often called the "C type" Merino because it is smooth bodied rather than wrinkled.

FIGURE 14-2.   The Debouillet is adapted for rugged, sparse grazing areas. (Courtesy of the Debouillet Sheep Breeders Association)

Rambouillet

France is considered the origin of this breed even though it was developed from Merino stock imported from Spain about 1850. The name is derived from the agricultural experiment station in Rambouillet, France, where the sheep were perfected through selective breeding.

FIGURE 14-3.   The Delaine Merino is a strain of the Spanish Merino. (Courtesy of Don Shugart Photography, Grapevine, Texas)

FIGURE 14-4. The Rambouillet is the largest of the fine wool breeds. (Courtesy of Don Shugart Photography, Grapevine, Texas)

This white, horned breed is noted for high-quality wool, adaptability to divergent and severe range conditions, and size, being the largest of the fine wool breeds. The Rambouillet ewe will breed year round. Figure 14-4 shows a good representative of the breed.

MEDIUM WOOL BREEDS

Many breeds in this category are selected primarily for mutton type; wool production is sacrificed when necessary.

Cheviot

The Cheviot Hills of Scotland is the origin of this relatively small-statured breed imported to the northern United States about 1830. As seen in Figure 14-5, the Cheviot has a distinctive look. The face and lower legs are white and

FIGURE 14-5. The face and lower legs of the Cheviot are free of wool and both rams and ewes are polled. (Photo by author, New Zealand)

free of wool. The nose is black and both sexes are polled. The head is carried high, the ears erect. Outstanding characteristics include vigor, good milking and mothering ability, quality carcasses, and adaptability to rugged grazing conditions.

Coopworth

Developed in the 1960s from Border Leicester and Romney breeds, this breed has increased to more than 11 million head in New Zealand, Australia, Eastern Europe, and the United States. The breed was developed by Ian Coop of Lincoln College in Canterbury, New Zealand. The Coopworth has replaced the Romney on wetter lowlands and easier hill country. Performance data emphasizes high lambing percentages, heavy fleece weights, physical soundness, rapid weight gains, easy care, and good mothering abilities.

This is a medium-sized to large sheep with white face and legs clear of wool. There is usually some wool on the poll. Equal emphasis is placed on meat and wool. The wool is used in heavier apparels and carpet (Figure 14-6).

Dorset

English imports about 1880 from Dorset, Somerset, and Wiltshire Counties in England marked the beginning of this medium-sized breed in the United States. The Dorset (Figure 14-7) is white faced. Both horned and polled strains are registered in the Continental Dorset Club. This breed is very prolific, lambs early, and is comprised of good milkers. Both sexes will breed year round, the rams being among the most active of any breed in hot weather. Because of this characteristic, the Dorset is widely known for its use in the production of hothouse lambs (lambs produced during a time of the year that most sheep do not normally reproduce).

FIGURE 14-6. The Coopworth is a dual-purpose breed with equal emphasis on meat and wool. Coarse, long, lustrious wool. (Courtesy of Raewyn Saville, Rotorua, New Zealand)

Chap. 14 SHEEP—GENERAL VIEW

FIGURE 14-7. The Dorset will breed out of season and is very prolific. (Photo by author, New Zealand)

Finnsheep (Finnish Landrace)

Brought to the United States about 1966, the Finnish Landrace (Figure 14-8) more commonly referred to as Finnsheep, is a native of Finland. It is widely used in this country today in crossbreeding programs because of its unusual prolificacy, often referred to as the sheep that "lambs in litters." It is not unusual for Finnsheep to have triplets, quadruplets, quintuplets, and even sex-

FIGURE 14-8. Finnsheep or Finnish Landrace is famous for multiple births. (Photo by author)

tuplets. They are good mothers and have among the highest ratings for ease of lambing. When a ewe has quadruplets or more, two or three are usually left on the ewe, the rest given colostrum and raised on lamb milk replacer.

Finnsheep in the United States have mostly been used for crossbreeding purposes by commercial sheep producers to increase the percentage of twins and triplets in the original stock. Up to one-eighth Finnsheep in a cross is reported to generate a statistically significant increase in total lamb crop.

Purebred Finnsheep should weigh 100 lb before breeding and be at least 125 lb at lambing. They can often be bred at 7 months of age and lamb before 1 year of age, although special feed and care during gestation will be needed to meet the nutritional requirements of multiple lambs.

The purebred Finnsheep has a large number of black lambs, a characteristic not discriminated against by the association. In spite of this negative feature, the high incidence of multiple lambs make it an increasingly popular breed for crossbreeding purposes in the United States. Finnsheep have a naturally short tail, which often does not need docking.

Hampshire

Imported to Hampshire County, England, about 1840, this breed has received wide acceptance and popularity in the United States. The black, woolless ears and nose and lack of horns distinguish the rugged Hampshire (Figure 14-9). This large, robust breed is known for its vigor and strength. It does well on the range, lambs easily, and has heavy, vigorous lambs.

FIGURE 14-9. The black, polled Hampshire is large and does well on rangelands.

Montadale

E. H. Mattingly of St. Louis, Missouri, in 1931 started crossing sheep to develop a bloodline composed of 40% Cheviot and 60% Columbia. He came up with a breed that has both a bare face and bare legs below the knee. White hair is required although black, but not brown, spots are acceptable. The Montadale is a polled, medium-sized, mutton-type sheep with good-quality wool.

North Country Cheviot

As might be expected, this breed comes from Scotland and is white, usually polled, and similar to the Cheviot discussed earlier. The outstanding characteristic is high-quality wool.

Oxford

About 1850, the Oxford was imported to the United States from Oxford County, England. Figure 14-10 shows a representative of the breed. It is polled and varies in color. A very large, mutton-type breed that produces good feeder lambs, the Oxford is best adapted to areas with good feed sources.

Perendale

The Perendale was developed (Figure 14-11) by Geoffrey Peren, Massey University, New Zealand, by interbreeding the Cheviot and Romney. A dual meat and wool breed, it is an "easy-care" breed and a classic hearty hill country forager. The Perendale is a small to medium-sized sheep with pointed ears, white face, and legs clear of wool. There is some wool on the poll. The nose is black. The wool has exceptional spring, which gives good shape retention to knitted apparel and a high insulation factor in garments. About 11 million

FIGURE 14-10. The Oxford, a mutton-type sheep, is very large. (Photo by author, New Zealand)

FIGURE 14-11. The dual-purpose Perendale breed, with equal emphasis on meat and wool, was developed for the hill country. (Courtesy of Raewyn Saville, Rotorua, New Zealand)

head are found, mostly in New Zealand, Victoria, and New South Wales in Australia.

Polypay

The Polypay medium wool breed of sheep (Figure 14-12) was developed by Clarence Hulet at the USDA sheep experiment station, Dubois, Idaho, in the late 1960s. The breed was established for the purpose of superior reproduction through twins and triplets.

FIGURE 14-12. Polypay ewe with twin lambs. Even triplets are not uncommon. (Courtesy of the American Polypay Sheep Association)

Bloodlines from Rambouillet, Targhee, and Polled Dorset were combined for the characteristics of size, long-breeding season, milking ability, herding instincts, and fleece characteristics. To this was added the early puberty and high multiple lambing characteristics of the Finnsheep. The result is a breed that produces impressive 200 to 300% lamb crops in well-managed flocks. Breeders are now found in several states and lines have been adapted to several areas, including hot, semiarid climates. A breed association was formed in 1980.

Ryeland

The Ryeland has been used as a dual-purpose breed for over 600 years (Figure 14-13). It takes its name from the rye-growing areas of southern Herefordshire, England. Originally developed by the monks of Leominster, the breed

FIGURE 14-13. The Ryeland is a short wool, meat breed used as a terminal crossing sire. (Courtesy of Raewyn Saville, Rotorua, New Zealand)

has a hardy constitution and thrives in most sheep countries, although numbers have steadily declined since its peak of popularity in the 1930s.

The Ryeland is moderate-sized, thick set, with a white face and legs which are usually covered with wool. The clean, white wool is suitable for textiles requiring a smooth finish and good resilience, mostly in high-quality tweeds and hosiery.

Shropshire

Shropshire and Stafford Counties, England, are considered the origin of the "Shrop." Importations to the United States came about 1850. Both sexes are polled and the face is black, but note that wool covering is complete from the tip of the nose to the hooves (Figure 14-14). Considered a dual-purpose breed (both wool and lamb), this medium-sized sheep is noted as a prolific producer with good mothering ability in the ewe.

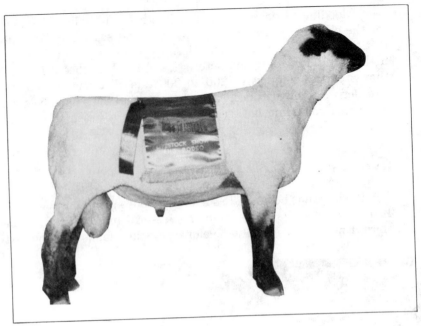

FIGURE 14-14. The Shropshire is considered dual purpose, being used for wool and lamb production.

Southdown

A cluster of hills in southeastern England, the South Downs, lend their name to the breed imported to the United States in the early nineteenth century. Figure 14-15 illustrates a typical Southdown. It is similar in appearance to the Shropshire although normally it has a white- or light colored nose. It is covered with wool on the face and legs, although there is seldom much wool below the knees. The Southdown is small to medium in size.

It is the breed most associated with carcass quality in lambs, having won the International Grand Champion Fat Lamb award in Chicago over 30 times.

FIGURE 14-15. The small to medium-sized Southdown is noted for producing a high-quality carcass. (Courtesy of Raewyn Saville, Rotorua, New Zealand)

Chap. 14 SHEEP—GENERAL VIEW

FIGURE 14-16. A popular meat breed, short wool with some pigmented fibers, is the Suffolk. (Courtesy of Raewyn Saville, Rotorua, New Zealand)

Carload lot winnings have been just as impressive. They are unexcelled in mutton type, conformation, and carcass quality.

Suffolk

The Suffolk had its origin in England; importations were first made to the United States about 1880. Black, bare face and legs mark the distinctive Suffolk (Figure 14-16). A very large breed, the Suffolk is heat tolerant and very hardy. It is a good rustler and adapts well to the western range conditions of the United States, being prolific and a good feeder. The ewe is an excellent milker and mother. One of the few breeds that has any aggressive characteristic at all, the Suffolk will sometimes "play rough."

LONG WOOL BREEDS

Bred chiefly for mutton and size, the long wool breeds are slow to mature and produce fatter carcasses—conditions that have led to a decline in their importance in the United States.

Cotswold

Originally from England, importations first came to the United States about 1830. The Cotswold (Figure 14-17) is polled and white, with long, ropy wool that usually falls over the eyes. The clean face and forelegs are white. This is a large breed used mostly for crossbreeding to improve size, fleshing, and length of fleece.

FIGURE 14-17. The Cotswold is a large, polled, and white long wool breed. (Courtesy of Vocational Instructional Services, Texas A&M University)

Leicester

Developed by the famous English breeder Robert Bakewell, the Leicester (Figure 14-18) was imported to the United States before the Revolutionary War. The breed never attracted much attention in this country although some popularity was gained in Canada. The chief use of the Leicester appears to have been in improving other breeds. It is well muscled and has excellent quality fleece for long wool. The Leicester ewe is a good mother.

FIGURE 14-18. Imported from England, the Leicester has achieved little popularity except in Canada. (Photo by author, New Zealand)

Chap. 14   SHEEP—GENERAL VIEW

FIGURE 14-19. The Lincoln is a dual-purpose breed used mainly for creating crossbred ewes. Long, coarse, lustrious wool. (Courtesy of Raewyn Saville, Rotorua, New Zealand)

Lincoln

Lincoln County, England, is the Lincoln's place of origin. The breed has not received much popularity in the United States but is widely used in New Zealand, Australia, and South America. The Lincoln is polled and white faced with fleece covering most of the face and legs (Figure 14-19). This breed holds the double distinction of being the largest breed of sheep and the heaviest fleeced of the mutton breeds. Because of heavy muscling it is used chiefly for cross-breeding purposes, particularly in South America and especially on Merinos.

Romney

An English breed referred to as the Kent or Romney Marsh breed originated in Kent County, formerly a marsh-like area that was drained ages ago. The Romney is white, open faced, carries a foretop above the eyes, and has little or no wool below the knees (Figure 14-20). It is very popular in New Zealand,

FIGURE 14-20. The large Romney is noted for high-quality carcass and fine fleece. (Photo by author, New Zealand)

a country that made perhaps the greatest improvement in the breed. U.S. imports came from England in 1904, later imports from New Zealand. In addition to large size, the Romney is noted for a rapid growth rate, a shorter, finer fleece than other long wool breeds, and a high quality carcass. The Romney ewe is a good milker.

CROSSBRED WOOL BREEDS

This category of sheep came about through the crossing of long wool and fine wool blood lines.

Columbia

This breed originated in Wyoming and Idaho by crossing Lincoln rams on Rambouillet ewes. A white, open-faced, polled head characterizes the Columbia (Figure 14-21). It produces a high-quality fleece, adapts to range areas, and is considered to have the qualities desired by the commercial sheep manager or breeder. Fancy details for show purposes receive little consideration.

FIGURE 14-21. The white, open-faced, polled Columbia is a favorite of many commercial sheepmen.

Chap. 14 SHEEP—GENERAL VIEW

FIGURE 14-22. The Corriedale is another good dual-purpose breed. (Photo by author, Rotorua, New Zealand)

Corriedale

The Corriedale estate in New Zealand is the origin of this respected producer of wool and mutton. Lincoln rams on Merino ewes served as the basic cross, although possibly some Romney and Leicester blood may have been infused in the early stages. The first importations came to Wyoming in 1914. A polled head and white, open face characterize the blocky Corriedale (Figure 14-22). It is medium sized, hardy, early maturing, and is considered a dual-purpose breed (very good producer of wool and mutton).

Panama

Idaho can claim the distinction for originating this breed. Rambouillet rams on Lincoln ewes served as foundation bloodlines. Because the same two breeds were used to produce the Columbia, characteristics and appearance are about the same.

Targhee

Another Idaho innovation is the Targhee (Figure 14-23). Developed by the U.S. sheep experiment station at Dubois, Idaho, the breed is ¾ Rambouillet and ¼ long wool breeding, mainly Lincoln. Some Leicester and Corriedale blood was also used.

The Targhee—named after Targhee National Forest, Idaho, where experimental animals were summer pastured—is white, polled, open-faced, rugged, adapted to range areas, very prolific, and intermediate in size. Perhaps its most outstanding characteristic is a heavy fleece yield of long staple wool.

FIGURE 14-23.   The Targhee is noted for a heavy fleece yielding long staple wool.
(Courtesy of the U.S. Targhee Sheep Association)

CARPET WOOL BREED

Black-Faced Highland

Except for New Zealand and Australia only one breed, the *Black-Faced Highland* from Scotland, makes any major contribution to the carpet industry. Wool from this sheep is coarse, tough, and long lasting, a necessity in the manufacture of carpets. The Highland is horned (both sexes), free from wool on the legs, and black or mottled colored.

Drysdale

Research by geneticist Francis Dry at Massey University in New Zealand in the 1930s led to the development of the Drysdale (Figure 14-24). In studying the inheritance of hairiness, it was found that certain Romneys carried a pow-

Chap. 14   SHEEP—GENERAL VIEW

FIGURE 14-24. The Drysdale orginated in New Zealand is unique because of its dual-purpose role in the production of carpet wool and meat. (Photo by author, New Zealand)

erful gene which produced large, primary fibers and horn growth on a normally hornless parent breed. In the 1950s this breed was commercially developed as a specialty carpet wool producer. The Drysdale has replaced the Scottish Blackface imports in New Zealand, producing the major component of carpet blends while maintaining a high-quality meat standard.

This breed is medium to large with a white face and legs usually clear of wool. Rams have heavy horns, while ewes have short horns. They are wide-

FIGURE 14-25. The Karakul is a fur sheep kept for production of lamb pelts. (Courtesy of Vocational Instructional Services, Texas A&M University)

spread throughout New Zealand and Australia. The Drysdale is unusual in that it is considered a dual-purpose breed, meat and carpet wool.

FUR-SHEEP BREED

The *Karakul* (Figure 14-25) was named for a village in central Asia where the breed originated. It is black or brown, only the rams are horned, and all are naturally short-tailed or broad-tailed. Few are produced in the United States. This breed is kept for the production of lamb pelts for fur production, producing a rather poor mutton type. Pelts vary in quality and are classified as *broadtail* (lambs killed within a few hours of birth), *Persian lamb* (lambs killed at 3 to 10 days of age), and *Caracul,* a different spelling (lambs killed up to 2 weeks).

STUDY QUESTIONS

1. A wild variety that modern sheep are thought to descend from is the _____ .
2. Nearly all wild sheep have short _____ .
3. Spain is the origin of the _____ wool breed of sheep called _____ .
4. Sheep are ruminants. (True, false) _____ .
5. Sheep belong to the hooved, even-toed order *Artiodactyla.* (True, false) _____ .
6. Sheep belong to the species _____ .
7. The most common classification of sheep is by _____ type.
8. The C-type Merino is _____ (smooth, wrinkled).
9. The largest of the fine wool breeds is the _____ , which originated in _____ .
10. The medium wool breeds were developed to produce more _____ .
11. The _____ breed is widely known for breeding out of season.
12. The breed most famous for production of carcass quality in fat lamb shows is the _____ .
13. The largest breed of sheep is the _____ , which is a _____ wool breed.
14. A _____ wool breed is developed by breeding long wool to fine wool lines.
15. The Black-Faced Highland is a _____ wool breed.
16. The Karakul breed of sheep produces _____ furs and _____ fur.

DISCUSSION QUESTIONS

1. What are the characteristics that all sheep have in common?
2. Name the types of Merino and breeds that were developed from them. Why has this breed had such a positive influence on the industry? What influence has it had on lamb and mutton production?
3. Name the breeds of sheep that are polled, bare legged, and open faced. What type of sheep is each?
4. Sheep may be broken down into how many use classifications? Name them.
5. If you were recommending a breed of sheep for your area, which would it be? Why?

Chapter Fifteen

REPRODUCTION, FEEDING, AND MANAGEMENT OF SHEEP

Sheep are ruminants. Therefore, many of the principles learned in the previous chapters on cattle can be applied, especially feeding and management. Reproduction principles are, however, considerably different. Sheep are seasonally polyestrus (*poly*—many, *estrus*—heat), meaning that ewes have numerous cycles during which conception can occur but only during a certain season, the fall. Principally, because of higher environmental temperatures and the wool coat of this species, both rams and ewes are not fertile during other seasons. A few breeds do not follow this general pattern (the Dorset is an example) and may produce out of season lambs. The males of all breeds are generally fertile all year but semen quality may be rather poor during the warm months.

The ewe also has a much higher incidence of twinning, a 110 to 150% lamb crop in a flock is not unusual. Unlike cattle that have the capability of being managed to calve in the spring, fall, or all year round, sheep normally lamb in the spring. Management and feeding practices prior to this time affect the efficiency of reproduction.

BREEDING CHARACTERISTICS OF EWES

Puberty

Ewe lambs in most mutton-type breeds reach puberty at 5 to 7 months of age, normally breeding the first fall at 8 or more months of age. Larger, slower-maturing wool breeds may be delayed to 16 months or more. Rams reach puberty about a month earlier than ewes in each case.

Estrus

This seasonally polyestrus mammal begins the estrous cycle triggered by cool temperatures in the fall. The cycle is repeated every 16 days on the average (actual range may be 14 to 20 days) until conception occurs. The length of heat averages 30 hours. Unlike females of other species, ewes show few external indications of estrus other than standing to be mounted.

Ovulation

The egg is released from the ovary about 24 to 30 hours after the onset of estrus. Therefore, conception is most likely when breeding occurs late in the heat period.

Gestation

The length of pregnancy averages 148 days (actual range may be 144 to 152 days).

BREEDING MANAGEMENT

Flushing (feeding extra grain or lush pasture) 2 or 3 weeks prior to the breeding season is recommended by all managers to increase the number of ova shed from the ovary and to enhance the incidence of twinning. Feeding ½ lb of oats or corn per head per day is a normal flushing ration. An increase in the lamb crop of 10 to 20% is often the result.

Tagging (shearing the locks of wool and dirt from the dock) the ewes makes service by the ram more certain. The ram is also trimmed around the sheath. The feet are also inspected at this opportune time and trimmed if necessary.

To obtain a visual picture of the progress of mating, the rams may be marked by painting their breast with a thick paste made for this particular purpose. A ewe that accepts a ram is therefore marked by a smear on the rump. The paste may be made by mixing lubricating oil and yellow ochre for the first application which may need applying every day or two. After 16 days, the color is changed by mixing the oil with venetian red and maintained for the next 16 days. The last 16 days, lamp black is mixed with the oil. This method shows which ewes settled (conceived) at the first, second, or third heat period. A ewe that has only a yellow mark bred early. A red mark over yellow means two services. A black mark over the other colors could mean that the ewe is not going to conceive and should possibly be culled from the flock. The colors should always progress from light to dark.

Where ewes are separated into flocks served only by one ram, this marking method can detect a sterile male easily because all ewes are repeat breeders. The problem can be detected in time to bring in a fertile male. Without the simple marking system, an entire lamb crop can be skipped and a year wasted.

The stocking rate for pasture mating is 25 to 35 ewes per yearling ram, 35 to 60 ewes per mature ram. Males are vigorously productive to 6 or 8 years of age.

FEEDING DURING GESTATION AND LACTATION

Ewes should be fed to gain 20 or 25 lb before lambing. This may be accomplished by feeding free-choice legume or mixed legume, hay, silage, or root crops such as turnips. A protein supplement and a mineral supplement may also be needed. The roughage rations and supplements previously suggested for cattle (Chapter 4) may be used. Consumption will be about 4 lb of hay or 12 lb of silage for the average 150-lb ewe.

Because of the rapidly developing fetus and the extra energy needed for strong lambs at birth, ½ lb of grain per ewe per day feeding 30 to 45 days before parturition is suggested. This grain allowance should be increased to one pound after birth to supply the energy needed for maximum lactation, particularly important because of the high incidence of twinning. If a legume hay is used, no protein supplement is generally needed. A salt–mineral mixture should always be available, as well as clean water. As soon as pastures become available, the ewes and lambs are switched gradually to them. Lambs may be creep fed but usually are not because unlike calves they attain sufficient finish on pasture to meet market demands. Complete nutrient requirements for sheep are contained in Tables 15-1 and 15-2.

GESTATION AND PARTURITION CARE

Sheep are relatively hardy but should have shelter to protect them from extended periods of soaking rain; a plain, open shed facing away from the wind will do. Exercise is very important to good blood circulation. If bad weather forces ewes to remain several days without activity near the hay bunks and feed troughs, they should be moderately walked or encouraged to move about by scattering roughage outside.

Tagging the ewes around the dock, flank, and udder as lambing time approaches is a worthwhile chore. This makes lambing more sanitary and nursing easier. Tagging is a term that refers to any trimming or cutting off of the wool (usually wool with manure and/or burrs in it) and may be done on either ewes or rams, near parturition, shearing, or market.

The practice of clipping the wool from the rear quarters of a ewe prior to lambing is called *crutching*. Tagging is also done at this time. The ewe is crutched by placing her on her rump and taking several strokes with the clippers across the belly just ahead of the udder. Then the wool on the inner side of each leg is sheared off. This is followed by placing the free hand over the two teats, making several strokes from the attachment of the udder to the dock. Crutching is a desirable practice; it ensures cleaner lambing and enables the new lamb to find the udder more readily. It is particularly desirable when the herdsman may not be on hand during lambing since, with uncrutched ewes, the lamb may start sucking on a tag and not find a teat before it is too weak to nurse.

Another procedure worthy of mention is *facing*. This is the practice of cutting the wool above and below the eyes of sheep to prevent wool-blindness. If this is not done in some breeds with heavy face covering, they are apt to be poor foragers and show poor mothering ability.

In the larger flocks in the West, lambing occurs on the open range with few problems (except for predators) and little assistance although a good shepherd watches closely and is on hand when needed. Smaller flocks are often confined and ewes "making bag" are moved to a lambing pen—consisting of four portable panels about 4 ft long. Fresh, clean straw or sawdust may be used for bedding. Normal presentation is the same as for cattle and assistance is offered only when obviously needed. When born in confinement, the lambs' navels should be painted with iodine as a precaution against tetanus, which is quite common among confined sheep. If the weather is very cold, it may be advisable to provide heat lamps to keep the lambs warm. Rubbing with old towels can also be effective in drying off the lambs and stimulating blood circulation.

If all goes well after a few days in the lambing pen, the ewes and their young are turned out. Occasionally a ewe will disown one or more lambs that require more attention and patience by the shepherd. Some methods employed to overcome this rejection are (1) milking the ewe and smearing the milk on the rump of the lamb and nose of the ewe, (2) smearing mucous membranes from the lamb's nose on the ewe's nose, (3) blindfolding the ewe, and (4) tying a dog near the lambing pen. The dog sometimes brings out protective maternal instincts and the other methods heighten the sense of smell to overcome rejection.

An orphan may be switched to a ewe having only one lamb using the same methods. A ewe that loses her lamb may be persuaded to accept an or-

TABLE 15-1. DAILY NUTRIENT REQUIREMENTS OF SHEEP (Daily Nutrients per Animal)

Body Weight [kg(lb)]	Dry Matter [kg(lb)]	TDN [kg(lb)]	DE[a] (Mcal)	Protein [kg(lb)]	DP[b] [kg(lb)]	Ca [g(lb)]	P [g(lb)]	Salt (g)	Carotene (mg)	Vit. A (μg)[c]	Vit. A (IU)	Vit. D (IU)
Ewes												
Nonlactating and first 15 weeks of gestation												
45(99)	1.08(2.38)	0.59(1.30)	2.6	0.095(0.21)	0.054(0.12)	3.2(0.007)	2.5(0.006)	9.0	1.7	280	935	250
54(119)	1.26(2.78)	0.68(1.50)	3.0	0.109(0.24)	0.059(0.13)	3.3(0.007)	2.6(0.006)	10.0	2.0	300	1100	300
64(1.41)	1.35(2.98)	0.77(1.70)	3.4	0.122(0.27)	0.068(0.15)	3.4(0.008)	2.7(0.006)	11.0	2.4	396	1320	350
73(161)	1.53(3.37)	0.86(1.80)	3.8	0.136(0.30)	0.073(0.16)	3.5(0.008)	2.8(0.006)	12.0	2.7	446	1485	400
Last 6 weeks of gestation												
45(99)	1.53(3.37)	0.91(2.00)	4.0	0.145(0.32)	0.082(0.18)	4.2(0.009)	3.1(1.007)	10.0	5.8	696	2320	250
54(119)	1.71(3.77)	1.00(2.20)	4.4	0.154(0.34)	0.086(0.19)	4.4(0.010)	3.3(0.007)	11.0	6.8	816	2720	300
64(141)	1.89(4.17)	1.09(2.40)	4.8	0.163(0.36)	0.091(0.20)	4.6(0.010)	3.5(0.008)	12.0	7.9	948	3160	350
73(161)	1.98(4.37)	1.13(2.49)	5.0	0.168(0.37)	0.091(0.20)	4.8(0.011)	3.7(0.008)	13.0	9.1	1092	3640	400
First 8 to 10 weeks of lactation												
45(99)	1.89(4.17)	1.24(2.73)	5.4	0.181(0.40)	0.100(0.22)	6.2(0.014)	4.6(0.010)	11.0	5.8	696	2320	250
54(119)	2.07(4.56)	1.33(2.93)	5.8	0.190(0.42)	0.104(0.23)	6.5(0.014)	4.8(0.011)	12.0	6.8	816	2720	300
64(141)	2.25(4.96)	1.40(3.09)	6.2	0.200(0.44)	0.109(0.24)	6.8(0.015)	5.0(0.011)	13.0	7.9	948	3160	350
73(161)	2.34(5.16)	1.43(3.15)	6.2	0.209(0.46)	0.113(0.25)	7.1(0.016)	5.2(0.012)	14.0	9.1	1092	3640	400
Last 12 to 14 weeks of lactation												
45(99)	1.53(3.37)	0.91(2.00)	4.0	0.145(0.32)	0.082(0.18)	4.6(0.010)	3.4(0.008)	10.0	5.8	696	2320	250
54(119)	1.71(3.77)	1.00(2.20)	4.4	0.154(0.34)	0.086(0.19)	4.8(0.011)	3.6(0.008)	11.0	6.8	816	2720	300
64(141)	1.89(4.17)	1.09(2.40)	4.8	0.163(0.36)	0.091(0.20)	5.0(0.011)	3.8(0.008)	12.0	7.9	948	3160	350
73(161)	1.98(4.37)	1.13(2.49)	5.0	0.168(0.37)	0.091(0.20)	5.2(0.012)	4.0(0.009)	13.0	9.1	1092	3640	400

Weight kg (lb)												
Replacement lambs and yearlings												
27(59)	1.08(2.38)	0.68(1.50)	3.0	0.136(0.30)	2.9(0.006)	0.073(0.16)	2.6(0.006)	8.0	1.7	230	765	150
36(79)	1.26(2.78)	0.73(1.61)	3.2	0.127(0.28)	3.0(0.007)	0.068(0.15)	2.7(0.006)	9.0	2.3	310	1065	200
45(99)	1.35(2.98)	0.77(1.70)	3.4	0.118(0.26)	3.1(0.007)	0.064(0.14)	2.8(0.006)	10.0	2.8	378	1260	250
54(119)	1.35(2.98)	0.77(1.70)	3.4	0.109(0.24)	3.2(0.007)	0.059(0.13)	2.9(0.007)	11.0	3.4	459	1530	300
Rams												
Lambs and yearlings												
36(79)	1.26(2.78)	0.91(2.01)	4.0	0.145(0.32)	3.0(0.007)	0.082(0.18)	2.7(0.006)	9.0	2.3	310	1035	200
45(99)	1.53(3.37)	0.95(2.09)	4.2	0.145(0.32)	3.1(0.007)	0.082(0.18)	2.8(0.006)	10.0	2.8	378	1260	250
54(119)	1.71(3.77)	0.95(2.09)	4.2	0.145(0.32)	3.2(0.007)	0.082(0.18)	2.9(0.007)	11.0	3.4	459	1530	300
64(141)	1.89(4.17)	1.04(2.29)	4.6	0.145(0.32)	3.3(0.007)	0.082(0.18)	3.0(0.007)	11.0	4.0	540	1800	350
73(161)	1.98(4.37)	1.09(2.40)	4.8	0.145(0.32)	3.4(0.008)	0.082(0.18)	3.1(1.007)	12.0	4.5	608	2025	400
Lambs												
Fattening (gaining 0.35 to 0.45 lb/hd/day)												
27(59)	1.08(2.38)	0.68(1.50)	3.0	0.145(0.32)	2.9(0.007)	0.082(0.18)	2.6(0.006)	8.0	1.0	165	500	150
32(71)	1.26(2.78)	0.82(1.81)	3.6	0.154(0.34)	2.9(0.007)	0.086(0.19)	2.6(0.006)	8.0	1.2	198	660	175
36(79)	1.35(2.98)	0.95(2.09)	4.2	0.163(0.36)	3.0(0.007)	0.091(0.20)	2.7(0.006)	9.0	1.4	231	770	200
41(90)	1.53(3.37)	1.04(2.29)	4.6	0.163(0.36)	3.0(0.007)	0.091(0.20)	2.7(0.006)	9.0	1.5	248	825	225
45(99)	1.62(3.57)	1.09(2.40)	4.8	0.163(0.36)	3.1(0.007)	0.091(0.20)	2.8(0.006)	10.0	1.7	280	935	250

Source: Committee on Animal Nutrition, National Academy of Sciences–National Research Council, *Nutrient Requirements of Sheep*, 4th revised edition, Publication 1693 (Washington, D.C.: NAS-NRC, 1968). Reprinted by permission.

[a] 1 kg TDN = 4.4 Mcal DE (digestible energy).

[b] DP, digestible protein.

[c] Vitamin A alcohol, 0.3 µg is equivalent to 1 IU of vitamin A activity. If vitamin A acetate is used, the vitamin A alcohol should be multiplied by 1.15 to obtain equivalent vitamin A activity. The comparable multiplier for vitamin A palmitate is 1.83.

TABLE 15-2. NUTRIENT REQUIREMENTS OF SHEEP (Percentage or Amount per kg of Total Ration Based on Air-Dry Feed Containing 90% Dry Matter)

Body Weight [kg(lb)]	Daily Gain or Loss [g(lb)]	Daily Feed		Percentage of or Amount per kg of Air-Dry Feed										
		Per Animal [kg(lb)]	Percent Live-weight	TDN (%)	DE[a] (Mcal)	Protein (%)	DP[b] (%)	Ca (%)	P (%)	Salt (%)	Carotene (mg)	Vit. A (µg)[c]	Vit. A (IU)	Vit. D (IU)
Ewes														
Nonlactating and first 15 weeks of gestation														
45(99)	32(0.07)	1.2(2.6)	2.6	50	2.2	8.0	4.4	0.27	0.21	0.8	0.7	108	360	96
54(119)	32(0.07)	1.4(3.1)	2.5	50	2.2	8.0	4.4	0.24	0.19	0.7	0.7	110	367	100
64(141)	32(0.07)	1.5(3.3)	2.4	50	2.2	8.0	4.4	0.22	0.17	0.7	0.7	116	388	103
73(161)	32(0.07)	1.7(3.7)	2.4	50	2.2	8.0	4.4	0.20	0.16	0.7	0.7	117	391	105
Last 6 weeks of gestation														
45(99)	168(0.37)	1.7(3.7)	3.8	52	2.3	8.4	4.6	0.24	0.18	0.6	1.5	183	610	66
54(119)	168(0.37)	1.9(4.2)	3.5	52	2.3	8.2	4.5	0.23	0.17	0.6	1.6	194	648	71
64(141)	168(0.37)	2.1(4.6)	3.3	52	2.3	8.0	4.4	0.22	0.16	0.6	1.7	206	687	76
73(161)	168(0.37)	2.2(4.9)	3.0	52	2.3	7.8	4.3	0.22	0.16	0.6	1.8	228	758	83
First 8 to 10 weeks of lactation														
45(99)	-36(-0.08)	2.1(4.6)	4.6	59	2.6	8.7	4.8	0.30	0.22	0.5	1.3	151	504	54
54(119)	-36(-0.08)	2.3(5.1)	4.2	58	2.6	8.4	4.6	0.28	0.21	0.5	1.4	163	544	60
64(141)	-36(-0.08)	2.5(5.5)	3.9	56	2.5	8.0	4.4	0.27	0.20	0.5	1.5	172	574	64
73(161)	-36(-0.08)	2.6(5.7)	3.6	55	2.4	8.0	4.4	0.27	0.20	0.5	1.6	192	638	70
Last 12 to 14 weeks of lactation														
45(99)	32(0.07)	1.7(3.7)	3.8	52	2.3	8.4	4.6	0.26	0.20	0.6	1.5	183	610	66
54(119)	32(0.07)	1.9(4.2)	3.5	52	2.3	8.2	4.5	0.25	0.19	0.6	1.6	194	648	71
64(141)	32(0.07)	2.1(4.6)	3.3	52	2.3	8.0	4.4	0.24	0.18	0.6	1.7	206	687	76
73(161)	32(0.07)	2.2(4.9)	3.0	52	2.3	7.8	4.3	0.24	0.18	0.6	1.9	228	758	83

Replacement lambs and yearlings

27(59)	136(0.30)	1.2(2.6)	4.5	55	2.4	11.0	6.0	0.21	0.19	0.6	85	283	50
36(79)	91(0.20)	1.4(3.1)	4.0	50	2.2	8.7	4.8	0.20	0.18	0.6	97	323	62
45(99)	64(0.14)	1.5(3.3)	3.4	50	2.2	7.6	4.2	0.20	0.18	0.6	111	370	74
54(119)	32(0.07)	1.5(3.3)	2.8	50	2.2	7.0	3.9	0.20	0.18	0.7	135	450	88

Rams

Lambs and yearlings

36(79)	181(0.40)	1.4(3.1)	4.0	62	2.7	10.0	5.5	0.20	0.18	0.6	97	323	62
45(99)	136(0.30)	1.7(3.7)	3.7	57	2.5	8.6	4.7	0.18	0.16	0.6	102	340	68
54(119)	91(0.20)	1.9(4.2)	3.5	50	2.2	7.6	4.2	0.17	0.15	0.6	109	364	71
64(141)	45(0.10)	2.1(4.6)	3.3	50	2.2	6.9	3.8	0.16	0.14	0.5	117	391	76
73(161)	45(0.10)	2.2(4.9)	3.0	50	2.2	6.6	3.6	0.15	0.14	0.5	127	422	83

Lambs

Fattening

27(59)	159(0.35)	1.2(2.6)	4.5	55	2.5	12.0	6.6	0.23	0.21	0.6	61	204	56
32(71)	181(0.40)	1.4(3.1)	4.4	58	2.6	11.0	6.1	0.21	0.18	0.6	64	213	57
36(79)	204(0.45)	1.5(3.3)	4.3	62	2.8	10.7	5.9	0.19	0.18	0.6	68	226	59
41(90)	204(0.45)	1.7(3.7)	4.2	62	2.7	9.5	5.3	0.18	0.16	0.6	70	230	61
45(99)	181(0.40)	1.8(4.0)	3.9	62	2.7	9.4	5.2	0.18	0.16	0.6	72	240	64

Source: Committee on Animal Nutrition, National Academy of Sciences–National Research Council, *Nutritional Requirements of Sheep,* 4th revised edition, Publication 1693 (Washington D.C.: NAS–NRC, 1968). Reprinted by permission.

[a] 1 kg TDN = 4.4 Mcal DE (digestible energy).

[b] DP, digestible protein.

[c] Vitamin A alcohol, 0.3 mcg is equivalent to 1 IU of vitamin A activity. If vitamin A acetate is used, the vitamin A alcohol should be multiplied by 1.15 to obtain equivalent vitamin A activity. The comparable multiplier for vitamin A palmitate is 1.83.

phan by rubbing the orphan's back with the dead lamb. A common practice is to remove the skin of the dead lamb and tie it to the back of the adopted lamb. This works surprisingly well. The skin may later be removed bit by bit.

One danger from this method is possible infection from the dead lamb if it died of disease. Also, in spite of all efforts, it is sometimes too time consuming or impossible to foster a "bummer" (orphan lamb) with another ewe. In large operations where there may be many bummers or a triplet or smaller twin that is not getting enough milk, it may be more advantageous to utilize an automatic lamb self-feeder using cow's milk or milk replacer. For the first few days, lambs should receive colostrum milked from a fresh ewe, goat, or cow.

Lambs are weaned at about 6 months of age, allowing ewes a short rest before beginning the next breeding season in the fall. Lambs may be marketed as grass fat spring lambs, enter the feedlot for further fattening and be classified simply as lambs when marketed, or kept as ewe lambs and ram lambs for breeding stock.

OUT-OF-SEASON LAMBS

The ewes of some breeds of sheep, principally the Dorset, Rambouillet, and Merino, will breed year round. Because males are fertile throughout the year, mutton-type rams are bred to these ewes to provide lambs for special markets.

Hothouse Lambs

This is a highly specialized business. Lambs are dropped in the early fall or winter and sold at the light weight of 30 to 60 lb live weight. Boston and New York are the markets demanding hothouse lambs, so called because they are kept in protected shelters and pushed to be ready for sale within 6 to 12 weeks from birth. The lambs are usually castrated but not docked because some buyers associate the docked condition with older lambs.

Easter Lambs

Some demand for light (20 to 30 lb live weight) lambs occurs each year at Easter, principally in the east. Heavier, more variable weights are also accepted, so this provides a market for early lambs, should the shepherd decide to take advantage of a favorable market.

STUDY QUESTIONS

1. Sheep are generally fertile only during the _____ season.
2. The fertility of the _____ is usually affected most by high temperatures.

3. The incidence of _____ is highest in sheep compared to other ruminants.

4. Puberty of ewes is reached at _____ months of age.

5. _____ weather initiates the seasonally _____ breeds.

6. The estrous cycle averages _____ days.

7. The length of estrus averages _____ hours.

8. Ovulation occurs _____ (before, after) estrus is over.

9. Gestation averages _____ days.

10. Feeding extra grain shortly before the breeding season is called _____ .

11. _____ the ewes insures more certain service by the ram.

12. The color scheme for marking ewes by ram breast paints should be in the color order of _____ , _____ , and _____ .

13. The colors are changed every _____ days.

14. A mature ram may be bred to _____ ewes.

15. A ewe should gain about _____ lb during gestation.

16. Shortly before and after parturition, a ewe should be fed extra _____ for _____ .

17. Navels of lambs born in confined quarters are treated with _____ to guard against _____ .

18. Lambs are weaned at about _____ months of age.

19. Hothouse lambs are usually not _____ .

20. Easter lambs will weigh about _____ live weight.

DISCUSSION QUESTIONS

1. Suppose you have just inherited a sheep ranch with 12,000 head of Rambouillet ewes. Your manager asks for instruction on buying rams for them and other breeding instructions. How many will you need? When should the ewes be bred? How many lambs should you expect at lambing time?

2. How much extra feed will be needed to flush the ewes in Question 1? Give answer in tons and specify when and for how long flushing should continue.

3. Discuss the color marking system to determine which ewes have bred. Give products, colors, and time factors.

4. Describe a system for producing hothouse lambs for eastern markets. Which breeds would be used? Why? When would ewes be bred? When would they lamb? How soon afterward would lambs be marketed?

Chapter Sixteen

SHEEP SELECTION

It seems that every species of animal has evolved with peculiar names describing certain parts of its anatomy. Sheep are no exception, as illustrated in Figure 16-1. The selection of sheep varies depending on the use to which the sheep will be applied. Breeding flock replacements have points not considered in feeder lamb prospects, and wool producers have requirements that are quite different. Selection is made using a variety of techniques that can be broken down into selection based on judging individuals, selection based on pedigree, selection based on animal performance, and selection based on production testing.

After becoming familiar with terminology, observations may be made for the purpose of selecting animals with the characteristics desired, most often put into practice by judging one animal against others.

JUDGING — MUTTON TYPE

Sheep produce two major products, meat and wool. However, in mutton-type breeds, the emphasis is on carcass traits; wool is generally a minor consideration. The procedure used in livestock shows or judging contests illustrates a

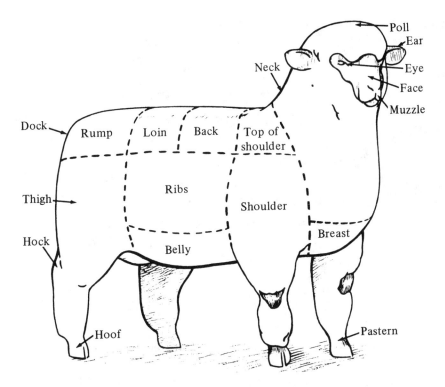

FIGURE 16-1. Parts of a sheep.

systematic, orderly procedure for evaluation and selection of stock. Although the practical application of selection may vary considerably, the show ring approach will be adhered to for the purposes of illustration. Several divisions are used to equalize comparisons between age, sex, or use; however, condensing these divisions to slaughter classes and breeding classes will simplify this discussion.

Slaughter Classes

In slaughter classes, wether lambs (males castrated before reaching sexual maturity) are most often compared, although ewes or other classes may be substituted. The principle to keep in mind is that the live animal should be mentally stripped of its pelt, feet, and head. Once a decision is made on this projected carcass, the finer points (wool, structural defects, color, etc.) can be evaluated in the event of a close decision.

As indicated in Figure 16-2, a class of slaughter sheep should be observed from a distance of 15 ft or more from the side (for depth and length of body, straightness of topline, depth of flank, size, and scale), front (depth and thickness of forequarter, strength of bone), and rear (thickness and depth of quarter, width of loin, and uniformity of width from dock to rack). Some idea of how the lambs may be placed can be determined from this visual appraisal, but

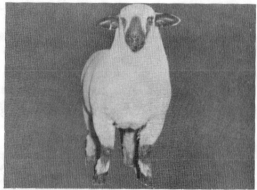
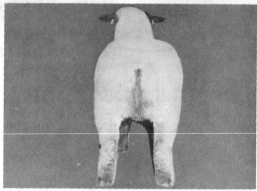

FIGURE 16-2. Slaughter sheep should be first judged from a distance, viewing them from the side, front, and rear. (Courtesy of the American Hampshire Sheep Association)

more than any other class of mammals, sheep must be judged by the hands. Thickness of wool and degree of fat can camouflage many defects. The most important step is moving in for a systematic, mental calibration of differences between pairs. Any system may be used in any combination. One common system is to measure first the loin (Figure 16-3) by placing all four fingers flat against the loin edge at a 90-degree angle to the topline. It is helpful to cross the thumbs for purposes of imagining this width. Second, the width of the rack is determined by using the edge of the arc made between the thumb and the index finger. Do not feel the rack with the palms. Next, feel the size of leg by placing the arc between thumb and index finger of one hand in the flank of the lamb. Place the other hand at the same level to complete a circle (Figure 16-4). A final check should be made for finish (fat covering) over the backbone and ribs. This is done with the finger tips as if feeling for the padding under a carpet.

It is essential that this procedure be conducted in the same order and with a minimum of time so that measurements are not "forgotten" before feeling of the next subject. An experienced judge may spend ten seconds or less

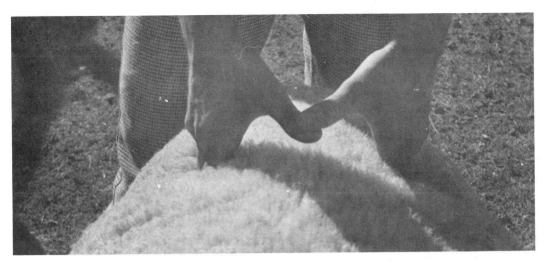

FIGURE 16-3. Measuring the loin.

feeling one lamb. However, the judge may return to repeat the process several times on one or all lambs before completing comparisons.

The visual observation combined with the physical handling determines the superior animals in the mind of the judge, and placings are made so that lambs may be ranked in order. Note that no mention of breed type, color, feet and legs, or wool was made. These are minor points in a carcass class and usually disregarded unless the lambs are very close in muscling and other characteristics.

FIGURE 16-4. Size of leg as determined in judging.

Breeding Classes

Ewes and rams that are to be kept for production purposes are judged the same as slaughter classes to determine muscling except that natural fleshing is looked for instead of fat. Sex classes are usually not mixed; rams are compared to rams, ewes compared to ewes. In addition to muscling, other characteristics become important, such as breed type, color, feet and legs, size, and wool.

Registered breeding stock must have the necessary characteristics of size, type, or color as required by the breed association, so these matters become all-important, whereas slaughter classes may disregard them almost completely. Perhaps the most important and most often overlooked characteristics of breeding animals are the feet and legs. Because sheep are range animals, they must have sound structures to carry them long distances without difficulty. In additon, rams or ewes with leg defects are reluctant or unable to breed.

Wool, even in mutton-type breeds, now becomes an important consideration. The fleece thickness can be judged by feeling with both hands on both sides at the britch. The fleece is also parted at these three points to visually estimate the length and quality of fiber. A heavy-fleeced animal with exceptional muscling is the goal of sheep breeders.

JUDGING — WOOL TYPE

When income from the sale of wool is the major objective, less emphasis is placed on body conformation. However, selection principles remain the same as previously discussed, the only difference being a higher value placed on the wool. The fleece must be judged for weight and quality. The fibers should be long, crimps and yolk (yellow color) should be noted, and black fibers should, ideally, not be present.

As a general rule, wool-type sheep are larger, more angular, less muscular in conformation but heavier fleeced. In selecting stock, the first consideration is given to the wool, then the meat. The opposite is true in selecting mutton-type animals.

PEDIGREE

The importance of bloodlines must not be overlooked or underestimated in selection of sheep. The pedigree is a written record of past appraisals and an estimate of potential performance. For instance, a ram that has produced lambs with high weaning weights, sired sons and daughters with heavy, quality fleeces, or produced champions of carcass shows might well be expected to pass on these qualities to his offspring. To the trained breeder reading the pedigree and making selections based on it, it is a blueprint of the foundation upon which future designs may be built.

ANIMAL PERFORMANCE

No matter how good an animal may look in the show ring or how illustrious its ancestors may be, it is without valid credentials until its own merit is proven. That measurement of merit is animal performance.

Measuring Techniques

Pencil and Pad.   The birth weight of lambs, regularity of lambing and incidence of twinning in ewes, weaning weight, fleece weight, and so on, are observations easily made. These data can be transferred to permanent records for lifetime histories of ewes or rams for use in aiding selection of flock replacements.

Scales.   Perhaps the most useful evaluation tool is the scales, which provide an accurate appraisal of wool, lamb, or mutton production. Weight gain and feed efficiency data over a specified period of time require their use. Yet many sheep breeders do not know the value of their product until the buyer's scales announce their findings. Scales are a useful measurement that should have a place in any measurement and selection system.

Complex Measuring Techniques.   Sheep readily lend themselves to sophisticated methods of evaluation such as ultrasonic measurement of loin-eye size and carcass cut out data, although these techniques are not as widespread in use as with other species. These findings are applied like those previously discussed for beef cattle (Chapter 5).

PRODUCTION TESTING

Adequate records demand the initiation of meaningful individual evaluation tests through performance testing or progeny testing. In order to make breed improvements, certain characteristics must be measured, recorded, and used so that intelligent selection principles may evolve. Table 16-1 gives the characteristics that are of greatest economic importance in production testing and the corresponding heritability range, indicating rate of improvement.

These characteristics are tested for the obvious reason of their strong economic importance. Prolificacy is important because sheep have the inclination and ability to twin regularly. Those ewes that are more productive, although this trait is not highly heritable, stand a better chance of passing on this ability to their daughters.

TABLE 16-1. HERITABILITY OF SOME CHARACTERISTICS IN SHEEP

Characteristic	Heritability
Prolificacy (rate of production, twinning)	Low
Birth weight	Medium
Weaning weight	Medium
Weaning conformation score	Low
Wrinkles or skin folds	Medium
Face covering	High
Fleece weight	Medium
Staple length	High

Birth weights, weaning weights, and conformation scores are easy to obtain, and improvement might be desired and expected at a somewhat faster pace because of slightly higher heritability.

Wrinkles and skin folds are important because shearing is difficult in their presence and fibers lack uniformity. Smoother-bodied sheep are generally preferred in the United States for ease of shearing.

Face covering may sound of minor importance but sheep people know it to be of major concern. Wool-blind ewes graze less, have high labor requirements if clipped around the eyes, and wean lighter lambs than open-faced ewes.

Fleece weight and staple length are an estimate of both quantity and quality of fiber produced. About two-thirds of the emphasis in sheep selection is placed on lamb or mutton production with the remaining one-third on wool production. A selection system will therefore be somewhat different from that previously discussed for cattle and swine.

The independent culling level method can be used if both wool and meat production minimums are established. Any individuals falling below these standards are removed from the flock.

The selection index, where each economically important trait is given a weighted value, is particularly adaptable to sheep selection. Extremely strong features tend to average out very weak faults to give a total score representative of all measurements. A minimum acceptable score is selected and only those animals meeting or exceeding that value are kept for breeding purposes.

A selection system for sheep, as with other species, varies with the changing times and market demands, but a combination of judging, pedigree, animal performance, production testing, and common sense makes improvement possible.

STUDY QUESTIONS

1. Sheep should be judged first from a _____ view.
2. Thickness of wool and the amount of _____ covering can camouflage poor muscling.

3. The width of the _____ and _____ should be judged over the back using the hands.

4. Leg size can best be determined by _____ rather than by sight.

5. Finish is checked by feeling over the _____ .

6. In judging the breeding classes, fat is discriminated against and _____ becomes important.

7. The most important characteristics often overlooked in judging breeding stock are _____ and _____ .

8. Wool is parted at the _____ , _____ , and _____ for observation.

9. The fleece is visually estimated for _____ and _____ of the fibers.

10. In judging wool-type sheep, less emphasis is placed on the _____ and more on the _____ .

DISCUSSION QUESTIONS

1. Discuss the differences in judging breeding classes and slaughter classes of mutton-type sheep; of wool-type.

2. How important is wool in judging mutton-type sheep? Why?

3. What qualities does a judge look for in fleece? Why?

4. Discuss the role of eye appraisal, handling, pedigree, production testing, and heritability estimates in practical sheep selection.

Chapter Seventeen

LAMB, MUTTON, AND FIBER

The slaughtering process of sheep is different from other species in several respects, and the end product, the carcass, is unique because of an age distinction that categorizes young carcasses as lamb and mature carcasses as mutton.

THE SLAUGHTERING PROCESS

Sheep are shackled by the hind leg and a double-edge knife is used to swiftly pierce the neck severing both jugulars. This allows rapid bleeding while the heart still beats to provide the necessary pressure to assure proper drainage. Note that the throat is not cut; it is pierced. The pointed, sharp knife is effective and humane, most sheep never showing signs of emotion or pain.

The foreleg is severed at the pastern above the foot, and the joint is broken with the hands. This is the factor that determines the lamb or mutton classification. Growing animals deposit calcium at the ends of the long bones at the epiphyseal cartilage which does not ossify (harden) until about one year of age. If this break occurs in the cartilage, it exposes a *lamb joint* (Figure 17-1) indicating a young sheep, and the carcass is classified as lamb. If ossifica-

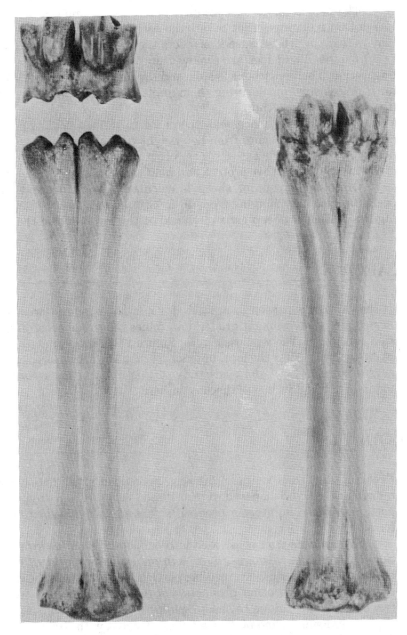

FIGURE 17-1. A "lamb joint" (left) indicates a young sheep, and the carcass is classified as lamb. A "spool joint" (right) indicates a more mature sheep with the carcass classified as mutton. (Courtesy of the National Livestock and Meat Board, *Meat Evaluation Handbook*)

tion has occurred, the joint has fused, growth is ended, and the *spool joint* (Figure 17-1) is exposed, indicating a mature mutton carcass.

The pelt is taken off carefully. Forcing the hands between the pelt and *fell* (a thin membrane that covers the carcass) is called *fisking*. Care is taken not to tear the fell because it protects the meat and reduces drying out.

The head and feet are removed and the offal (intestines) taken out for inspection. The breastbone is split and the forelegs folded back with a wooden or stainless steel pin securing them in place to shape the carcass in an attractive style for marketing.

Washing takes place next, then a 24- to 72-hour cooling period. Some 1% shrinkage usually occurs from hot weight to cool weight because of moisture loss. The better-grade carcasses are often shrouded (wrapped in cheesecloth or light canvas) to further protect and hold them in shape.

Dressing percent of sheep is calculated from live and cooled carcass weights, generally ranging from 46 to 53%. Fat influences higher percentage figures and older sheep generally yield at the lower end of the scale.

THE CARCASS

If the carcass is passed as suitable for human consumption after a thorough inspection, a government grader determines the grade according to conformation and finish. The wholesale cuts (Figure 17-2) are produced by the packing plants and will enter the local butcher market for further breakdown into retail cuts (Figure 17-2). A detailed study of this information is suggested for the inquisitive mind in lieu of discussion here.

BY-PRODUCTS

Recalling that the maximum dressing percent of sheep is usually 53%, there remains 47% to be disposed of or utilized. Very little, if any, is wasted—this portion of the slaughtering process is utilized as by-products in a variety of forms.

By-products may be divided into edible and inedible groups. Edible portions include heart, tongue, liver, and similar items previously discussed in beef cattle (Chapter 6). The inedible portion, in addition to the customary glue, soap, fertilizer, and other products discussed in Chapter 6, contains at least two by-products unique to sheep—pulled wool and chamois skins.

After being fisked off, the skins are treated with a chemical to loosen the follicles that hold the wool fibers. This wool, even though often very short, is removed from the skin by hand and enters the fiber market. The wool-free skin was once tanned and college diplomas were printed on it, giving rise to the adage, "awarding a sheepskin." Its most common use today is as the chamois

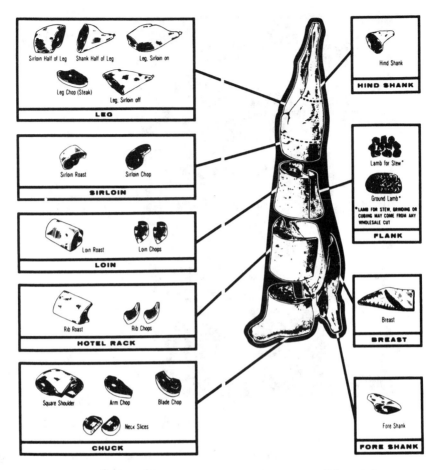

FIGURE 17-2. Mutton wholesale and retail cuts. (Courtesy of the U.S. Department of Agriculture)

skin seen in gas stations and car washes. Recently tanned skins with the wool have become popular as throw rugs, automobile seat covers, and bedding covers for convalescing patients (Figure 17-3).

FIBER

More than 40% of the total income derived from sheep can come from the fleece. Farm flocks usually produce less than 40%, but income from range flocks in the western United States is often higher. The ancient art of weaving wool fibers for apparel, carpets, and novelty fabrics continues despite the development of synthetics. Because of a worldwide shortage of oil and consequently its synthetic derivatives, wool will continue as an important and necessary fiber to meet world market demands.

FIGURE 17-3. Tanned sheepskin, used for throw rugs and car seat covers, accounts for over $3 billion annually to the New Zealand economy. (Photo by author, New Zealand)

The United States produces about 10% of the world's wool. Australia leads all nations and recognizes the importance of sheep to its economy in word and song. A good example is the well-known song "Waltzing Matilda," which is the Australian expression for rustling sheep. It tells of a scoundrel who waits near a billabong (watering place) in order to steal a sheep. Many prayers through the centuries have been uttered by the shepherd to protect his flock and bless his billabong. The production of wool means warm clothing, protection from the elements, and cash income.

Characteristics of Wool

The wool fiber is similar to human hair except that it is smaller in diameter and crimped (like the bellows of an accordian) rather than curly or straight, has a scaly exterior, and stretches more readily. The sawtooth edges of the fibers when woven together interlock to give body and strength; the crimped characteristic gives elasticity. Fine wool is more crimped than coarse wool and demands a higher price because of this characteristic.

Wool will absorb up to 18% of its own weight in moisture without feeling damp and as much as 50% of its weight before being completely saturated, an important characteristic in protecting the body in cold, damp weather. It is

Chap. 17   LAMB, MUTTON, AND FIBER

also an excellent insulator in protecting the body from the sun's rays. For this reason, wool is as common in the Sahara as in the frozen north. Wool is also almost fireproof—fibers will burn only as long as they are exposed to direct flame from another, combustible product—a great safety feature in clothing. Reports of wooden warehouses burning to the ground with only minimum losses of the stored wool are not uncommon. The fiber can stretch 30% of its length and return to shape repeatedly, and in comparison to a strand the same diameter, it is stronger than steel.

Shearing

Shearing is done when the weather begins to turn warm because the yolk softens at this time easing the chore. Depending on the area in the United States, shearing is done from March to July. Removal of the fleece (Figure 17-4) by electrically driven clippers is the method employed by most flock owners, although this job can be done by small, electric clippers or hand shears. Tagging (removal of stained wool and dung locks) is completed prior to actual shearing. Tags are kept in separate bags to be cleaned and processed. Black sheep are separated, shorn last and their wool bagged separately, as are rams. The fleece must be dry when shorn to prevent gumming of the clippers. Care is taken to remove the fleece in one piece as if unwrapping a child in a blanket. Second cuts (going over the same surface a second time) are objectionable because they create short fibers difficult to weave.

To create the best impression on buyers, the unbroken fleece is always rolled the same way, flesh side down on the floor, sides folded to the center, neck folded to the shoulder and rolled starting at the britch. This exposes only

FIGURE 17-4. Wool shearers like world champion Paul Bowen, Rotorua, New Zealand, can shear over 400 sheep per day. (Courtesy of Raewyn Saville, Rotorua, New Zealand)

the best parts of the fleece. A special paper twine is used to tie the bundles, and similar bundles are sorted into regulation jute bags for shipment.

Since shearing is a skill that must be learned by actual demonstration and practice, the serious student of sheep should attend a sheepshearing school or get experience with professional shearers. If unable to do either, a logical alternative is to contact makers of shearing equipment. Some companies maintain a supply of self-teaching charts.

Grease wool is the term given to the raw product. *Scouring* the wool (treating with soap and soda solution) removes the soluble impurities which include yolk (yellow coloring), suint (salts caused from sweating that cause the characteristic odor), and gland secretions that serve to protect the fibers while on the sheep. *Lanolin,* a common product used in cosmetics, is refined from this grease. The weight loss of these impurities is called *shrinkage.* Most wools shrink 50 to 65%, which has an influence on their grade. Insoluble impurities, such as dirt, straw, burrs, insects, paint, and tar, must be removed by hand, by a chemical process called *carbonizing,* or by *carding* on huge machines built for this specific purpose.

Grading Wool

An expert grader examines the scoured wool for length, diameter, fineness, crimp, pliability, color, luster, and other qualities before giving each bundle a grade. Grading can be based on the English or American system. The English method or spinning count system estimates the number of hanks of yarn that can be spun from one pound of scoured wool. A hank is 560 yards. Grades range from 36 to 80. For example, a grade of 40 would mean "40 hanks could be spun from one pound of scoured wool. The American or "blood" system is based on a comparison of quality to the Merino wool. The grades are Fine (representing full blood Merino), Half blood, Three-eights blood, Quarter blood, Low quarter blood, and Common and braid. Figure 17-5 illustrates the American grades of wool.

Wool Classes

Two types of fabrics are made from wool, *worsted* (fibers 2 in. long or more) and *woolens* (short fibers woven to give a fuzzy look). Fibers are also classed according to length and use (apparel or carpet).

Combing Wool.   The longest, finest, highest priced wool, 2 in. or more in length, is used mostly for worsted fabrics. The name comes from the manufacturers' combing process of mechanically combing to separate longer fibers from short ones. These long fibers are woven parallel to each other giving a smooth product.

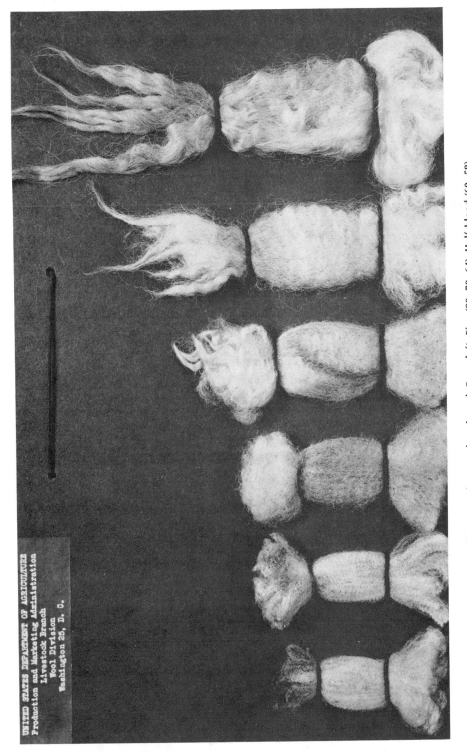

FIGURE 17-5. American grades of wool. From left, Fine (80, 70, 64), Half blood (60, 58), Three-eights blood (56), Quarter blood (50, 48), Low quarter blood (46), and Common and braid (44, 40, 36). (Courtesy of the U.S. Department of Agriculture)

Clothing wool.   The shorter fibers from the combing process fall in this category. The fibers are laid in all directions giving rise to the fuzzy look of woolens, but are also used to make felt, hats, and other products.

French Combing Wool.   The French manufacturers have developed a process that can utilize fibers longer than clothing wool but shorter than combing wool to make worsted fabrics, thereby expanding wool utilization even further.

Carpet Wool.   This wool can vary in length but must be coarse and elastic to withstand heavy use. Most carpet is produced outside the United States.

STUDY QUESTIONS

1. A break joint distinguishes _____ from _____ carcasses.
2. A mutton carcass exposes a _____ joint.
3. The _____ is a membrane that covers the carcass.
4. The fell should be left intact to prevent _____ .
5. A 100-lb sheep will produce _____ lbs of meat.
6. Fat will _____ dressing percent.
7. _____ wool is an inedible by-product of sheep slaughter.
8. Elasticity of wool fibers comes from the _____ characteristic.
9. A great safety feature of wool clothing is that it is almost _____ proof.
10. Shearing is usually done during the months of _____ .
11. Raw fleeces are referred to as _____ wool.
12. Removal of yolk, suint, and other impurities is called _____ .
13. Scouring may shrink a fleece by _____ %.
14. The American or _____ system compares the quality of wool using the Merino as a standard.
15. The English system measures wool by _____ that are _____ yards long.
16. The longest wool fibers are woven into _____ fabrics.
17. Woolens use _____ fibers.
18. _____ wool is used to make felt and hats.

DISCUSSION QUESTIONS

1. Define:

a. lamb joint	**f.** crimp	**k.** carding
b. spool joint	**g.** tagging	**l.** blood system
c. fell	**h.** grease wool	**m.** hank
d. offal	**i.** scouring	**n.** worsted
e. fisking	**j.** shrinkage	**o.** woolens

2. What is the difference between the English and American system of grading wool? Explain and give terms used in each.

3. Discuss the manufacturers' methods and standards for classifying wool.

4. Why is wool considered such an excellent clothing fiber?

5. Name the mutton wholesale and retail cuts.

Chapter Eighteen

DISEASES OF SHEEP

The sheep is a very gentle species of animal, almost devoid of self-protective instincts because of its dependence on man. Shepherds, thousands of years ago, watched over their flocks to keep them from harm. At that time, little management could be done except to frighten off predators. Sometimes a wolf or wild dog would attack and be driven off before seriously wounding an animal, but because sheep are so susceptible to shock, death often claimed a victim. This implies the importance of early recognition of problems and preventative measures. More than other species, sheep have a tendency to give up their hope for life once they are stressed. Diseases, of course, cause stress, and treatment should include not only modern methods and drugs, but also the gentle touch of a compassionate handler.

DISEASES OF SHEEP

As with other classes of livestock, body temperature is often the first signal of developing disorders. The normal temperature range should be from 100.9 to 103.8 °F.

Blackleg

Although sheep are affected to a lesser extent, signs, treatment, and prevention are the same as for cattle. See "Blackleg" in Chapter 9.

Bloat

Signs, treatment, and prevention are the same as for cattle. See "Bloat" in Chapter 9.

Bluebag (Mastitis)

Known as bluebag in sheep, this is the same disease as mastitis in cattle. See "Mastitis" in Chapter 9.

Bluetongue

Depressed appetite, inflammation of the inside of the mouth and nose which may turn blue, frothing at the mouth, and labored breathing are major signs. A red band at the top of the hoof may appear. Although fatalities are not high (10 to 30%), the disease is dreaded in the southwestern part of the United States where it is found. Insects transmit the virus that causes the disease. Treatment has largely been ineffective. Prevention through the use of a vaccine is recommended.

Circling Disease (Listerellosis, Encephalitis)

Principal signs are awkward staggering, walking in circles, and paralysis. Usually fatal, this disease is caused by a bacterial infection affecting the brain and producing incoordination. In ewes, the disease is sometimes confused with pregnancy disease (ketosis), but may be distinguished by moving the head, which is usually turned to one side in circling disease. The head will always return to the same position but remain straightened out in pregnancy disease. There is no prevention other than proper sanitation. Treatment is not very effective, although some response has been noted if antibiotics are given early enough.

Enterotoxemia (Overeating Disease)

Very common in feedlots, enterotoxemia is responsible for the largest death losses in feedlot lambs. Signs are sudden loss of appetite, staggering, convulsions, and death. The cause is a bacterial development brought on by a high scale of feeding, whether in feedlot or lush pasture. Treatment using antitoxins has been effective when supervised by a veterinarian. Vaccines (toxoids) are available and are used as a preventative.

Foot Rot, Founder, Grass Tetany

Signs, prevention, and treatment are the same as for cattle. See Chapter 9.

Lamb Dysentery (Scours)

The principal signs are diarrhea and fever the first few days after birth. Death losses are very high. The cause is a bacterium usually found where there are high concentrations of sheep in close confinement. The problem is seldom seen on the open range or pasture. Antibiotics may reduce death losses, but no treatment has proved very effective. Prevention through good sanitation and well-sheltered, dry quarters is recommended.

Pinkeye

This is the same condition that affects cattle. See Chapter 9.

Pregnancy Disease (Ketosis, Acetonemia)

Called pregnancy disease in sheep, this metabolic disorder is the same as ketosis (Chapter 9) in beef cattle. The condition usually strikes ewes during the last two weeks of gestation. Trembling when exercised, weakness, and collapse are characteristic signs. As with cattle, the feeding of a high-energy feed such as molasses at the time of stress is beneficial as a preventative and treatment.

Scrapie

The name is derived from the chief sign of scraping wool off by rubbing against fences and other objects because of intense itching (Figure 18-1). There is usually no fever, but an unsteady, uncoordinated gait precedes paralysis and death. The cause is a transmissible virus that has defied destruction and is not fully understood. No treatment is known. Destruction of infected flocks has been conducted to prevent spread of this highly fatal disease.

Soremouth (Contagious Ecthyma, Contagious Pustular Dermatitis)

Lambs are most commonly affected. Small blisters appear on the mouth, nose, lips, and tongue. A scab soon develops and pus drains from infected areas.

FIGURE 18-1. Scrapie is an infectious disease that attacks the nervous system of sheep. The name scrapie describes a main symptom of the disease—an infected animal scrapes off patches of wool as it rubs against objects to relieve intense itching. (Courtesy of the U.S. Department of Agriculture)

The condition is sometimes misdiagnosed as bluetongue. Death losses are low, but economic losses because of failure to eat and gain are tremendous. The cause is a virus. Treatment consists of isolation, the only cure being time. Vaccines are available for control where serious outbreaks are likely.

Stiff Lamb Disease (White Muscle Disease, Muscular Dystrophy)

Stiff rear legs and a "humped back" are chief signs. Usually, only young lambs less than a month old are affected. Lambs may die or remain stunted. The cause has been linked to a deficiency of selenium and/or vitamin E, but it is not well understood. Treatment with injections of selenium or vitamin E has proved useful. Prevention by feeding ingredients such as linseed meal, which contains adequate vitamin E, appears the most logical course of action.

Tetanus (Lockjaw)

Sheep are quite commonly affected by *Clostridium tetani* following shearing, docking, castration, and even vaccination. The use of elastrators (rubber bands) for castration or docking is known to predispose the lamb to the infectious disease. One of the first signs is localized stiffness followed by muscle spasms, rigidity, convulsions, and death. Treatment is seldom effective and recovery unlikely. Prevention is the key management practice to avoid tetanus. Tetanus toxoid vaccinations are used in flocks having a history of tentanus. Vaccinations should be given well in advance of docking, castration, parturition, and so on.

Urinary Calculi

Stone-like pebbles composed chiefly of calcium develop in the urinary tract of rams or wethers, blocking passage. Males have difficulty urinating, or complete blockage occurs, resulting in death because of uremic poisoning. Females are seldom affected because stones are easier to pass. The cause is nutritional, occurring mostly under feedlot conditions brought on by an improper calcium/phosphorus ratio.

When the condition arises, an increase in the salt content of the diet by 1.0 to 1.5% (normally 0.5%), a calcium/phosphorus ratio of 2:1, or use of 0.5% ammonium chloride have been of benefit in reducing incidence. The use of 20% alfalfa in rations may also be valuable as a preventative.

Vibriosis

The signs, prevention, and treatment are the same as for cattle (Chapter 9).

PARASITES OF SHEEP

Internal Parasites

Although sheep are less susceptible to health problems than other farm animals, internal parasites are a constant threat to economical production. Sheep have a cleft upper lip allowing them to graze very close to the ground. This characteristic provides for close contact with eggs and larvae of internal parasites that also can live in the soil.

Few sheep are killed by internal parasites—damage comes mostly from poor growth and unthriftiness. *Tapeworms, lungworms, nodular worms,* and *stomach worms* are the most common infestations. Low-lying areas may also harbor *liver flukes* because snails are intermediate hosts that inhabit wet localities.

Signs of parasites in sheep are paleness of eyelids (anemia), poor growth, potbellies, and "bottle jaw" or "poverty jaw" (swellings under the jaw). Treatment with a drench prescribed by a veterinarian is quite effective in restoring optimum health. New drugs are appearing on the market almost annually making any further specific recommendations difficult for this text.

External Parasites

Because of their thick wool covering, sheep are both protected and victimized by external parasites. While providing warmth and protection from the elements, the wool also provides a favorable environment for *maggots, ticks, mites,* and other parasites.

Wool maggots are the larvae of some kinds of blowflies. They are especially a nuisance when unsanitary conditions of the wool prevail because of hot, wet, muddy weather. Contamination by urine and manure in close confinement is also a chief cause. Commercial chemical sprays and dips are very effective as a control.

Ticks are a problem in some areas. A regular schedule utilizing the dipping vat has been found effective to prevent a buildup of this parasite.

Mites cause mange or sheep scab. At least five types that create problems in specific body areas have been identified—those that attack the top of the skin, burrow into the skin, live on the lower extremities, attack the hair follicles, or just cause a mild form of itching. The most common symptoms are itching and loss of patches of wool and a skin condition similar to severely chapped hands with blood sometimes appearing in the cracks. The mites can be seen with the naked eye by scraping some infected skin on a piece of black paper. Sheep dip is a practical control method.

PREDATOR CONTROL

When the Environmental Protection Agency (EPA) banned the use of some potentially environmentally harmful chemicals, predators increased at an alarming rate and losses forced many sheep and goat producers out of the business. However, some producers were determined to find a way to control losses to predators, and several breeds of dogs were imported into the United States for this purpose. One such breed, the Anatolian (Figure 18-2), was imported from Turkey, where for centuries they have been used to guard Angora goats against predators. Dogs live with the flocks and share little human contact except for daily feedings, instinctively guarding and protecting without training.

In many sheep and goat raising areas, losses of 30% or more to predators was not uncommon. Introduction of the Anatolian (and several other equally effective breeds) decreased losses to predators to a level comparable to, or better than, pre-EPA banning of poisonous chemicals.

Another fascinating possibility of predator control among sheep and goats, although still experimental, is the use of a guard mule (Figure 18-3). Reports first came from a few ranchers in western states of the United States that there were few if any losses in pastures where a mule was allowed to range with a flock of sheep. This obviously works best when there are no other mules or horses on the same range. Apparently, the mule stays with the flock for

FIGURE 18-2. Anatolian guard dogs watch over a flock of registered Angora goats. Annual predator losses of 30% or more are often cut to zero through the use of such dogs. Sheep and goat raisers are resorting to this effective measure in the United States. (Courtesy of Sam Harris and *Livestock Weekly*)

companionship but takes on the role of aggressor if any other animals approach the flock.

Although ranchers have never found a dead predator, they have seen mules "worrying" a stray dog, badger, and other animals that get too close to the mule's "family." With its keen eyesight, ability to see almost 360° without turning its head, and its natural curiosity, the mule apparently spots an intruder quickly and chases it until it leaves.

Although these reports have not been verified by university experiments, there are sufficient reports from sheep raisers to indicate that the system has possibilities. Some have even indicated that not all mules share the same enthusiasm for guarding sheep, but that a mule produced from a Shetland stallion, especially one with an aggressive temperament, is almost always successful in reducing predator losses. Some reports have indicated that a guard mule reduced predator losses from 40% down to zero, making commercial lamb production possible in the face of potential economic chaos because of the predator population explosion due to banning of poison control measures.

The stocking rate, because of the flocking instinct of sheep, is quite attractive. Apparently a mule with the guard instinct can protect a flock of 100 or more. Some experimentation may be necessary to find the best ratio.

FIGURE 18-3. This mule stands guard over a flock of sheep. No losses to predators have been reported since the accidental symbiotic relationship between the mule and sheep was discovered.

STUDY QUESTIONS

1. The normal maximum body temperature for sheep is _____.
2. A red band at the top of the hoof is a symptom of _____.
3. The largest death losses in feedlot lambs is from _____.
4. Lamb dysentery usually occurs under _____ conditions.
5. Ketosis or _____ in sheep strikes during the last weeks of gestation.
6. A common itching-type disease is called _____.
7. Soremouth is often confused with _____.
8. Vitamin E or selenium has been suggested as a treatment for

   _____.

9. Urinary calculi usually does not affect _____.
10. Urinary calculi results from an imbalance of _____ and

    _____.

11. The use of 20% _____ rations may prevent development of calculi stones.
12. Sheep are susceptible to internal parasites because of a cleft

    _____.

13. Wet low-lying areas can be expected to harbor _____ which damage the _____ .

14. Paleness of the inside of the eyelid indicates _____ parasites.

15. Other indications of internal infestations are swollen _____ and swellings under the _____ .

16. The most common symptoms of external parasites are _____ and losses of _____ .

17. A practical control method for external parasites is use of a recommended _____ .

DISCUSSION QUESTIONS

1. Describe the signs of bloat, blackleg, and bluebag in sheep. In what order would these three problems be easiest to treat?

2. What disease results in the highest death losses among fat lambs? Describe signs, prevention, and possible treatment.

3. How can encephalitis be distinguished from ketosis?

4. What precautions should be taken at docking, shearing, and castration time? What disease is commonly associated with these jobs, and what signs would indicate a problem?

5. What diseases and parasites directly cause damage to fleece?

Chapter Nineteen

GOATS—MILK, MEAT, AND MOHAIR PRODUCTION

The over 400 million goats in the world have a unique place among farm livestock. The playful and loving goat, kept as a pet by many, produces milk, meat, mohair (goat's hair), and skins for a large part of the world's human population. Ninety percent of the goats in the world are in Asia and Africa, with many a developing nation's population deriving the majority of its milk and meat from goats.

Goats are unique because they are easy to keep, require only a small acreage of land, and are very hardy. Goat diets consist of items not usually consumed by other animals. The bulk of a goat's diet is browse (brush, twigs, weeds). In addition, the goat is an efficient converter of these low quality forages to valuable products. Thus the goat produces agricultural products from otherwise nonproductive lands.

As we shall discuss later, goat's milk has various advantages over cow's milk and has been long recognized by doctors for use by persons with digestive disorders. Goat's milk is also used to process soft cheeses.

Angora goat's hair (mohair) is smoother than wool and fits in the textile industry between wool and the rarer specialty furs such as cashmere and camel's hair (See Figure 19-1). Goat meat is used by many nations in the world (Table 19-1).

FIGURE 19-1. An Angora buck like the ones represented in this photo sold to a New Zealand breeder for a world record $250,000.

HISTORY OF THE GOAT

The goat was among the first animals to be domesticated. The goat originated as the wild goat (*Capra hircus aegagrus*), which dwelled in very rough and rocky terrain. It is thought that early hunters brought young goats (kids) back from their hunts. These kids were raised in the village as pets, later used for milk, meat, and skins. The playful goats probably quickly found a place in these primitive villages. Biblical references are noted for goat hair and goat

TABLE 19-1. DISTRIBUTION OF GOATS IN THE WORLD USED FOR MILK AND MEAT

Country	Distribution of the Goat Population (%)	
	Used for Meat	Used for Milk
Asia	62.5	46.2
Africa	24.7	19.4
South America	3.7	1.9
North and Central America	1.3	3.6
Europe	5.4	22.8
USSR	2.3	6.1

milk. Goat milk has been recommended by doctors and through home remedies throughout history.

In the United States, goats were among the animals brought along with the Virginia settlers. These goats were used for milk and meat. The first purebred dairy goats were imported to the United States in 1893. Today, dairy goats are scattered throughout the United States, with most kept close to the larger metropolitan areas where goat's milk is marketed.

Angora goats kept for mohair were imported in 1849. Angora goats spread westward after the Civil War when Texas became the leading state in mohair production. Texas still maintains this rating with 90% of the Angora goats in the United States kept in the hill country of Texas. Other states with mohair production are Arkansas, New Mexico, Mississippi, Oregon, California, and Utah.

BREED OF GOATS

Goats used in the United States are grouped as dairy, mohair, and meat breeds. Keep in mind that the dairy breeds are also used for meat, as is the mohair breed (Angora) (Table 19-2).

TABLE 19-2. BREEDS OF GOATS CLASSIFIED BY USAGE

Dairy	Mohair and Meat	Meat
Alpines	Angora	Spanish
Nubians		
Toggenburg		
Saanens		
La Mancha		

Dairy Breeds

Alpines.　　These originated in the French Alps (Figure 19-2). They have a wide variation in color, including white, gray, brown, and black. They are large, rugged animals with alert eyes and erect ears. They have very few kidding (giving birth) problems. The doe mature size is 125 lb. They are noted for good milking ability.

Nubians.　　The Nubian goat originated in Africa. They have a short, sleek coat of mostly black and tan colors with long drooping ears. This is a very gentle breed, which gives less milk than the Swiss originated breeds but gives milk high in percent milkfat. The mature doe is about 130 lb. Nubians tend to be more meaty than the other dairy breeds (Figure 19-3).

FIGURE 19-2. The Alpine breed of dairy goat originated in the French Alps. Note the alert, erect ears. (Courtesy of Don Shugart Photography, Grapevine, Texas)

FIGURE 19-3. The Nubian is a gentle breed with drooping ears noted for a higher percentage of milkfat and is meatier than other dairy breeds. (Courtesy of Don Shugart Photography, Grapevine, Texas)

La Mancha. This American breed developed in 1959 from Spanish goats (see Figure 19-4). It is extremely hardy and adapted to adverse conditions. The La Mancha goat has short or no ears, which distinguishes it from other breeds. They are noted for milk high in milkfat content.

Saanen. The Saanens originated from the Saanen Valley in Switzerland. They are a very popular goat, white with long or short hair. Ears are erect and alert. This is the largest of the Swiss breeds with does weighing over 140 lb at maturity. They are noted for giving large quantities of milk, but milk lower in milkfat content.

Toggenburg. The Toggenburg or Togg breed (Figure 19-5) orginated in the Swiss Alps. This is a small breed with short and compact bodies. The does weigh about 100 lb at maturity. Toggs have a shade of brown to the body with white on the lower legs, the base of the tail, and down each side of the

FIGURE 19-4. This champion La Mancha dairy goat was derived from the Spanish meat-type goat. It has short or no ears and is noted for high milkfat content.

face. They can be long or medium haired with beards. Toggenburg goats are good milkers.

Mohair Breed (Angora Goats)

The Angora goat originated in Ankara, Turkey, a mountainous area with a dry climate and extreme temperatures. Both sexes are horned and open faced with long locks of hair over the rest of the body. The hair of the Angora goat is called mohair. Mature bucks weigh from 125 to 175 lb; mature does 80 to 90 lb.

FIGURE 19-5. The Toggenburg came from the Swiss Alps. Good milkers, the Toggs are small with short, compact bodies. (Courtesy of Don Shugart Photography, Grapevine, Texas)

Spanish Goats (Meat Goats)

Spanish goats are larger than Angora goats and have less hair. They come in a variety of colors. They are used primarily for meat production. Spanish goats are very hardy and take a minimum of management and labor. The unique feature of Spanish goats is their reproductive physiology. Unlike other goats who only breed in the fall to winter of the year, Spanish goats can breed throughout the year. This allows for yearlong kidding and yearlong meat production.

SELECTION OF GOATS

Like all animals, goats are selected based on performance, and those body characteristics that indicate performance.

Selecting Dairy Goats

Selection of a dairy doe is based on the same characteristics as selecting a dairy cow. The dairy goat judging card (Table 19-3) is divided into general appearance, dairy character, body capacity, and mammary system. This is just like that for the dairy cow. Description of the ideal animal closely resembles that of the dairy cow with emphasis on feminity; strong back and topline; straight legs and feet; angularity of body; ribs wide apart; large body capacity with a deep barrel, large heart girth, and wide chest; and a capacious, strongly attached, well-carried udder.

Selecting Angora Goats

Angora goats are selected for mohair and meat production. The evaluation for Angora goats is divided into body characteristics and fleece quality. Body characteristics include size and weight for age; constitution and vigor; confor-

TABLE 19-3. THE MILKING DOE SCORE CARD

1. General appearance (30 points)
   Attractive individuality, revealing vigor, feminity with blending of parts. Style impressive. Walk graceful.
2. Dairy character (20 points)
   Animation, angularity, general openness, free of excess tissue. Neck long, lean, withers well defined. Ribs wide apart, flat and long. Thighs flat and lean. Skin loose and pliable. Hair fine.
3. Body capacity (20 points)
   Relatively large in proportion to the animal, providing ample digestive capacity, strength, and vigor. Deep barrel. Large heart girth. Wide chest.
4. Mammary system. (30 points)
   A capacious, strongly attached, well-carried udder of good quality indicating heavy production and long period of usefulness.

mation; amount of bone; and breed type. Fleece characteristics are freedom from kemp (large, white, chalky hair); uniformity and completeness of fleece covering; luster and oil in the fleece; density of fleece (number of fibers per unit area of the goat); length of the fleece (about 1 inch per month of growth); fineness of fleece (the fleece diameter); and character of fleece (uniformity of fleece type).

Selecting Meat Goats

Selection of meat goats (as the Spanish goat) is based on characteristics for meat type animals. This includes size and weight for age, body constitution and conformation, amount of bone, and dressing percentage.

STUDY QUESTIONS

1. Ninety percent of the goats in the world are in _____ and _____ .
2. Goat _____ and _____ are of major importance in feeding humans in many developing countries.
3. In the United States, Angora goats are located mostly in the state of _____ .
4. _____ goats are usually located close to large cities.
5. The _____ and _____ dairy goat breeds give milk high in milkfat content.
6. The _____ breed is popular, giving larger a quantity of milk which is lower in milkfat content than other goats.
7. The Angora is raised for _____ and _____ production.

DISCUSSION QUESTIONS

1. Why are goats unique among farm livestock?
2. List and describe the five breeds of dairy goats used in the United States.
3. Describe the ideal dairy goat.
4. List the characteristics used in judging dairy goats.
5. Describe the desirable body characteristics of a goat used for meat production; for mohair production.

Chapter Twenty

DAIRY GOAT
MANAGEMENT

The dairy goat is often thought of as a miniature dairy cow. Although there are many similarities, distinct differences exist that make goat dairying an important part of today's agriculture. In the United States, dairy goats are second to cattle in milk production for human consumption, yet over the world dairy goats supply the majority of milk for humans in many countries.

Like the dairy cow, the dairy goat has been bred and selected throughout history to produce large amounts of milk. Desirable body conformation of the dairy cow and goat are similar. Individual glandular structure of the goat's udder with the alveoli, milk ducts, gland cistern, and teat anatomy and function in milk production is like that of cattle. Conversion of feed nutrients to milk is very similar in both. Lactation of 305 days with a 60-day dry period is also the norm for both species.

Characteristics unique to the goat for milk production make it a different enterprise. The most obvious is the udder of the goat. Instead of the four teats and udder divisions of the cow, the dairy goat's udder has only two. Second, the dairy goat is very efficient in milk production. In general, seven goats would produce as much milk as one cow, yet 10 goats can be fed on one cow's ration. It is not uncommon for a 125-lb doe to produce over 4000 lb of milk in one 305-day lactation.

TABLE 20-1. AVERAGE COMPOSITION (%) OF MILK FROM THE COW AND GOAT

Species	Water	Fat	Protein	Lactose	Ash	Solids Not Fat	Total Solids
Goat	87.0	4.25	3.52	4.27	0.86	8.75	13.00
Cow	87.2	3.70	3.50	4.90	0.70	9.10	12.80

The dairy goat is only one-tenth the size of the cow, hence it is much easier to maintain. As we shall see, nutrient requirements are less, and the goat will eat a wide variety of feeds, converting them into milk. It is easy to see why goat dairies range from a few goats in the backyard to large commercial dairies with hundreds of does.

CHARACTERISTICS OF GOAT MILK

Goat's milk has been known for its nutritional and medical value since Biblical times. Compared to cow's milk (see Table 20-1), goat milk has the following characteristics:

1. Goat's milk is whiter in color.
2. The milkfat globules are smaller and stay in emulsion with the milk. Fat must be separated by a mechanical separator since it will not rise to the surface over time.
3. The goat milkfat is easier to digest.
4. The protein curd is softer, which enables special cheeses to be produced.
5. Goat's milk is higher in the minerals calcium and phosphorus, and the vitamins A, E, and various B vitamins.
6. Goat's milk is used for persons allergic to cow's milk and for persons with various digestive disorders.

FEEDING THE MILK GOAT

A goat is a ruminant with a digestive tract very similar in size, anatomy and function to that of sheep. Goats, however, are more efficient ruminants than sheep or cattle. They are able to consume more dry matter for their body size (5 to 7% of body weight). This compares to only 2 to 3% of the body weight of a cow in dry matter intake. Goats are also more efficient in digesting coarse fibrous feeds than cattle or sheep. Hence a goat will consume and use roughages that cattle or sheep will not touch.

As with other milking animals, the nutritive requirement of the lactating doe is much larger than that of the dry doe or growing animal. Thus the diet for does in milk should be based on high-quality roughages plus excellent pasture or a concentrate feed. Two to three pounds of leafy, immature hay per day or good-quality pasture per goat is required, plus ½ lb of a 16% concentrate ration per quart of milk produced. Excellent pasture can replace up to half of the concentrate required per day. An example of a concentrate mixture for goats is 40% corn, 20% oats or barley, 25% wheat bran, and 15% soybean or cottonseed meal. To this add 1% salt and 1% calcium/phosphorus supplement. The concentrate mixture should be coarsely ground because goats dislike finely ground, dusty feeds. Plenty of clean water and a salt/mineral supplement should be given free choice to all goats.

Feeding does during the dry period is important for development of the unborn kids and obtaining proper body condition of the doe for maximum milk production. Seventy percent of the birth weight of the unborn kids is developed during the dry period. Feeding the doe during this time is done by supplying good-quality pastures alone, or 1 to 2 lb of concentrates per day on poor- to fair-quality pastures. The concentrate mixture fed to milking does will also suffice for the dry doe.

REPRODUCTION IN THE GOAT

Puberty

The buck kid (male offspring) will reach puberty at 3 months of age but is used only occasionally during this time. Goats are seasonal breeders and will not come into heat until the fall or winter season. A doe kid (young female offspring) born in the spring will go into heat the following breeding season (fall to winter). Whether she should be bred at this time depends upon her growth or size at breeding. Breeding size for most breeds is 85 to 90 lb at 9 to 10 months of age. A well-grown doe will be of adequate size for breeding the first season. If she is not, she must wait and be bred at the next seasonal breeding, or at about 18 months of age.

Breeding Management

As stated, goats are seasonal breeders and will cycle in August through January with the months of September, October, and November the prime months for breeding. During the breeding season, does will come into heat every 21 days with the heat period lasting 2 to 3 days. Heat signs are similar to those of cattle (restlessness, frequent urination, constant bleating, swollen vulva, riding other does, and standing for does to ride her). Does bred on the second day of heat realize high pregnancy rates. A healthy buck can serve at least 30 does. Mating is accomplished by taking the doe to the buck for a single service.

Bucks are kept separate from milking does because their strong odor can be taken up by the milk. Artificial insemination can be used but is not widespread.

Gestation

The gestation period of the goat is from 147 to 155 days (or 5 months). Does must be kept in good condition during gestation for normal kid development. Goats in poor condition during gestation often abort their young.

Parturition (Kidding)

Prior to expected kidding (giving birth), the does are put into a clean, dry, quiet, draft free quarter with about 30 ft^2 allowed per doe. Goats usually have easy kidding, with few complications. Normal kidding occurs in about 30 minutes. Multiple births are normal for goats with twins and triplets common. Multiple births can be encouraged through selection and proper nutrition during gestation.

Care at Parturition

After kidding, the doe should expel the afterbirth in about 4 hours. If this does not happen, consult your local veterinarian. Feed hay free choice to the doe and begin grain feeding the day following kidding. Start with only 1 to 2 lb of concentrate per day and increase it gradually until the doe is consuming $\frac{1}{2}$ lb of concentrate per quart of milk produced.

At kidding, clean the mucous from the noses of the young and dip their navels in iodine to prevent infection. Be sure the newborn kids get colostrum either by sucking or through a bottle.

Milking and Lactation Period

Like dairy cows, goats need to be milked at least twice daily to remove the back pressure within the udder. The lactation period for a goat is 7 to 10 months with a 2-month dry period. Milk yield per day will increase after kidding to a maximum at about 2 to 3 months postpartum. Milk production then declines gradually during the rest of the lactation. As with dairy cows, maximum milk yield per year is obtained from goats with good persistence (slower decline in milk production after the peak).

Although normal lactation is 10 months long, outstanding does are often not rebred as soon as possible. This delay in rebreeding prolongs the lactation up to 20 months to provide a milk supply during the off season when other goats, bred during the breeding season, are dry. Such goats can be managed to kid every other year.

Management of Kids from Birth to Puberty

Dairy goats give colostrum milk for 3 to 5 days after kidding. This must be given to the young kids fed four to five times a day for the first 3 to 5 days. This is done by nursing or by feeding from a bottle, depending upon the management of the owner. Kids raised on bottles should receive about 8% of their body weight in milk per day, fed in two to three portions. As with dairy calves, kids can be successfully raised on milk replacers. Kids are weaned from milk or milk replacer at 6 to 12 weeks of age, depending upon the amount of concentrate mix eaten per day. Kids should be encouraged to eat a good 20% protein starter ration (such as the calf starter ration in Chapter 11).

High-quality hay and fresh water should be made available to the kids. Early consumption of hay and grains promotes rumen development. When the kid is eating hay and grain well, milk can be discontinued. Kids are dehorned, and bucks not kept for breeding are castrated at 2 to 5 days of age. Buck kids kept for replacements should be removed to separate pens at 2 to 3 months of age to prevent premature pregnancies.

Replacement kids should be fed so that they reach breeding size at 9 to 10 months of age. This is accomplished by providing good-quality pastures or hay plus one pound of concentrate mixture per day. The concentrate mixture can be eliminated when high-quality pastures are available.

DISEASES OF GOATS

Goats are subject to many of the diseases and parasites discussed in the chapters on diseases of cattle and sheep. These include blackleg, bloat, brucellosis, encephalitis, foot rot, Johne's disease, ketosis, mastitis, milk fever, pneumonia, soremouth, and tetanus. Also, control of the external and internal parasites as discussed for sheep is important in goat management.

STUDY QUESTIONS

1. The goat's udder has _____ divisions, each with its glandular structure and teat.
2. The goat has a _____ digestive system.
3. Goats can consume _____ % of their body weight in dry matter per day.
4. _____ % of the unborn kid's birth weight is developed during the dry period.
5. Puberty of bucks is at _____ months of age.
6. Goats are bred during the months of _____ thru _____ .
7. Gestation in the goat is about _____ months.

8. Multiple births like twins and triplets are common in goats. (True, false) _____ .

9. The lactation period of the goat is normally _____ .

10. Kids are weaned from milk or milk replacer at _____ to _____ weeks of age.

DISCUSSION QUESTIONS

1. How is the dairy goat different from the dairy cow?
2. Compare goat's milk and cow's milk.
3. How is the reproduction of the goat different from other farm animals?
4. Describe a management practice which would allow does to produce milk when other does are in their dry period.

Chapter Twenty-One

GOATS FOR MOHAIR AND MEAT PRODUCTION

Goats are unique grazers and are able to produce mohair (Angora goat's hair) and meat from otherwise nonproductive lands. Goats are excellent climbers, will travel longer distances in search of preferred forages than cattle or sheep, and are thus well suited to mountainous or hilly, rough ranges. Goats eat a wide range of plants and prefer brushy browse. They are used to help control brushy and weedy plant species in many ranges. Due to their different diet, they are very compatible with sheep and cattle, and combining the two species will increase total range production. Goats are very hardy and require little or no protection from the climate, except after shearing when sheds may be required in cold, wet weather. Angora and Spanish goats do not need extra feed if good pasture is provided. Supplemental feeding is advised only when pastures are short due to drought or cold weather.

MANAGEMENT OF ANGORA GOATS FOR MOHAIR PRODUCTION

Angora goats are kept for mohair and meat production (Figure 21-1). Eighty-five percent of the mohair produced in the world comes from the United States and Turkey. In the United States, 90% of the Angora goat population is kept

FIGURE 21-1. Champion Angora buck. Texas leads the nation in mohair production.

in the hilly ranges of central Texas, where a variety of browse and range vegetation exists.

Angora goats have long locks of mohair, which can be of three types—
based on the type of lock. These are ringlet (C type), flat (B type), or web locks.
Mohair grows 6 to 12 in. per year, and goats are clipped twice yearly. The
average production of mohair is 6.5 lb of mohair per goat per clipping.

Mohair is distinguished from wool in that it is smooth and straight without crimps. The diameter of mohair is similar to that of long wool. The finer
mohair is more valuable. Mohair from young goats is finer than the coarse
mohair of older goats. Mohair is used in making sweaters, dresses, coats, suits,
hats, gloves, scarves, and blankets. Industry uses of mohair include uphol-
stery goods and paintbrushes. The quality of mohair is based on fineness or
diameter of the fiber, fiber length, tensile strength, elasticity, grease content,
condition of the fiber, and freedom from kemp (large, white, chalky hairs).

Management of Angora goats is similar to that of the dairy goat. Seasonal breeding, age of breeding, care at kidding, and basic nutrition are similar. Since they do not give as much milk, however, they do not need
supplemental feeding during lactation if pastures are adequate. Kids are allowed to nurse the doe until weaning. At weaning, kids are either kept for
replacement does and bucks, or slaughtered for meat production.

GOAT MEAT

Angora goats and Spanish goats are raised for goat meat production. Spanish
goats differ from other goats in that they are not seasonal breeders. Hence
they are bred year round to supply a constant supply of meat.

TABLE 21-1. CHEVON CLASSIFICATION

Class	Age at Slaughter
Kid	A few weeks to 9 months of age
Mutton	Castrated males 9 to 18 months of age
Billy	Uncastrated males 9 to 18 months of age
Older goat	Mature goats

Goat's meat is unique in flavor, palatability, and tenderness. It is leaner than other meat and usually less tender. Its leanness has a place in today's demand for less fattening meats. Goat's meat is termed either cabrito or chevon, depending on the age of the goat at slaughter. Cabrito (Spanish for "little goat") is from kids slaughtered after they receive colostrum in the first few days of life. Its main use is for barbecue meat. Chevon is from goats slaughtered at weaning or older. It is classified by age according to Table 21-1. Of these meats, the kid is the most tender, with the meat from the older goat less tender. Older goat meat is used primarily in processed meats such as goat sausage, frankfurters, bologna, and chili con carne.

STUDY QUESTIONS

1. Angora goats are raised for _____ and _____ production.
2. Most mohair is produced in the countries of _____ and _____ .
3. Mohair grows _____ in. per year.
4. A goat will produce about _____ lb of mohair per year.
5. Goat's meat is less lean than other meats. (True, false) _____ .

DISCUSSION QUESTIONS

1. List some grazing characteristics of goats that enable them to utilize otherwise nonproductive lands.
2. Distinguish between mohair and wool.
3. List the quality characteristics of mohair.
4. How is the Spanish goat's reproduction different from other goats?
5. Define cabrito, chevon, and mutton.
6. List the uses of cabrito and chevon.

Chapter Twenty-Two

THE SWINE INDUSTRY

Swine production has always been noted for its ups and downs. Hog markets throughout history have increased and dropped rapidly, often without warning. However, profits have been made throughout history. It has been said that a producer is . . .

absolutely assured of a good profit . . .

if the boar doesn't go "bad,"
if the sow receives proper nutrition,
if the sow doesn't abort from bangs or lepto,
if more than half of the pigs aren't stillborn,
if the dow doesn't have MMA,
if diseases don't strike,
if the gilts don't eat their pigs,
if the pigs don't develop scours,
if the edema isn't too severe . . .

Swine producers have their problems—yet through production knowledge and proper techniques, pork production is a sound investment.

SWINE HISTORY

Swine belong to the phylum *Chordata* (vertebrates), class *Mammalia* (milk giving), order *Artiodactyla* (even-toed, hoofed animals), family *Suidae* (nonruminants), genus *Sus*. The hog that evolved from the wild hog of Europe is of the species *Sus scrofa*; those derived from the wild hog of East India are *Sus vittatus*. The European wild hog had coarser hair (much like a mane along the neck) than present swine. It had large legs and feet, a long head and tusks, a narrow body, and great ability to run and fight. Wild descendants of the European wild hog are found today with many of the same characteristics. The East Indian pig was smaller and more refined than the European wild hog. Instead of solid black, the East Indian pig had a white streak along its side. Both gave rise to present day swine. Swine revert very quickly to the wild state, and razorbacks and other wild hogs are found in the United States.

Early Domestication

Because swine could not be driven by moving tribes like cattle, horses, and sheep, they were domesticated by herdsmen. These people were often low in social standing, and this, coupled with the odor of confined swine, led to dislike and contempt of the early swine producers.

Swine in the United States

Columbus is credited with introducing swine to the New World on his second voyage in 1493. Eight head were purchased in the Canary Islands and released on the island of Haiti where they became foundation stock and spread to other islands. By 1506, the wild hogs had multiplied to the point that they were killing cattle and many had to be hunted down with dogs and destroyed to preserve the balance of nature.

Hernando de Soto came to Haiti in 1538, launched an expedition, and landed in Florida in 1539. Thirteen hogs were reported to have been taken on the expedition. As they were driven through the midwest, many strayed off and reproduced, populating the land with a new species that rapidly adapted. When de Soto died in 1542, the original 13 hogs had multiplied to 700 and continued to provide pork for the diet of the explorers. Occasionally, de Soto, who prized swine, offered pork to the Indians. One recorded incident indicates just how prized they were. Three Indians who had tasted the new meat at the invitation of de Soto were so impressed that they returned during the night to steal a few head; de Soto gave orders to capture the Indians. Two of the Indians were killed, and the third had his hands cut off and was sent back as a warning to others.

Some wild hogs from de Soto's days introduced swine to the United States, but the first importation of good-quality hogs came in 1609 at Jamestown (500 to 600 head). Settlers moving westward later spread them with their movement. By 1790, over 6 million pounds of pork and lard per year were being exported. The center of production shifted to the corn growing areas of the central United States about 1840, where it remains today, although swine are an important part of the economy in every state.

PORK PRODUCTION — TODAY AND TOMORROW

In the midwest, hogs are on one out of every three farms. The Corn Belt and north central states lead in pork production. Iowa leads the nation with 25% of all hogs on inventory. Illinois, Indiana, Missouri, and Minnesota join Iowa to make up the traditional top five hog-producing states.

Throughout the United States, the trend in pork production is toward the larger and more specialized. Improved production techniques, better business management, and improved pork quality keep the industry increasing. The trend is toward an increase in the number of hogs on farms and a decrease in the number of farms producing hogs. Average consumption of pork is remaining fairly constant at 70 lb of pork consumed per person per year in the United States.

BREEDS OF SWINE

Throughout swine production history in the United States, breeds have been developed and maintained to fulfill a demand for a certain type of retail product. The demand has moved from a lard-type hog (carcass high in fat) to the present meat-type hog (carcass high in meat compared to lard). The lower demand and profit for lard as a final product of pork have brought about this switch in hog types. Today, only the meat-type breeds as described below are maintained.

TRADITIONAL BREEDS

American Landrace

The Landrace breed was developed in Denmark and is responsible for the Danes' pork producing name for which they are still famous today. The Danish government protected this breed and did not allow importation of it until an agreement was signed in the 1940s, allowing shipment of surplus stock. The American Landrace Association, Inc., was organized in 1950. The breed is white in color (Figure 22-1) and is known for bacon, carcasses with very long sides, square hams, trim jowls, and short legs. It is noted for good feed conver-

FIGURE 22-1.   The white American Landrace is noted for a long, bacon-type carcass. (Courtesy of the American Landrace Association, Inc.)

sion and prolific reproduction. The American Landrace is longer than other breeds because of an extra vertebra.

Berkshire

The county of Berkshire, England, gave rise to a breed that is black with white feet, face, and tail switch (Figure 22-2). The Berkshire's face is dished and its ears are erect. Desirable characteristics include excellent meat type, long body, and high-quality carcass. Its reputation throughout the years has been for a meaty, well-balanced carcass with a high cut-out value.

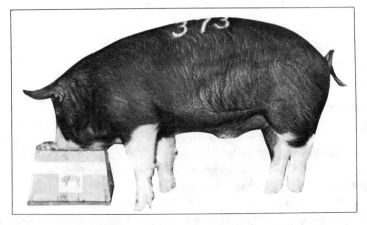

FIGURE 22-2.   The Berkshire is black with white feet and white in the face and tail switch.

FIGURE 22-3. The Chester White breed is noted for large litters, mothering ability, and a carcass with large hams.

Chester White

This white hog originated in Chester and Delaware counties of Pennsylvania (Figure 22-3), and the breed was recognized in 1848. The Chester White hog is prolific; the sow has large litters and is noted for being a good mother. The carcass is high quality, lean, and has large hams and a high dressing percent.

Duroc

A blending of two breeds (Jersey Reds and Durocs of New York) in the northeastern United States gave rise to one of the most popular breeds, the Duroc. It is red with shades from light to dark with a cherry red color preferred (Figure 22-4). The Duroc is noted for excellent rate of gain and feed efficiency.

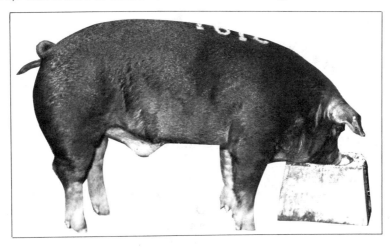

FIGURE 22-4. The red Duroc has an excellent rate of gain.

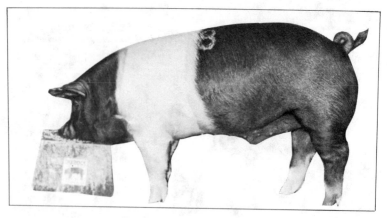

FIGURE 22-5. The white-belted, black Hampshire produces a long carcass which is high in muscling and low in backfat.

Maturing early, the Duroc sow has large litters and is a good mother. The carcass is considered a good meat type.

Hampshire

The New Hampshire is a well-known black hog with a white belt around the shoulders and front legs that originated in Boone County, Kentucky (Figure 22-5). The Hampshire has a refined head and body and is known for producing a good, long carcass, high in muscle and low in backfat. It is a good gaining breed of swine and is known for good mothering ability.

Hereford

The Hereford is a breed whose color is similar to Hereford cattle—white face, red body with white on at least two feet, the underline, and switch (Figure 22-6). This breed was developed in 1830 in La Plata, Missouri. A well-finished carcass and heavy shoulders with compact type are desirable characteristics for the Hereford.

OIC (Ohio Improved Chester)

The OIC breed was developed by L. B. Silver of Salem, Ohio, in the 1860s (Figure 22-7). The breed was founded to improve the size of the original Chester White, a feat claimed by OIC breeders. Controversy persists over their existence separate from the parent breed, but the name is still recognized.

FIGURE 22-6. The Hereford swine, like Hereford cattle, is red with a white face and white on at least two feet, the underline, and the tail switch. (Courtesy of Vocational Instructional Services, Texas A&M University)

FIGURE 22-7. The Ohio Improved Chester (OIC) is larger but similar to the Chester White. (Courtesy of Vocational Instructional Services, Texas A&M University)

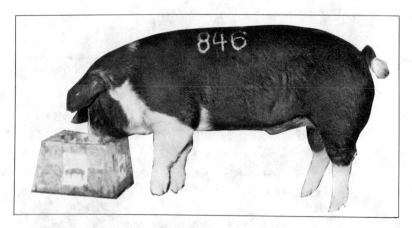

FIGURE 22-8. The Poland China is known for producing a heavy ham carcass and for its gainability.

Poland China

The Poland China originated in the Miami Valley of Ohio. No swine from Poland were used in the foundation—the name Poland China was given to the breed by a Polish farmer, Mr. Asher. Under his direction, the association voted to name the new breed Poland China at an early breed association meeting. The Poland China is black with white feet, face, and tip of tail (Figure 22-8). The breed was developed for and is known for a heavy ham carcass and gainability, reaching the maximum weight at a given age.

Spotted Swine

Spotted swine, previously known as spotted Poland China, originated in Indiana from the Poland China breed that is often spotted at birth. The color pattern consists of a spotted black and white pattern with about 50% of each color (Figure 22-9). It is similar in size and carcass cut-out to the breed from which it originated.

Tamworth

A breed originating in England, the Tamworth was imported in 1822. It is red in color, varying from light to dark (Figure 22-10). The breed is noted for its bacon type with long, smooth sides of the carcass. The Tamworth sow is an excellent mother, and all Tamworths have long legs, making them adaptable to rough country.

FIGURE 22-9.   The Spotted Swine breed developed from the Poland China breed, which is often spotted at birth.

Yorkshire

The Yorkshire originated in England where it was known as the Large White. It is white with a dished face and erect ears (Figure 22-11). The Yorkshire carries the title of "The Mother Breed" because the Yorkshire sow is known for large litters and good mothering ability. The Yorkshire is a good feed converter and yields a carcass of high dressing percent.

FIGURE 22-10.   The red Tamworth is a bacon-type hog having trim long sides of the carcass. (Courtesy of Vocational Instructional Services, Texas A&M University)

FIGURE 22-11. The Yorkshire is referred to as "The Mother Breed." (Courtesy of the American Yorkshire Club)

NEWER BREEDS

Newer breeds have been developed from a crossbred foundation and usually are intended to fulfill a demand for a specific type of hog. Characteristics of each new breed usually combine the desirable characteristics of the breeds used in its development. Numerous hybrids have been developed by swine breeding companies to provide the necessary quality control. Because so many of the new breeds today are being replaced by swine breeding companies, the names of the breeds are of little interest. Details of specific crosses to produce hybrid breeds are kept by the Inbred Livestock Registry Association founded in Noblesville, Indiana.

STUDY QUESTIONS

1. Swine belong to the genus _____ .
2. Like cattle, swine were introduced to the United States by _____ .
3. The largest population of swine is in the _____ states of the United States.
4. The state of _____ is the leader in swine production in the United States.

5. The average yearly consumption of pork per person in the United States is _____ lb.

6. Today, only the _____ types of breeds are maintained in the United States for commercial pork production.

7. The _____ breed is white, and is noted for its long sides, bacon carcass, and length of body because of an extra vertebra.

8. The Berkshire has a _____ face, erect ears, and is black with white feet, face, and tail switch.

9. The _____ breed is a white hog developed in Chester County, Pennsylvania, and is noted for large litters, mothering ability, and a high-quality carcass with large hams.

10. The Duroc is various shades of _____ in color and is noted for rate of gain and feed efficiency.

11. The _____ breed is black with a white belt around the shoulders and front legs.

12. Like their cattle counterparts, _____ swine have a white face and red body.

13. A larger-bodied Chester White breed was developed by L. B. Silver of Ohio and is called _____ .

14. The Poland China breed was developed for and known for heavy _____ carcasses and gainability.

15. The _____ is black and white spotted and resembles the original Poland China breed.

16. The _____ is red in color, noted for its bacon type, mothering ability, and ruggedness.

17. The _____ carries the title "The Mother Breed."

18. The Yorkshire is _____ in color and is noted for large litters, good feed conversion, and good mothering ability.

19. Many new breeds have been developed in the United States by _____ two or more existing breeds.

20. The outstanding characteristics of a newer breed are the outstanding characteristics of the breeds that went into its development. (True, false) _____ .

DISCUSSION QUESTIONS

1. Where are hogs most heavily concentrated in the world, and why?

2. Name and give the outstanding characteristic of every breed of white-colored hog of importance in the United States.

3. Two solid red breeds are popular in the United States. Name them, give their outstanding characteristic, and describe how to tell them apart.

4. Name every breed of swine you can that has some black on the body and some white on the legs.
5. Discuss the zoological classification of swine.
6. What species are responsible for United States swine, and how did they get here?

Chapter Twenty-Three

SWINE REPRODUCTION

Swine are one of the most prolific animals raised for farm market. Litter bearing capacity and short generation intervals make large numbers of marketed animals possible. An outstanding sow can easily produce two litters of 10 market pigs or 4000 lb of pork per year. This allows swine to be an "in and out" business requiring about as short a time to get into the swine business as to get out.

Although differences exist, the reproductive systems of the boar (Figure 23-1) and sow (Figure 23-2) are essentially the same in physiology and function as the bull and cow. For specific details and a review of reproduction see Chapter 3. The boar, like the bull, has the function of producing viable sperm (male sex cells) in the testes and depositing sperm in the sow's reproductive system at the proper time for fertilization of the ova. The sow produces ova from the ovaries similar to those of the cow. The ova are fertilized and the embryos carried until the farrowing.

Swine do have some reproductive characteristics that are unique compared to cattle, sheep, and horses. The most important difference is that swine are *polytocous* (litter bearing) animals producing multiple ova for multiple births (Figure 23-3). It is not unusual for sows to ovulate 15 to 20 ova during an estrous period. However, in average herds only 8 to 9 pigs are normal. Some sows' ova are never fertilized; others that are fertilized die in early develop-

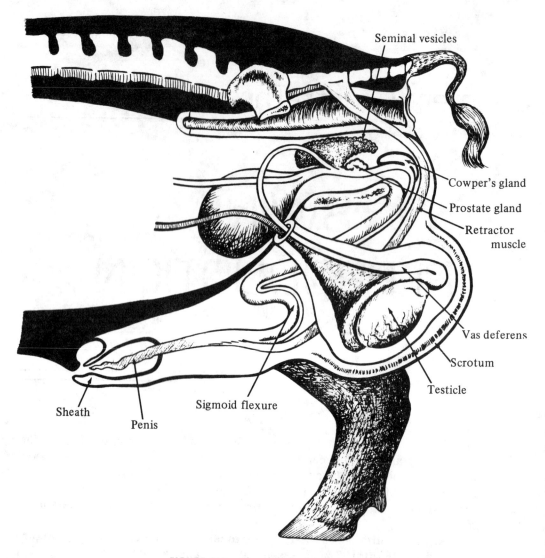

FIGURE 23-1. The reproductive system of the boar.

ment or after birth. Thus, few 15- to 20-pig litters are ever weaned. Knowledge of swine reproduction and proper breeding managment may help increase litter size.

PUBERTY AND ESTRUS IN SWINE

Puberty occurs in gilts at 4 to 7 months of age (Table 23-1), compared to 12 to 18 months in cattle. Thus offspring from swine are obtained in 1 year. Size of gilts at puberty varies from 150 to 250 lb, depending on management and

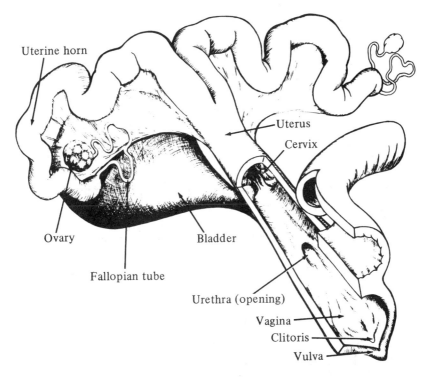

FIGURE 23-2. The reproductive system of the sow.

Labels in figure: Uterine horn, Uterus, Cervix, Ovary, Bladder, Fallopian tube, Urethra (opening), Vagina, Clitoris, Vulva

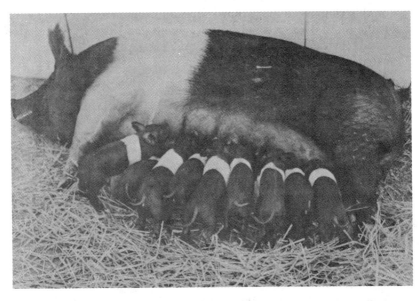

FIGURE 23-3. Swine produce multiple ova for multiple births. (Courtesy of the Hampshire Swine Registry)

TABLE 23-1. PUBERTY, OVULATION, AND GESTATION IN SWINE

Phenomena	Range	Average
Age at puberty (month)	4–7	6
Weight at estrus (pounds)	150–250	200
Duration of estrus (days)	1–5	2–3
Length of the estrous cycle (days)	18–24	21
Time of ovulation (hours after onset of estrus)	12–48	24–36
Best time to breed	Second day of estrus	
Gestation period (days)	111–115	114

nutrition. Boars are somewhat slower and reach puberty at 5 to 8 months of age and 175 to 250 lb.

Estrus in swine varies from 1 to 5 days in length with a normal range of 2 to 3 days. Unlike the cow, which ovulates one egg after the heat period, swine ovulate an average of 10 to 20 ova on the second day of heat. If these ova are not fertilized, the estrus will reoccur every 21 days.

BREEDING FOR MAXIMUM LITTER SIZE

Age to Breed Gilts

Gilts (young female hogs), although reaching puberty at 4 to 7 months of age, are usually not bred until they are 8 months of age or on their third heat period (Figure 23-4). This allows gilts to farrow at about 1 year of age. Gilts farrowing at this time have more longevity and produce larger litters throughout their lives. More ova are shed on the second and third heat periods than the first, and an average of two to three extra pigs are farrowed when bred on the third estrus rather than the first. Approximately one extra pig per litter results from breeding on the second heat period rather than the first.

Time of Breeding

For maximum fertilization of the ova, breeding is usually done 12 to 24 hours after the onset of heat. Because ova are ovulated on the second day of heat, sperm must be present to fertilize the ova before they die (within 12 hours of ovulation). A practice of allowing two services during mating, at 12 and 24 hours after heat starts, results in higher conception rates, more ova fertilized, and thus larger litters.

Hand Mating Versus Lot Mating

The most common method of mating is by *lot mating*. This involves penning sows or gilts and mating one or more boars to them during expected heat

FIGURE 23-4. Gilts are usually not bred until they are 8 months old. (Courtesy of the United Duroc Swine Registry)

periods. This reduces labor requirements. *Hand mating* (putting a gilt or sow and boar in a pen, allowing service, and removing) has been used to regulate breeding, to increase litter size, and to keep accurate breeding records for registration purposes.

Flushing of Gilts

Flushing (increasing feed at mating of gilts is describe in Chapter 24) has increased the number of ova ovulated by gilts and thus has increased litter size.

GESTATION OF SWINE

Compared to cattle, swine have a short gestation period (114 days). This is easily remembered as 3 months, 3 weeks, and 3 days. The most critical times during gestation are the first 30 to 35 days and just prior to farrowing. Most embryonic death (loss of developing embryo) occurs during the first 30 to 35 days of gestation. Control of extreme temperatures, proper nutrition, and control of diseases are important for all fertilized ova to develop into healthy pigs.

Fetal development is quite similar to that of the cow, yet has some variation. Because multiple ova are shed and fertilized, multiple embryos develop in separate embryonic membranes within the uterus. Each embryonic unit functions separately without effect on the others.

FIGURE 23-5. Approaching parturition is characterized by restlessness, rearranging of bedding into a nest, and enlargement of external reproductive organs and mammary glands. (Courtesy of the American Landrace Association, Inc.)

SIGNS OF APPROACHING PARTURITION

Approaching parturition in swine is characterized by restlessness, rearranging of bedding into a nest, and enlargement of external reproductive organs and mammary glands (Figure 23-5). Milk is present in the teats 12 to 48 hours preceding farrowing. Accurate breeding records are also helpful in determining approaching parturition.

CARE AT FARROWING

The sow is usually moved to farrowing quarters 3 to 5 days prior to farrowing. The sow is cleaned, placed in well-sanitized quarters, and kept comfortable. The good swine handler is always present to ensure pigs and sow are healthy at farrowing. Swine usually farrow completely within 2 to 6 hours. The young pigs should be freed from their embryonic membranes, dried, warmed if chilled, and ensured of getting colostrum from the sow.

REBREEDING AFTER FARROWING

Sows will show a heat period 2 to 7 days after farrowing. This first estrus is sterile (infertile). Most sows will not come into estrus again until 3 to 5 days after pigs are weaned or removed. Thus rebreeding depends on the age of weaning pigs. To speed up the period between litters, some producers have gone to early weaning.

CARE OF THE BOAR

Although boars reach puberty sooner, they should not be used for breeding purposes before 8 to 9 months of age. Often, young boars are tested for fertility by laboratory examination or by breeding them to two or three market gilts. If the gilts do not return into heat, the boar is assumed fertile. If the gilts are slaughtered 4 to 5 weeks after breeding, the presence of 8 to 10 developing embryos in the gilts' uteri ensures high boar fertility.

Boars do not have as high a sexual capacity as bulls and rams and must be managed to keep libido (sex drive) high for good service. It is often desirable to rotate boars in the sow herd. A boar placed with one sow group in lot mating is left for 24 hours and moved to another group and replaced with another boar. This system may be more difficult in a hand mating breeding system. Care should be taken that boars are not overworked. Recommended mating of a yearling boar is 7 to 8 gilts or sows per week. A mature boar can service up to 12 gilts or sows per week.

ARTIFICIAL INSEMINATION IN SWINE

In recent years, a lot of interest in the use of artificial insemination in swine has developed. Semen can be easily collected from the boar and inseminated in the sow. Fresh boar semen is viable for about 40 hours when cooled and can be used in this state. When properly diluted, boar semen is viable for use for 3 days without freezing. These two methods have been the chief insemination procedures in swine. With recent advancements in boar semen freezing techniques, commercial semen is now available. The use of frozen boar semen could become the most important method used in artificial insemination of swine in the near future. The advantages of AI for swine are like those for beef and dairy animals: reducing the spread of diseases, making use of outstanding boars, reducing the investment of boars and facilities, and allowing for testing and identification of outstanding boars. The difference in the acceptance of AI in swine is due to the low concentration of sperm for high volume of semen of the boar. Large volumes of extended semen are needed per insemination and greatly reduces the number of animals that can be bred per insemination.

In artificial insemination, semen is collected by use of an artificial vagina and a boar trained to mount a dummy. Semen ejaculated from the boar is in three factions: presperm, sperm, and fluids. The semen or sperm faction is usually collected for use in AI. The semen is strained through cheesecloth and can be used without dilution immediately, diluted and used within 3 days, or frozen. Freezing is done with pellets on dry ice. Pellets are stored in liquid nitrogen. Ten cubic centimeters of pellets is thawed, and the solution is added to correct the volume of liquid needed and used for insemination of the sow.

Detection of sows in heat is usually accomplished by using a teaser boar, observing sows walking fences near boars, observing restlessness in sows, and by knowing sows come into heat 3 and 5 days after weaning. Sows are inseminated with 30 to 50 ml of extended semen (fresh or frozen). This large volume needed for artificial insemination in swine compared to cattle (1 ml) limits the number of sows that can be bred from one collection of semen. Usually, only 6 to 12 sows can be bred from one collection of semen through AI. Sows are usually inseminated two times—at 12 hours after the onset of estrus and 12 hours later.

STUDY QUESTIONS

1. Swine are unique because they are _____ (litter bearing).
2. Sows generally ovulate _____ ova in the estrous period.
3. Puberty in swine occurs at _____ months of age.
4. Estrus in swine varies from _____ days.
5. Sows ovulate the ova on the _____ day of estrus.
6. Gilts are usually bred on the _____ heat period.
7. Sows are usually bred _____ hours after the onset of estrus for maximum fertilization of ova.
8. Gestation in swine is _____ days or _____ months _____ weeks and _____ days.
9. Multiple embryos develop in swine in separate units, each developing separate from the others. (True, false) _____ .
10. Care of sows at farrowing usually begins _____ days prior to farrowing.
11. Sows usually farrow within _____ hours.
12. Rebreeding of a fertile estrus does not occur until after _____ .
13. A mature boar can mate up to _____ sows per week.

DISCUSSION QUESTIONS

1. What are the latest conclusions on the practicality of AI in swine? Describe details of collection, storage, and insemination.
2. Starting with one boar and 10 gilts of sexually mature age and size, calculate the numbers of swine you could conceivably have in 3 years, provided that all young males were sold and all sows were kept.
3. How can you detect when a sow is about to farrow?

4. What special handling techniques are needed to assure a high percent livability in baby pigs at farrowing time?
5. What is the function of the following?
   a. ovary
   b. retractor muscle
   c. secondary sex glands
   d. ova
   e. estrus
   f. placenta

Chapter Twenty-Four

FEEDING SWINE

A 1923 bulletin, "A History of Hogs," by Ashton, University of Missouri, referred to the age-old observation that led to the phrase "Don't make a pig of yourself." Ashton wrote:

> The hog not only eats to live but lives to eat and next to eating, derives his greatest pleasure from sleeping. But this merry life is short while that of his master is prolonged, and in some cases, is more sad than happy.

A hog's desire to eat is so intense that some breeders have questioned the use of a mixed ration, preferring to give free choice ingredients and let the hog balance its own ration. The answer that is often given when the validity of this practice arises is "If the hog is smarter than you are, let it balance the ration. If you are smarter than it is, then you balance it. But don't underestimate the hog." In other words, one must have knowledge and know-how to apply it in order to do a better job than the hog following its natural instincts. Feeding swine greatly affects the success of a swine operation. Feed is the greatest expense in swine production, normally representing from 65 to 80% of the total cost of producing pork.

THE SWINE DIGESTIVE SYSTEM

Because the digestive anatomy of the pig (a monogastric animal) is considerably different from the ruminant (cow, sheep), the feed used and the way swine digest it are not the same (Figure 24-1). Following feed through the digestive system of a monogastric animal will illustrate digestion, absorption, and excretion.

The pig takes in food, chews it well, mixes it with saliva before swallowing. Saliva serves the same function in the pig as in the cow (lubrication primarily) but differs in swine because it contains enzymes that begin to break down feed to its "building blocks." Also, no rumination of feed occurs in the pig because all feed is well chewed prior to swallowing.

Food, when swallowed, moves down the esophagus into the stomach. Unlike the cow, the pig's stomach is one compartment and is often called a *simple stomach*. Because the pig does not have a rumen (fermentation vat) for bacterial digestion of fiber, the level of roughage that can be digested by the pig is low. The stomach of the pig must also function as a storage place of ingested

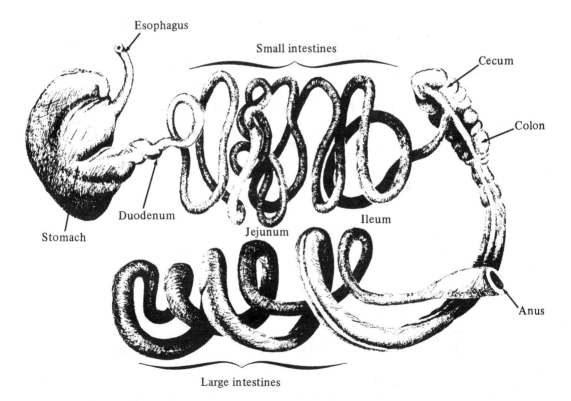

FIGURE 24-1. The monogastric digestive system.

feed. The one compartment stomach of the pig is small compared to the four compartment ruminant digestive system of the cow and holds only about 2 gallons. Most digestion occurs in the stomach and the rest in the small intestine.

The small intestine, made up of duodenum, jejunum, and ileum, is the place maximum absorption of nutrients occurs. The feed that is undigested or not absorbed moves from the small intestine to the cecum (appendix) and into the large intestine. Here water is absorbed and the remaining wastage from the digestive processes is excreted through the anus.

SPECIAL SWINE NUTRIENT REQUIREMENTS

Swine, like all species, require the same six nutrients covered in Chapter 4 (water, protein, carbohydrates, fats, minerals, and vitamins). The functions of each in the pig's body is essentially the same as in the cow. The pig requires nutrients from different feed sources and often in different ratios and quantities than the cow.

Because the simple stomach animal does not have a rumen, it lacks the benefits of microorganism digestion such as creation of high-quality protein, manufacture of B vitamins, and utilization of roughages. Therefore, the fiber content in swine rations must be below 5% and of very high quality (alfalfa leaf meal, legume, or grazing cereal grains). Roughages in the ration serve as vitamin supplements, add bulk to control weight in sow rations, and have a laxative effect. The protein in a swine ration must be of sufficiently high quality to furnish the necessary amino acids (protein building blocks) in proper amounts and proportions for correct swine protein synthesis. Therefore, the amino acid level and content of a feed source is important in swine feeding. Additional supplementation of the B vitamins stems from the inability of the swine digestive system to synthesize quantities of B vitamins for proper swine nutrition.

Protein and Amino Acids

Protein should be provided in terms of both quantity and quality. A variety or mixture of several sources (such as soybean meal, tankage, and dried milk) can provide a high-quality supplement premix.

Protein quality refers to the amino acid content—the higher the level and the more the variety, the higher the quality is. Ten essential amino acids are required by pigs for normal growth. According to research, they are arginine, histidine, isoleucine, leucine, lycine, methionine, phenylalanine, threonine, tryptophan, and valine.

Fat

Although practical swine rations are always thought to contain adequate fat, it is generally accepted that a level of 1.0 to 1.5% fat in the diet will ensure adequate essential fatty acids. Adding fat up to 20% has increased growth rate but often produces a fatty carcass or "soft pork." The demand is now for lean pork so the fat in the ration seldom exceeds 5%.

Carbohydrates, TDN, Net Energy, Metabolizable Energy

The carbohydrate content of a ration makes up most, although not all of the energy requirements. Because some energy is derived from fat and in some instances from protein, rations balanced for energy are common. *Total digestible nutrients* (TDN) is used as an estimate of energy requirements.

The only loss of nutrients deducted from total energy consumed by an animal is the loss that occurs in the undigested portion of the feces to arrive at a figure for TDN. Although generally assumed by the layperson that the remainder is all assimilated and used in the body, it is not.

The total energy of a feed (gross energy) is not all available. After the energy lost through the feces, combustible gases, urine, and heat increment or work of digestion is deducted from gross energy, *net energy* remains. Net energy is often used in place of TDN as a more accurate approach to meeting requirements.

Metabolizable energy, or *available energy,* may also be argued as a better figure on which to base calculations. Only the loss of energy through the feces, urine, and combustible gases is the total loss to the animal. The heat increment or work of digestion produces heat to keep the body warm, a useful and necessary function when environmental temperatures are sufficiently low. For this reason, metabolizable energy values and requirements can be used as the most accurate method of energy calculations. However, because TDN values are so readily available and widely used, they are likely to remain popular or in use as a base for calculating net energy or metabolizable energy figures.

Minerals

Next to common salt, calcium is probably the most needed mineral. Calcium deficiencies in swine are rather gradual in appearance. Lameness is among the first observable symptoms with a gradual deterioration ending in paralysis of the hindquarters. After experiencing three or four months of a deficient ration, the paralysis will have set in. Steamed bone meal, ground limestone, and ground oyster shell flour are common sources of calcium.

Iodine is needed for growth, gestation, and lactation. A deficiency results

FIGURE 24-2. Iodine deficiency. The litter of hairless pigs shown at the top was stillborn. The hairless pigs shown at the bottom were born alive. [Reprinted by permission from Committee on Animal Nutrition, National Academy of Sciences–National Research Council, *Nutrient Requirements of Swine,* 7th revised edition, Publication ISBN 0-309-02140-5 (Washington, D.C.: NAS–NRC, 1973)]

in goiter (a swelling of the neck area due to enlargement of the thyroid gland) and loss of hair (Figure 24-2). Iodine is used by the thyroid gland to manufacture the hormone thyroxine, which regulates body temperature, among other things. Stabilized iodized salt containing 0.007% iodine added as 0.5% of the ration or fed free choice will usually act as a preventative.

Iron and copper are needed for the formation of hemoglobin and for the prevention of nutritional anemia. Anemia occurs in pigs during the suckling period when they are kept on dry lot. The inability of hemoglobin to carry sufficient oxygen through the bloodstream creates an accelerated heart beat that compensates by increasing the blood flow. This results in a readily detectable pounding of the heart known as *baby pig thumps* or just *thumps*. The most effective anemia preventative, in addition to iron and copper in the ration, is two injections of iron dextran into baby pigs at 3 and 21 days of age. Injections of 100 mg per pig for the first injection and 50 mg for the second are recommended (Figure 24-3).

Magnesium deficiency (Figure 24-4) in the pig includes weakness of the pasterns, uncontrollable twitching, a staggering gait, and tetany and leads to death 50% of the time. Magnesium is necessary for normal muscle control, contraction, and equilibrium. Practical rations usually have sufficient magnesium making the possibility of deficiency highly improbable.

Manganese-deficient symptoms in swine have been produced experimentally by feeding unnaturally low levels (Figure 24-5). The main symptom in adults is increased fat deposition, while the young pigs are born weak with a poor sense of balance. While it is known that 18 mg per pound of feed appears to be the optimum level for growth, the occurence of deficiency symptoms under normal circumstances is rare.

Phosphorus is also needed for proper skeletal growth and development. Typical deficient symptoms are weak legs and crooked leg bones (Figure 24-

FIGURE 24-3. Iron deficiency. Note the listlessness and wrinkled skin of this anemic pig. [Reprinted by permission from Committee on Animal Nutrition, National Academy of Sciences–National Research Council, *Nutrient Requirements of Swine*, 7th revised edition, Publication ISBN 0-309-02140-5 (Washington, D.C.: NAS–NRC, 1973)]

FIGURE 24-4. Magnesium deficiency. The pig at the left was fed 413 ppm of magnesium for 3 weeks; the pig at the right was fed 70 ppm for the same period. Not the extreme leg weakness, arched back, and general unthriftiness of the pig on the right. [Reprinted by permission from Committee on Animal Nutrition, National Academy of Sciences–National Research Council, *Nutrient Requirements of Swine*, 7th revised edition, Publication ISBN 0-309-02140-5 (Washington, D.C.: NAS–NRC, 1973)]

6). Steamed bone meal and dicalcium phosphate are good sources and also supply calcium.

Salt (sodium chloride) is the most commonly needed mineral by most animals. Swine have a minimum requirement of 0.2% of their ration, but as a practical course 0.5% is the accepted rule. This upper level should not normally be exceeded because of swine susceptibility to mild, ill effects known as *salt poisoning syndrome*. A deficiency of salt, on the other hand, may result in a depraved appetite and poor growth.

Zinc is needed for proper skin health, growth, and feed conversion. A condition known as *swine dermatosis* or *parakeratosis* (a sloughing off of the skin) has been cured or prevented by zinc addition to the ration (Figure 24-7). The functions of zinc are not fully understood, but it is known that high levels of calcium and possibly phosphorus bring on parakeratosis necessitating higher levels of zinc also. This indicates a relationship that has led to the recommended level of 23 mg of zinc per pound of feed, which includes a safety margin should calcium and phosphorus levels exceed the normal. Even higher levels of zinc are considered quite safe because as much as 450 mg per pound has been fed without toxic effects.

Vitamins

Vitamin A is necessary for growth, to prevent lameness or stiffness of the joints, and for normal reproduction (Figure 24-8). A deficiency causes incoordination of movement, lameness, swollen joints, and night blindness because of constriction of the optic nerve. Pregnant, deficient sows often bear pigs that

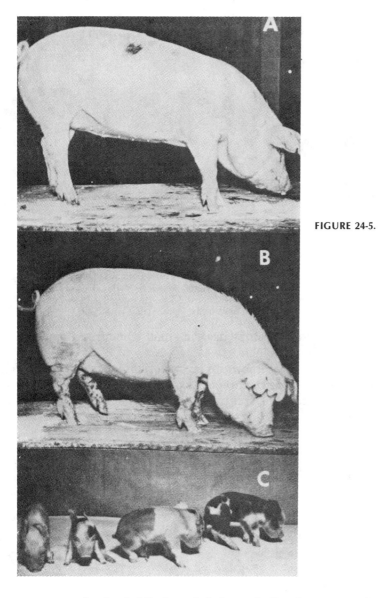

FIGURE 24-5. Manganese deficiency. (A) Gilt, 132 days old, that has been fed 40 ppm of manganese since she weighed 3.9 kg. (B) Littermate of the gilt shown in (A). This gilt was started at the same weight and has been fed 0.5 ppm of manganese. Note the increased fat deposit due to low manganese diet. (C) Litter from a sow that was fed 0.5 ppm of manganese. The pigs showed weakness and poor sense of balance at birth. [Reprinted by permission from Committee on Animal Nutrition, National Academy of Sciences–National Research Council, *Nutrient Requirements of Swine*, 7th revised edition, Publication ISBN 0-309-02140-5 (Washington, D.C.: NAS–NRC, 1973)]

are weak, dead, blind, and deformed. Good sources of vitamin A are alfalfa leaf meal and synthetic vitamin A, both of which are readily available.

Vitamin D is necessary for normal growth and bone development. Swine exposed to direct rays of the sun, which converts 7-dehydrocholesterol in the skin to vitamin D, seldom develop problems. However, confinement in housing where little or no sunshine falls directly can create conditions favorable for a shortage. Deficiency symptoms are rickets, enlarged joints, and weak bones. Vitamin D is associated with the utilization of the minerals calcium and phosphorus, which produce similar deficiency symptoms (Figure 24-9). Any sun-

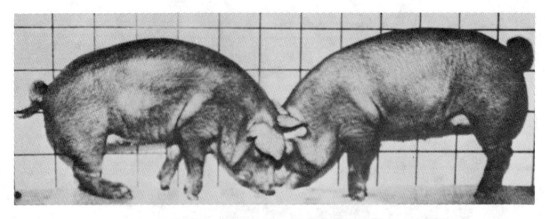

FIGURE 24-6. Phosphorus deficiency. On the left is a typical phosphorus-deficient pig in an advanced state of deficiency. Leg bones are weak and crooked. The pig on the right received the same ration as the one on the left, except that the ration was adequate in available phosphorus. [Reprinted by permission from Committee on Animal Nutrition, National Academy of Sciences–National Research Council, *Nutrient Requirements of Swine*, 7th revised edition, Publication ISBN 0-309-02140-5 (Washington, D.C.: NAS–NRC, 1973)]

cured forage, such as alfalfa, can be used in small amounts as a supplement and synthetic sources are available.

Vitamin E (tocopherol) is known to be essential for normal reproduction in swine. However, no definite requirements have been proven, making a supplementary source questionable. It is unlikely that swine rations would be deficient in vitamin E.

FIGURE 24-7. Zinc deficiency. The pig on the left received 17 ppm of zinc and gained 1.4 kg in 74 days. Note the severe dermatosis and parakeratosis. The pig on the right received the same diet as the pig on the left except that the diet contained 67 ppm of zinc. [Reprinted by permission from Committee on Animal Nutrition, National Academy of Sciences–National Research Council, *Nutrient Requirements of Swine*, 7th revised edition, Publication ISBN 0-309-02140-5 (Washington, D.C.: NAS–NRC, 1973)]

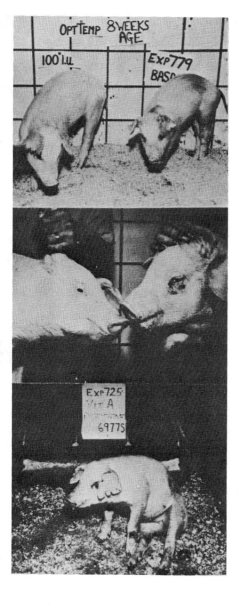

FIGURE 24-8. Vitamin A deficiency. The pigs shown at the top are 8 weeks old. The one on the right is deficient in vitamin A and shows a slow rate of growth. The pig on the right in the middle illustration is deficient in vitamin A and has xerophthalmia. The pig at the bottom is paralyzed in its hind legs. [Reprinted by permission from Committee on Animal Nutrition, National Academy of Sciences–National Research Council, *Nutrient Requirements of Swine*, 7th revised edition, Publication ISBN 0-309-02140-5 (Washington, D.C.: NAS–NRC, 1973)]

Vitamin K is needed mainly as a necessary blood clotting factor when the hog is cut or sustains a similar injury. It appears that microorganisms synthesize sufficient amounts in the intestine to meet daily requirements, making a supplement unnecessary.

Biotin, although needed by the pig, has not proven to be of enough concern to consider supplementing the normal ration. A biotin deficiency has been produced experimentally by feeding desiccated raw egg white. A substance, *avidin,* in the egg white has an antagonisitic effect on biotin creating the deficiency symptoms consisting primarily of loss of hair, cracks in the feet, and

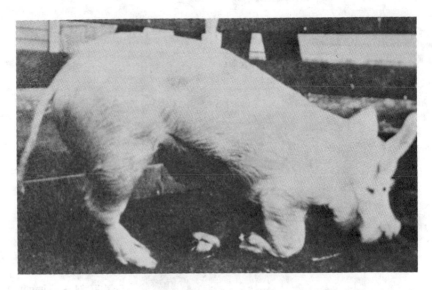

FIGURE 24-9.　Vitamin D deficiency. This pig has advanced rickets resulting from lack of vitamin D. The pig was fed indoors. Because of leg abnormalities, it was unable to walk. It later responded to supplementary vitamin D. [Reprinted by permission from Committee on Animal Nutrition, National Academy of Sciences–National Research Council, *Nutrient Requirements of Swine*, 7th revised edition, Publication ISBN 0-309-02140-5 (Washington, D.C.: NAS–NRC, 1973)]

dermatosis of the skin. Avidin is inactivated by heat treatment and thereby made safe for feeding.

Choline, inositol, *para*-aminobenzoic acid, and pteroylglutamic acid (folic acid) are vitamins that play a minor role in swine rations but are not thought to be in short supply naturally.

Niacin is necessary for normal hair and skin health. A deficiency may result in slow growth, occasional vomiting, pig pellagra (skin lesions, gastrointestinal disturbance, and nervous disorders), and impaired reproduction (Figure 24-10). Natural sources that have been used for niacin supplement are distiller's dried solubles, condensed fish solubles, and alfalfa leaf meal.

Pantothenic acid is necessary for proper muscle and nerve function. The most obvious deficiency symptom is the classic goose stepping gait (Figure 24-11) that is manifest in severe ration shortages. The hog walks with an uncoordinated, wobbly motion, and kicks upward with the hind legs because of impairment of proper nervous function. Dried brewer's yeast, dried whey, and alfalfa leaf meal are all good insurance as natural supplements.

Pyridoxine is needed by swine for proper growth and normal development. Deficiencies are usually exhibited by convulsions or epileptic-like fits and slow growth. Although deficiencies are uncommon and requirements have not been clearly defined, dried brewer's grains and condensed fish solubles are good natural sources often added to ensure proper nutrition.

Riboflavin is an important vitamin needed in many functions of growth and reproduction. Dietary deficiency results in the most obvious symptom of

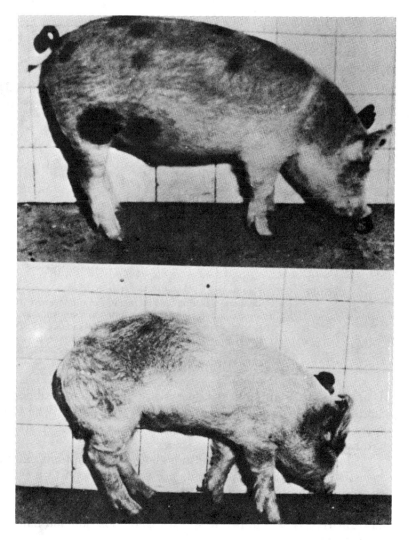

FIGURE 24-10. Niacin deficiency. The pig shown at the top has received adequate niacin, the pig at the bottom has not. The difference in growth and condition is due to the addition of niacin in the diet containing 80% ground yellow corn. [Reprinted by permission from Committee on Animal Nutrition, National Academy of Sciences–National Research Council, *Nutrient Requirements of Swine*, 7th revised edition, Publication ISBN 0-309-02140-5 (Washington, D.C.: NAS–NRC, 1973)]

crooked legs. Other symptoms include slow growth, lameness, diarrhea, anemia, and impaired reproduction (Figure 24-12). Any milk product such as dried skim milk is usually considered a good natural source.

Thiamine is associated with growth (Figure 24-13), normal functions of the heart, and temperature regulation. A deficiency results in slow pulse rate and low body temperature, and autopsies have revealed flabby hearts.

Vitamin B_{12} is important for growth and normal appetite. Major vitamin

FIGURE 24-11. Pantothenic acid deficiency. Locomotor incoordination (goose stepping) was produced by feeding a ration (corn–soybean meal) low in pantothenic acid. [Reprinted by permission from Committee on Animal Nutrition, National Academy of Sciences–National Research Council, *Nutrient Requirements of Swine*, 7th revised edition, Publication ISBN 0-309-02140-5 (Washington, D.C.: NAS–NRC, 1973)]

B_{12} deficiency symptoms are hyper-irritability and incoordination of the posterior. The mineral cobalt is a part of the B_{12} molecule. The value of tankage, meat scraps, and dairy by-products for swine rations is related in part to their vitamin B_{12} content.

It should be noted that although known symptoms exist for many deficiencies, not all dietary problems are simple enough to be traced to a single cause and symptoms do not always appear obvious. Also, a complex shortage of several dietary essentials may exist making clinical identification less important than a careful review of ration composition and balance.

FEEDING SWINE CLASSES

Swine are classified according to the purpose and age for which they are fed. Each class has unique nutrient requirements and is fed in a different way (Table 24-1). In some cases, rations are simplified to three categories—a starter ration, a grower-finisher ration that is also fed to lactating sows, and a gestating ration that is also fed to boars.

Feeding the Young Pigs

Young pigs are given small amounts of creep feed in the farrowing stall at 7 to 10 days of age. Very small amounts are given two or three times daily to

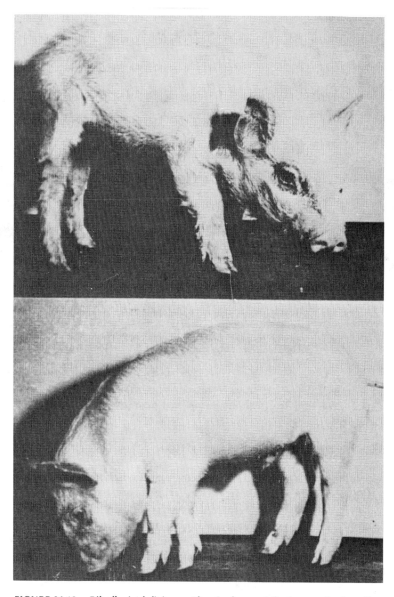

FIGURE 24-12. Riboflavin deficiency. The pig shown at the top received no riboflavin. Note the rough hair coat, poor growth, and dermatitis. The
pig at the bottom received adequate riboflavin. [Reprinted by
permission from Committee on Animal Nutrition, National Academy of Sciences-National Research Council, *Nutrient Requirements of Swine*, 7th revised edition, Publication ISBN 0-309-
02140-5 (Washington, D.C.: NAS–NRC, 1973)]

ensure that fresh feed is available. When pigs begin eating (about 2 weeks of
age), enough starter is left to last 2 or 3 days at a time. A fresh water source
other than that for the sow should be provided. Only very palatable, high-
quality feed sources are used in formulating the starter ration. Iron in the form

FIGURE 24-13. Thiamine deficiency. The pig on the right received no thiamine; the pig on the left received the equivalent of 2 mg of thiamine per 45 kg of live weight. Otherwise, their diets were the same. The pigs were littermates. [Reprinted by permission from Committee on Animal Nutrition, National Academy of Sciences–National Research Council, *Nutrient Requirements of Swine*, 7th revised edition, Publication ISBN 0-309-02140-5 (Washington, D.C.: NAS–NRC, 1973)]

of iron sulfate is included to prevent anemia. Antibiotics are added to starter rations for disease prevention and to promote growth.

Growing and Finishing Pigs

Growing and finishing of swine are usually accomplished on a full feed of high quality growing–finishing ration. A different ration is usually used for growing from 40 to 120 lb than that used in the finishing stage of 120 lb to market weight. The grower ration is higher in protein for formation and growth of body tissue and bone, while the finishing ration is high in energy for fattening and finishing for market. Full feed requires some method to limit wastage—such as pelleting, using feed troughs with lids on them, and maintaining a low

TABLE 24-1. RATIONS FOR VARIOUS CLASSES OF SWINE

Class	Ration	Period or Weight	Percent Protein	Daily Intake
Birth to weaning	Starter	Birth to 40 lb	18	Free choice
Weaning to market	Grower	40 to 120 lb	16	Full feed
Weaning to market	Finisher	120 to 240 lb	14	Full feed
Breeding herd	Gestation	Weaning to farrow	12–14	5 lb
Breeding herd	Lactation	Farrow to weaning	16	12 lb or full feed
Breeding herd	Boar		14	6 lb

level of feed in the feeders. Many producers hand feed (give a certain amount each day) rather than self-feed finishing swine to maximize utilization of feed and control wastage.

Feeding Replacement Gilts (Flushing)

Replacement gilts are usually selected at 5 to 6 months of age during the finishing period and placed on a different feeding system. Feed should be reduced from full feed to 5 or 6 lb per day. This would keep them in good condition until breeding time at 7 or 8 months of age. Litter size in gilts has been increased by flush feeding at breeding (increasing feed at breeding to provide additional energy). The conventional method of flushing is to increase feed to about 8 lb per day, 7 to 10 days prior to breeding. This feed level is maintained 4 to 8 days after service (breeding) and reduced to 5 to 6 lb until farrowing. The newer method of flushing is increase feed one day prior to breeding and double feed on the first day of mating. The objective here is the improvement of physiological condition at breeding. When weaning occurs, sows are relieved of the strain of sucking pigs and undergo a similar physiological boost at breeding.

Feeding Gestating Sows

Limited feeding of gestating sows and gilts has proved to ensure large litters. Sows receiving too much energy and putting on too much weight during gestation have higher embryonic mortality rates (dying of unborn pigs). The amount of gain desired during gestation is 60 to 80 lb for a sow and 75 to 100 lb for a gilt. The amount of feed required under dry lot conditions is 1.5 to 2% of the body weight in gilts (about 5 lb) and 1 to 1.5% of the body weight in sows (about 4 lb). Many producers save feed cost by using legume and cereal grain pastures for gestating sows. Good pastures supplemented with 0.5 lb of protein supplement and 2 to 3 lb of grain can cut feed cost of gestating sows by 30%. The stocking rate on pasture is 8 to 10 sows per acre or 10 to 12 gilts per acre.

To eliminate labor involved in hand feeding gestating sows, a skip-day feeding method has been developed. In this type of feeding, gestating sow groups are given access to self feeders for 6 to 12 hours every third day. Thus they receive their pasture supplement with less investment and less labor, but more management is required to keep sows and gilts in desired condition.

Feeding Breeding Boars

Boars in the breeding herd are fed the same ration as gestating sows. In drylot, feeding 3 to 4 lb when the boar is not used for breeding and 6 to 7 lb per day when used will keep the boar in excellent condition. The boar should be kept on legume or cereal grain pastures requiring 1% of its body weight of the concentrate ration.

TABLE 24-2. RATION SUGGESTIONS (Rations Using Premixed Commercial Supplements)

	18% Starter		16% Grower		14% Finisher	
Grain (corn or milo)	50	60	42.5	38.5	66.5	62.5
Cereal grain (oats, barley, wheat)	16.5	12	35	35	15	15
35% supplement	33.5	28		21.5		17.5
40% supplement			17.5		13.5	
Alfalfa leaf meal			5.0	5.0	5	5
	100 lb	100 lb	100 lb	100 lb	100 lb	100 lb

	14% Gestation				16% Lactation	
	Drylot		On Pasture			
Grain (corn or milo)	78.5	76.5	78.5	82.5	42.5	38.5
Cereal grain (oats, barley, wheat)					35	38
35% supplement		18.5	21.5			21.5
40% supplement	16.5			17.5	17.5	
Alfalfa leaf meal	5	5			5	5
	100 lb	100 lb	100 lb	100 lb	100 lb	100 lb

Feeding Lactating Sows

To ensure adequate milk production by the lactating sow, a high-quality ration is full fed throughout the nursing period. The only exception to this rule comes into play when small litters occur and sows gain excessive weight during lactation. The amount of feed will then rest on the judgment of the breeder.

RATION SUGGESTIONS

Although experienced herdsmen may have the background necessary to completely formulate rations using raw ingredients, most of them prefer to use a premix. Commerical premix supplements contain antibiotics, protein vitamins, and minerals in the proportions required by swine. All that remains then is to add the small premix to the other ingredients as illustrated in Table 24-2.

STUDY QUESTIONS

1. _____ cost represents the greatest expense in swine production.
2. Swine are _____ animals.
3. Swine saliva differs from that of the cow because it contains _____ that begin the digestion process.

Chap. 24 FEEDING SWINE

4. The level of _____ that can be digested by swine is low because the swine's simple stomach does not contain a place for bacterial digestion.

5. Like the cow, the place of maximum absorption of feed nutrients in swine is the _____ .

6. Protein in swine rations must be of _____ quality for the essential amino acids to be supplied.

7. Swine are unable to synthesize vitamins as a cow does; thus the vitamin _____ in addition to vitamins A, D, and K must be included in swine feed.

8. Different swine weight groups and the breeding herd should be fed the same ration. (True, false) _____ .

9. Rations for small pigs and lactating sows are _____ in protein than those for finishing pigs and boars.

10. Young pigs should start receiving starter feed at _____ to _____ days of age.

11. Grower rations are high in _____ while finishing rations are high in _____ for fattening.

12. _____ gilts, increasing feed before estrus, increases litter size.

13. During gestation, sows and gilts should be given _____ feed.

14. Gilts should gain _____ to _____ lb and sows _____ to _____ lb during gestation.

15. Lactating sows should be given _____ feed, except where small litters exist.

DISCUSSION QUESTIONS

1. Compare the anatomy, function, and limitations of the hog's digestive system as compared to sheep.

2. What is meant by "quality" of protein supplements for swine? Do swine actually need this? Why?

3. Describe deficiency signs for the following and give the nutrients responsible for correcting the condition.
   a. anemia
   b. goiter
   c. parakerotosis
   d. xerophthalmia
   e. rickets

Chapter Twenty-Five

SWINE SELECTION

Selection criteria for swine follow similar lines, with some minor variations because of the different species, to those discussed previously for beef cattle (Chapter 5). The broad selection bases are judging, pedigree, performance, and production testing.

EYE APPRAISAL — JUDGING

A thrilling sight is the action of the grand champion barrow auction at the local county fair (a barrow is a castrated male hog lacking sex characteristics). The exhibitor proudly guides the barrow with a cane to show the animal's strong points that won it the grand champion banner. How did the barrow become grand champion? A judge had to exercise judgment on the field of candidates, no matter how many were entered for the contest. However, judging is never more than comparing two animals at a time. Thus, through comparing two animals with an ideal-type hog in mind and selecting one over the other, animals are eliminated until a grand champion emerges. The desirable, ideal hog is described in this chapter.

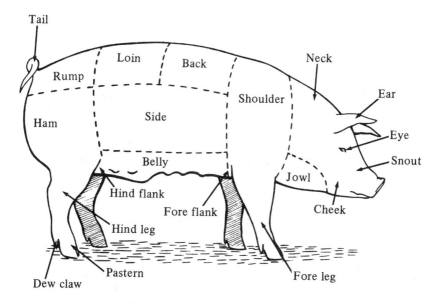

FIGURE 25-1. The parts of a market hog.

Parts of a Hog

As explained in Chapter 5, a judge must speak in terms of the class of live-stock being judged. Figure 25-1 illustrates the locations and names of the different parts of a market hog.

Description of the Ideal Meat-Type Hog

At one time, hog types were referred to as *meat type, bacon type,* or *lard type.* Because the lard type (overly fat) is no longer popular or necessary, all breeds have the goal of producing a lean, muscular carcass with little waste. Most writers describe this ideal as the meat-type hog.

Knowledge of the parts of a hog is helpful only when the animal scientist can use the terminology to describe faults or strong points of one hog as compared to the ideal meat-type hog. In viewing the swine from the side, back, and rear, the following desirable points should be noted.

Side View. When judging a hog from the side, the judge looks at the size for age, balance, length of side, depth of body, topline, trimness of under-line and jaw, legs, size of bone, finish, breed type, and teats in breeding classes. Because rapid gains are usually economical gains, a large, early maturing hog is desired (Figure 25-2). A moderately long body that is fairly deep gives a balanced appearance to the hog. An evenly arched back with high tail setting is desired. The belly and jowl should be trim, the flanks deep. The ideal hog is

FIGURE 25-2. The side view should show a lengthy, moderately arched back, straight legs, and smoothness throughout showing moderate finish. (Courtesy of T. D. Tanskley, Texas A&M University)

smooth over the shoulders, free from wrinkles, and moderately finished. Legs are moderately long, straight, true, and squarely set with strong pasterns. Ample bone size is desired. Breed type and teats, well developed and numbering twelve or more, are desired for breeding swine.

Rear View. The rear observation is an important view showing muscling and finish. Width from front to rear is noted with good, uniform width desired (Figure 25-3). Smoothness of the back of the shoulders with the shoulders "well laid in" (blending in the uniform thickness) should be present.

The ideal hog will be wide and full throughout the loin area of the back. The rear of the hog should show a plump, full, trim, and firm ham with muscling well down to the hocks (Figure 25-4). The hind legs should be set well apart and be sound and straight. This side stance is referred to as *pegging down;* the wider the stance, the heavier the muscle along the backbone.

Front View. A shapely, trim head showing sex characteristics and having wide-set eyes is best seen from the front view of the ideal hog (Figure 25-5). In breeding classes, the head should show breed characteristics as well as sex. The front legs should be medium in length, straight, and correctly set.

Feel for Proper Finish. Under the current trend of producing meat-type hogs rather than lard-type hogs, proper finish with a minimum of fat is desired. The ideal barrow possesses proper conformation, finish, and quality for maximum development of the high-priced cuts and a minimum of lardiness.

Chap. 25 SWINE SELECTION

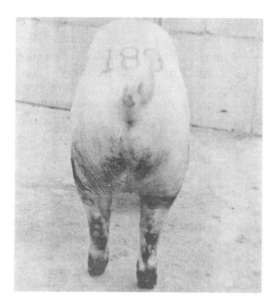

FIGURE 25-3. The width along the topline, muscling, and finish are noted from the rear view. (Courtesy of T. D. Tanskley, Texas A&M University)

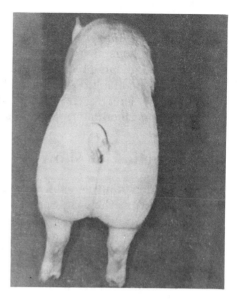

FIGURE 25-4. The rear of the hog should show a plump, full, trim, and firm ham with muscling down to the hocks. (Courtesy of T. D. Tanskley, Texas A&M University)

FIGURE 25-5. A trim head showing sex and breed characteristics is desirable from the front view. (Courtesy of T. D. Tanskley, Texas A&M University)

Although some fat is essential for the quality and cooking properties of the meat, large layers indicate waste, over-finish, and poor conformation. Well-finished hogs have a moderately thick, firm covering of flesh all over the body that is indicated by width of body, plumpness of hams, and firmness of jowls, belly, and flanks. A firm, uniform layer of back fat that is 1.5 in. or less is desired. This backfat can be measured on the chilled carcass as illustrated in Chapter 26 or determined on the live animal by backfat probing (Figure 25-6).

Selection of Show Animals

Selection of future show animals in either the market or breeding class is made throughout the year. Selection traits are kept in mind at the time of breeding, and sows and boars are carefully selected to produce the next grand champion. Farrowing dates are often arranged so that market hogs are at the optimum stage of development for show. Actual selection of show animals must be made far enough in advance to allow time for fitting and training. Often, many animals are selected and fitted, but only the best are used at show time.

FIGURE 25-6. Determining backfat on a live animal using a probe. [Reprinted by permission from John F. Lasley, *Genetics of Livestock Improvement* (Englewood Cliffs, N.J.: Prentice-Hall, Inc., 1963), p. 242]

Animals of the breeding herd are selected by pedigree, show-ring record, or individual merit. The pedigree of an animal (record of the animal's ancestry) can be used to select animals used for breeding. Pedigrees that give the production records of closely related animals (parents and grandparents of the animal) can serve as an indication of the breeding value of a young prospect.

Swine breeders have always recognized show ring winners as superior animals. Show auctions for the breeding classes are always filled with spirited bidding for the top hogs, and this often leads to paying a premium price for the purchased hog. The amateur buyer often prefers this method of selection, thereby allowing an acknowledged judge to select the best animals. Experienced swine buyers can judge breeding stock on the farm and can often obtain breeding stock of equal quality at more reasonable prices.

Pedigrees are especially useful in swine breeding because the generation interval is very short compared to other species and because more than one or two offspring is produced at one parturition. This makes possible a large number of observations of a particular bloodline or results of crossing bloodlines to achieve desired results. Similar results might be expected by selecting animals related to participants of a desirable production record. The pedigree offers an excellent starting place. As in most other species, the more recent individuals such as parents or grandparents are recommended because each generation may reduce potential improvement by one-half.

Fads and promotional advertising are always possible fallacies in selection by pedigree. Swine are generally felt to be less susceptible to serious manipulation because they have rarely demanded the extremes in high prices as compared to cattle and horses that attract that type of speculative breeding. Therefore, demands for change are dictated at the market, and breeders scan their pedigree books for genetic material to produce quality and quantity, knowing that the latter is the key to profitable returns.

The highest and best use of pedigree information is a basis for selecting young animals before their performance or that of their offspring is known. Because the generation interval is so short, traits such as longevity are not as critical, but many others, such as soundness, milk production, rate of gain, feed efficiency, and so on, are of definite concern. If these fast developing traits can be predicted with some degree of probability based on other bloodline observations, then the value of the pedigree as a selection tool becomes obvious. Also, one line that is weak in a particular trait, such as weak pasterns, may be bred to a line strong in the same area. Each may overcome a fault of the other.

Pedigrees serve as a useful tool, but their limitations must be realized. Once an individual's performance is measured or recognized, the pedigree may be relegated to obscurity or, at best, receive less weight in an individual's evaluation.

ANIMAL PERFORMANCE

Measuring the individual's ability to perform may be as simple as memory of past experiences or as complex as use of atomic age tools like radioactive isotopes.

Practical measurements are at the disposal of nearly all swine breeders. Gain, feed conversion, backfat thickness, carcass cut-out data, and loin-eye measurement are examples of valuable data that supplement other selection methods.

Measuring Techniques

Pencil and Pad. A pocket notepad can be used for recording such data as litter size at birth, litter size weaned, conformation score, and general remarks too valuable to be left to memory. Transfer to permanent records for lifetime evaluation of bloodlines may be made for useful records.

Scales. Litter weight at birth and at weaning is useful information obtained from a simple scale. Weighing of feed consumed over a period of time can also be used along with weight gains to calculate feed efficiency. Rapid, efficient gains would be difficult to select for were it not for the scale. Some breeders feel this is the most important tool at their disposal for developing the framework of a selection program.

Swine Testing Stations. The Agricultural Extension Service and other interested organizations have devised centers in many states to evaluate individual swine performance. Most stations screen proposed applicants for inherited defects such as swirls and hernia. Only good, genetically sound swine are desired for test and later breed improvement. Insofar as possible, nutrition and environment are kept identical so that accurate appraisals may be made. In many instances, a boar and a barrow that are full brothers are tested together so that the barrow may be slaughtered and data collected to estimate the boar's potential breeding capabilities for passing on a similar trait.

Rate and economy of gain are important measurements. The goal is attainment of a weight of 200 lb to be reached during the 150 days from weaning to slaughter, utilizing 320 lb of feed per 100 lb of gain. The net carcass value is also a valuable measurement based on backfat thickness, length of side, area of the loin eye, and the percent of four lean cuts of the chilled carcass. Females may also be on test. Prolificacy, the number of pigs over a long life, and weaning weight of the litter are the criteria of chief interest.

As pointed out in Chapter 5, the effectiveness of selection can be meaningful provided that traits of reasonably high heritability are tested for and used in evaluation of potential breeding stock. Breeding animals can only transmit unfailingly those genes that they themselves have been found to possess in relatively homozygous (pure) form. Higher heritability estimates indi-

TABLE 25-1. HERITABILITY OF SOME ECONOMICALLY IMPORTANT CHARACTERISTICS IN SWINE

Characteristic	Heritability
Litter size at birth	Low
Litter size at weaning	Low
Birth weight of pigs	Very low
Litter weight at weaning	Low
Rate of gain	Medium
Feed efficiency	Medium
Conformation score	Medium
Length of side	Very high
Backfat thickness	High
Loin-eye area	High

cate the most rapid improvement to be expected in a trait. Some traits commonly used in production testing and their heritability ratings are given in Table 25-1. Faster progress can be expected when dealing with traits of medium to very high heritability.

Ultrasonic devices may be used to determine loin-eye area in swine using a method similar to that described for cattle in Chapter 5. Other complex techniques such as cut-out data may be obtained if facilities are available to the testing station.

PRODUCTION TESTING

In developing a production testing program, permanent markings are essential. Ear notches at birth or shortly thereafter are the traditional and effective system used on swine. Observations are made from birth to slaughter or determination of useful life, and this information is recorded on lifetime records and used as an aid in selecting breeding stock.

Production Registry

Some record associations of the United States have a system of production testing known as *production registry*. The associations are generally open only to registered purebreds with no outstanding defects. The swine must be ear notched and spring from or produce a minimum of eight or nine (depending on the association) pigs in a litter. Minimum litter weights are also specified and must be witnessed by a person acceptable to the registry association. To qualify, a sow must produce at least one litter that meets requirements. Some associations require two litters before certification. A boar becomes certified when it produces at least five such litters (some associations require as high as 15) or at least two certified daughters (some require as high as 10).

Certified Meat Hogs

The National Association of Swine Records adopted a program that adds carcass evaluation to production registry to designate certified meat hogs. A certified litter must meet certain minimum standards to qualify. The first step is qualifying for production registry. Then two gilts or barrows from the litter are slaughtered, provided they are between 200 to 220 lb equivalent weight at 175 days of age (equivalent weights may be calculated by adding 2 lb for each day under 175 days or by subtracting 2 lb for each day over 175 days). The carcasses of both hogs must have at least 4 in.^{2} of loin-eye area and 29 in. of length (measured from the front of the first rib to the front of the aitch bone). No more than 1.6 in. average backfat thickness is allowed. If these standards are met, the entire litter is certified.

A certified boar is one that has produced five certified litters. However, each litter must be from a different sow, not more than two of which can be full sisters or direct descendants. A certified mating means the rebreeding of a sow and boar that have produced a certified litter.

A SYSTEM OF SELECTION

Finally, using a combination or all of the previously described techniques, the swine breeder needs to follow a system of selection that will result in maximum progress (Figure 25-7). Three common systems follow: tandem selection, independent culling level, and selection index.

Tandem Selection

Using tandem selection, one trait is selected for a time until maximum progress has been achieved; then another trait is selected for improvement. A good example of this method was the change from the lard-type hog to the long, lean variety. Because heritability of length of side and backfat thickness is higher than most other traits, rapid progress was made. Then it was discovered that muscle, loin eye, and other traits were often substandard so they become the target for improvement. The limitations of this system are obviously that many traits are interrelated and selection for one may not always result in improvement of another.

Independent Culling Level

Production testing fits well with the independent culling level's systematic procedure. As already discussed, minimum standards are set for several traits. A failure to meet these standards by any individual results in slaughter or removal from the herd. Even with high standards, this system works well and

FIGURE 25-7. Selection has changed the ideal hog of the nineteenth century depicted here to our modern meat-type swine.

is the most used with swine. Large litters give many numbers to work with, and even if only a few animals remain, the reestablishment of a herd is relatively rapid. For this reason, the standards are usually high and progress has been the envy of cattlebreeders and others working with a less flexible species.

Selection Index

By using a selection index all important traits can be easily combined into one figure or score. Higher scores mean more valuable animals for breeding purposes. The weight assigned to each trait will depend on its economic importance, heritability, and genetic linkage to other traits.

STUDY QUESTIONS

1. In viewing swine from the side, a _____ , _____ maturing hog is desired.
2. An even, arched _____ with high tail head and trim underline is indication of a meaty, sound-boned hog.
3. From the rear view, finish and _____ can be seen.
4. A well-muscled hog has good width throughout the back with a plump, full, trim, and firm _____ with muscling well down to the hocks.
5. A firm, uniform layer of fat over the hog of less than _____ in. in thickness is desired.
6. In selecting breeding stock, performance testing and progeny testing are the best criteria to use as they indicate the performance of a hog's _____ as well as its performance.

DISCUSSION QUESTIONS

1. Name the parts of a meat-type hog. Why is it necessary to know these terms?
2. Why do judges look for length, trimness, and heavy bone?
3. What are the various ways of detecting finish on a hog?
4. What is the chief value of pedigree information once show ring evaluation has been determined?
5. What are the economically important traits that have a high probability of improvement through selection and breeding for these features?
6. Define the following terms:
   a. production registry
   b. certified boar
   c. swine testing
   d. ultrasonic measurements

Chapter Twenty-Six

THE PORK CARCASS

The production of pork products that meet consumer demands keeps swine markets strong. No other livestock carcass is marketed to the consumer in such a variety of products varying all the way from sow belly to pigs' feet. The marketing, slaughtering, inspecting, grading, and retailing of pork carcasses fulfills the producers' goal of producing a high-quality product for a ready consumer market.

MARKET CLASSES AND MARKETING PORK

Most hogs are marketed either by direct sale (sale directly to a packer) or by route of a local auction. Market hogs are classified and sold in four market classes based on sex, the use to which the animal is best suited, and weight.

Barrows and Gilts

Barrows (castrated male hogs showing no sex characteristics) and *gilts* (young female hogs) form the class of most *finished market hogs*. Weights of barrows and gilts usually vary from 120 to 300 lb with the most desirable weight being

200 to 250 lb. Most pork for human consumption is derived from this class, depending on the U.S. Carcass Grade of the animal in the market class.

Sows

Sows (either pregnant or having had at least one litter) represent the second market class of swine. Usually they represent culled sows or bred gilts ranging in weight upwards of 220 lb. Pregnancy reduces dressing percent and results in a carcass that carries more fat than do barrows and gilts. Pork from sows is mainly marketed for human consumption as cured pork, and the carcasses are graded according to U.S. Carcass Grades.

Stags

Stags (castrated males showing sexual development) as a class are usually not marketed in sufficient numbers to warrant U.S. grading and are usually used for nonhuman consumption. Culled boars are generally castrated after culling and fed for a short time, and thus fit into this class.

Boars

Boars (uncastrated males) are low in market value because a large portion of the carcass is condemned for human consumption because of odor. Manufactured by-products such as inedible grease, fertilizers, hides, and others are the main use of this class.

SLAUGHTERING AND DRESSING HOGS

Upon arrival at the packing plant, hogs are penned in small pens, showered, and kept until slaughter (Figure 26-1). The slaughtering process is a set pattern of stunning, hoisting, sticking, scalding, dehairing, and dressing to yield a cooled pork carcass ready for inspection, grading, and further processing.

MEASUREMENTS OF CARCASS QUALITY

Carcass quality is measured in various ways, depending on the use of the carcass. Carcasses used primarily for bacon are measured by length and percent of trimmed belly, while those used for other purposes are measured by percent lean cuts, minimum backfat thickness, and loin-eye area. In all evaluations, the emphasis is on the meat-type hog.

FIGURE 26-1.   Veterinarian examining swine before slaughter. (Courtesy of the U.S. Department of Agriculture)

Carcass Yield or Dressing Percent

One of the first calculations made in slaughtering hogs is the dressing percent or carcass yield (the percentage yield of chilled carcass in relation to live weight). For example, a hog slaughtered at 200 lb live weight yields a 140-lb chilled carcass. The dressing percent would be calculated as

$$\frac{140 \text{ lb}}{200 \text{ lb}} \times 100 = 70\%$$

Thus, 70% of the live weight resulted in a chilled carcass. Normal dressing percents vary from 60 to 70%. The other 30 to 40% of the live weight is by-products of pork slaughter. The dressing percent of swine is higher than that of cattle or sheep because hogs do not have the large chest cavity and four-compartment stomach.

Fat- or lard-type hogs have a higher dressing percent than meat type. Yet because lard is low in retail price, the carcass value of lard-type hogs is less. Thus most packers use the cut-out value of the carcass, especially on the maximum yield of the more important primal cuts.

Once weighed, the chilled carcass is halved and other measurements are taken. The first of these is carcass length, which is measured from the front of the aitch bone to the front of the first rib (Figure 26-2). Desired length is 29 in. or more.

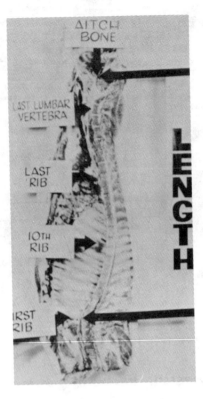

FIGURE 26-2. Carcass length is from the front of the aitch bone to the front of the last rib. Location of backfat measurements is also shown here. (Courtesy of the National Association of Swine Records)

Backfat of the chilled carcass is measured at three locations: the first rib, the last rib, and the last lumbar (Figure 26-3). The average of these three measurements is reported as carcass backfat. Standards for backfat differ among sex and market classes, with the general rule of 1.5 in. or less desired.

Loin-Eye Area

Because of the importance of pork chops in retailing pork, the lean meat area of the loin eye is often measured. For this measurement, the carcass is split between the tenth and eleventh ribs, and the loin-eye muscle is traced and measured by use of grid paper (Figure 26-4). The loin-eye area should be 4.5 in.^{2} or more in meat-type hogs.

Percent of Lean Cuts

A measure of the retail value of the carcass is the percent of the four major lean cuts (ham, loin, picnics, and Boston butts). These cuts, although only 45 to 50% of the hog's live weight, represent 75% of the carcass retail value. These four cuts are cut out, trimmed, and weighed. The percent lean cuts are determined similarly to dressing percent.

Chap. 26 THE PORK CARCASS

FIGURE 26-3. Backfat on a carcass is the average of backfat readings measured at the first rib, last rib, and last lumbar as shown in Figure 26-2. Shown here is the measurement at the first rib. (Courtesy of the National Association of Swine Records)

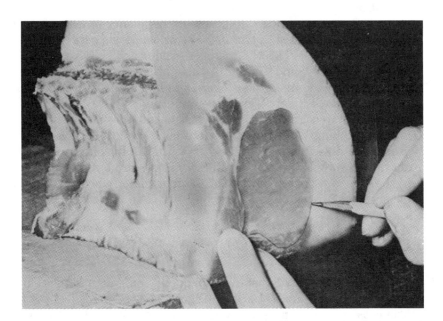

FIGURE 26-4. Loin-eye area is measured between the tenth and eleventh ribs. Only the one large muscle is included in the measurement. The muscle is traced on a clear piece of acetate, with the area later determined using a planimeter. (Courtesy of the National Association of Swine Records)

Ham-Loin Index

The ham-loin index is often used to evaluate carcasses in barrow shows. The score is based on the percent of trimmed ham and loin-eye area. The score is determined as follows:

Percent ham above 10% × 10 = _____
Square inch loin-eye area × 10 = _____
Total score

For example, a hog yielding a 19.6% trimmed ham that had a 4.6-in.² loin-eye would have the following ham-loin score:

$$9.6 \times 10 = 96$$
$$4.6 \times 10 = \underline{46}$$
$$142 \text{ ham-loin index}$$

CARCASS GRADES AND GRADING

Chilled carcasses are inspected for human consumption. Carcass grades are an indication of the excellence and quality of an animal in its specific market class. U.S. carcass grades are based on the quality of lean, the amount of fat, and the expected yield of trimmed major wholesale cuts (hams, loins, picnics, and Boston butts). The U.S. carcass grades are *U.S. No. 1, U.S. No. 2, U.S. No. 3, U.S. No. 4,* and *Utility.* Although live grades are not covered in this chapter, the same five grades exist, and live grades are based on the correlation of the live animal to yield the corresponding carcass grades. The skilled swine buyer recognizes this correlation and through visual evaluation and backfat probing can accurately determine resulting carcass grade. Carcass grades are determined by inspecting both the fat and lean for quality (firmness, color, actual backfat thickness, fatness around the belly, and marbling of the loin) and the percentage of the four major wholesale cuts (hams, loins, picnics, and Boston butts).

U.S. No. 1

Carcasses in this grade have a high-quality of lean, a high yield of lean cuts, and a low percentage of backfat (Figure 26-5). The yield of the four principal lean cuts is 53% or more; the maximum backfat is 1.3 to 1.6 in. with loin eye slightly firm, a slight amount of marbling, and pink to red coloring. The carcass will have an intermediate degree of finish, be well muscled, and have good length.

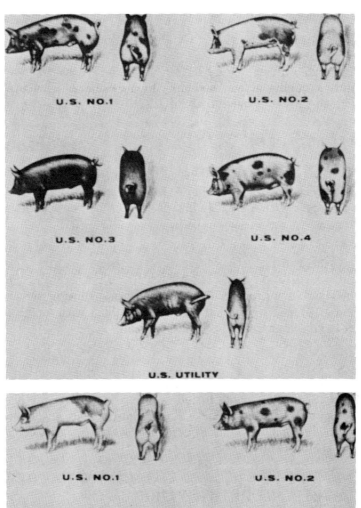

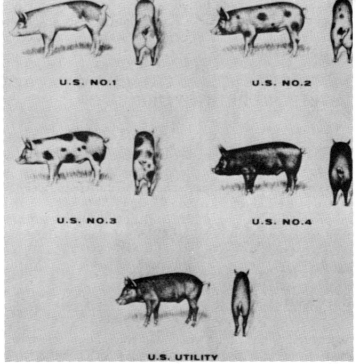

FIGURE 26-5.   USDA grades of swine. Shown at the top are slaughter swine; below are feeder pigs. (Courtesy of the U.S. Department of Agriculture)

U.S. No. 2

Figure 26-5 illustrates a U.S. No. 2 grade carcass. Such a carcass has an acceptable quality of lean, a slightly high number of lean cuts (50 to 52.9%), and more thickness of backfat than U.S. No. 1 carcasses (1.6 to 1.9 in. backfat).

U.S. No. 3

Carcasses yielding 47 to 49.9% of the four lean cuts, having 1.9 to 2.2 in. backfat, and an acceptable quality of lean are graded U.S. No. 3 (Figure 26-5). This is a carcass with a slightly low percent of lean cuts and a high percent of fat cuts.

U.S. No. 4

The U.S. No. 4 grade is assigned to carcasses low in percent of lean cuts (less than 47%) and high in backfat (above 2.2 in.) with a high percent of fat cuts, yet having an acceptable quality of lean.

U.S. Utility

The U.S. Utility grade includes carcasses with some lean quality but not enough to be graded U.S. No. 1 through U.S. No. 4. Carcasses with unacceptable belly thickness and all carcasses which are soft or oily are also graded U.S. Utility (Figure 26-5).

WHOLESALE AND RETAIL PORK CUTS

Once carcasses are graded, they are usually cut up and sold by the packing plant in the form of wholesale cuts (Figure 26-6). The most common wholesale cuts are ham, bacon, loin, picnic shoulder, Boston butt, jowl, spareribs, and feet. Only about 30% (loins, shoulders, and spare-ribs) of the wholesale cuts are sold as fresh meat with the remaining 70% sold as cured pork. The average lard production for hogs is 15% of the live body weight. Wholesale cuts are further processed into the retail cuts found at the common meat market (Figure 26-6).

PORK BY-PRODUCTS

All products from slaughtering other than the carcass meat and lard are designated as by-products. This usually represents about 30 to 40% of the animal's live weight. The successful use of all by-products is a goal for which all packers

Retail Cuts | **Wholesale Cuts** | **Retail Cuts**

Boneless Loin Roast
Tenderloin Frenched and Whole
Canadian-Style Bacon
Loin Chop
Rib Chop
Frenched Rib Chop
Butterfly Chop
Sirloin Roast
Loin Roast Center Cut
Blade Loin Roast
Crown Roast
Fat Back
Lard
Blade Steaks
Smoked Shoulder Butt
Boston Butt
Rolled Boston Butt

HAM
SIDE
LOIN
SPARE RIBS
BOSTON BUTT
PICNIC
JOWL

Jowl Bacon Square

Ham (Butt Half)
Ham (Shank Half)
Ham Butt Slice
Center Ham Slice
Fresh Ham Roast
Rolled Fresh Ham Roast
Bacon
Salt Pork
Spareribs
Fresh Picnic Shoulder
Smoked Picnic Shoulder
Cushion Picnic Shoulder
Rolled Fresh Picnic Shoulder
Fresh Shoulder Hock
Arm Steak

FIGURE 26-6. Wholesale and retail cuts of swine. (Courtesy of the U.S. Department of Agriculture)

aim, and a saying that there is no waste and the packer saves "everything but the squeal" has developed. By-products vary from blood to glycerin for explosives, meat scraps to medicine, glue to leather, and violin strings to fertilizers.

STUDY QUESTIONS

1. Market classes of swine are based on _____ , _____ , and _____ .

2. _____ are castrated male hogs showing no sex characteristics.

3. Most pork for human consumption comes from the _____ and _____ market classes.

4. _____ are condemned for human consumption because of odor and must be slaughtered for by-products.

5. Castrated males which show sexual development are sold in the _____ market class.

6. Dressing percents of swine vary from _____ to _____ %.

7. Length of swine carcass is measured from the front of the _____ bone to the _____ rib.

8. The desired carcass length is _____ in. or more.

9. Carcass _____ is measured at the first rib, last rib, and last lumbar.

10. Backfat on chilled carcasses should be below _____ in.

11. Loin-eye area is determined between the _____ and _____ ribs with 4.5 in.^{2} or above desired.

12. Percent yield of the major cuts: _____ , _____ , _____ , and _____ should be 45 to 50%.

13. Ham-loin index has been used in many barrow _____ shows.

14. U.S. Carcass Grades are U.S. No. 1, U.S. No. 2, U.S. No. 3, U.S. No. 4, and _____ .

15. The average lard production for hogs is _____ % of the live body weight.

16. From _____ to _____ % of the hog's live weight is manufactured as by-products.

DISCUSSION QUESTIONS

1. What are the hog market and sex classifications? Name them.

2. A trim, young 215-lb barrow is sent to slaughter. What would be the expected offal weight? How long (inches) should the carcass be? Where is this measurement taken? How much backfat (inches) is desirable?

3. If the hog in Question 2 were divided into the four lean cuts, how many pounds would remain?

4. If the hog in Question 2 yielded 20.5% ham and measured 4.9 in.^{2} loin eye, what would be the ham-loin index?

5. Discuss the carcass grades and the relationship of each to lean cuts and backfat.

6. Give the wholesale cut from which each of the following come:
   a. Canadian bacon
   b. blade steaks
   c. bacon
   d. salt pork
   e. picnic ham

Chapter Twenty-Seven

SWINE MANAGEMENT

Being of a different species, swine require different methods of handling; and methods found successful in cattle, sheep, or horses may not work. Swine management techniques vary somewhat among swine breeders. All swine operations, however, will fall into one of the following production systems. The type of production system that should be followed by a producer depends on management skill, available capital, feed supply, available labor, and personal preference.

FEEDER PIG PRODUCTION

Feeder pig production systems produce weaned pigs that are sold to be grown and finished out on other farms (Figure 27-1). Such a system has a comparatively low feed requirement but requires a high management level for success. Management is involved in selection and care of the breeding herd, at breeding time, at farrowing, and in care of the young pigs. Proper management ensures large litters that can be marketed as outstanding feeder pigs. Above-average feeder pig producers market 2.2 litters per sow per year, with 8.5 pigs or more marketed per litter at an average market weight of 45 lb or more.

FIGURE 27-1.   Feeder pig production systems produce weaned pigs. (Courtesy of the American Landrace Association, Inc.)

Management of the Breeding Herd

The quality of the breeding herd determines the quality of the marketed feeder pigs. Selection of replacement gilts and boars, therefore, must be based on visual appraisal, records of parents, and predetermined criteria. The desired gilts and boars should be from large litters and selected on number of nipples, growth weight at 154 days of age (200 lb or more for gilts, 220 lb or more for boars), backfat probe (1 inch or less), and litter mate carcass, cut-out data (60% or more lean cuts, 40% or more ham and loin, 1.2 inches or less backfat, and 29 inches or more length). It should be noted that selection for all these factors is almost impossible, but the goal should be to approach these standards in as many areas as is practical. Because litter size has a low heritability, it cannot be improved upon very much by selection. Therefore, management becomes as important, or more so, than selection.

Success depends on litter size so a critical management time is during breeding and gestation. Maximum litter size results when sows are bred on their second or third heat period, when gilts are flushed, and when sows and gilts are bred with more than one service and are bred to a tested, well-cared-for boar. More eggs are ovulated at the second and third estrous cycles, and maximum fertilization is achieved when breeding occurs 12 to 24 hours after the onset of estrus (see Chapter 23). Flushing of gilts increases the number of eggs ovulated. Feed is increased 7 to 10 days prior to and 4 to 8 days after breeding to flush a gilt by conventional methods. A newer method that shows some improvement in litter size is flushing gilts by increasing feed one day prior to breeding and double feeding the first day of breeding. Proper nutrition, disease prevention, and control of extreme environmental conditions during gestation ensure large litters.

Care of the Sow at Farrowing

Sows are generally moved into the farrowing crates (stalls) 5 days prior to farrowing date to allow proper adjustment to new conditions before farrowing (Figure 27-2). The efficient swine manager is present at farrowing—no matter what time of day or night it may be. The manager makes sure that newly born pigs are freed from their embryonic membranes so they will not suffocate and helps sows that do not farrow in 6 to 8 hours by an injection of *oxytocin* (a hormone that causes uterus contractions and milk letdown). After farrowing, the sows must be properly cleaned out and be giving milk for the young pigs. Sows that do not give milk may require an oxytocin injection for proper milk letdown.

Management of the Young Pigs after Birth

Most feeder pig producers follow a set schedule for management of young pigs. Although the order varies, the following management duties are generally performed. The day following farrowing, the needle teeth are clipped and the litters are evened up (splitting up litters of sows farrowed at the same time). To prevent anemia, pigs are injected with iron dextran two to three days

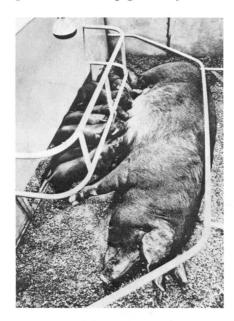

FIGURE 27-2. Farrowing crates are used at farrowing to allow for proper care of both sow and young pigs. (Courtesy of the U.S. Department of Agriculture)

after birth. Ear marking for proper identification of pigs is done at the same time. Starter rations are made available 7 to 10 days after birth. Male pigs are castrated at about two weeks of age. Iron dextran injections are again given and all pigs weaned at about 3 to 5 weeks of age. The pigs should weigh about 25 lb at weaning. Pigs are then grouped according to their size and sex and moved to the nursery where they are grown to about 40 lb at the time of marketing as feeder pigs. In many areas, feeder pigs are the major income producer in swine herds.

Use of Crossbreeding in Commercial Swine Operations

The advantages of crossbreeding are well known, and approximately 90% of the sows are crossbred in commercial herds. Crossbred pigs are stronger at birth and grow faster; thus more and heavier pigs are marketed. Crossbred sows produce more pigs. Most crossbreeding is not a single cross of two purebred parents, but some type of crossing using crossbred sows. This is accomplished by using two breeds in a crisscross or backcross system or using more than two breeds in a rotation crossbreeding program (Figure 27-3). By rotating breeds without backcrossing to one of the original breeds, high levels of hybrid vigor are maintained.

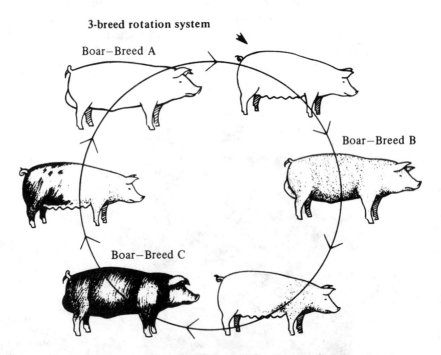

FIGURE 27-3. A rotational crossbreeding program for swine.

Multiple Farrowing

In recent years, swine producers have gone from twice a year seasonal farrowing to multiple farrowing (farrowing pigs throughout the year). This has allowed for maximum use of buildings and equipment and provides a steady supply of feeder pigs to a ready market.

Multiple farrowing requires a planned management schedule for rotating sow groups to fit the farrowing house schedule. Sows that will farrow within 7 to 10 days are rotated together throughout the year. The group is moved into the farrowing house about 3 to 5 days before the farrowing date. An example of such a schedule using three sow groups which farrow 48 days apart follows in Table 27-1 (two litters per year). Thus sows in group A are in the farrowing house April 19 to May 29, group B June 7 to July 17, and group C July 25 to September 3. Groups A, B, and C farrow their second litter in the same house as indicated in the schedule.

TABLE 27-1 48-DAY FARROWING SCHEDULE

Sow Group	Date Bred	Date Moved to Farrowing House	Farrowing Date	Date Weaned
A	Jan. 1	Apr. 19	Apr. 24	May 29
	June 3	Sept. 20	Sept. 25	Oct. 30
B	Feb. 12	June 7	June 12	July 17
	July 22	Nov. 7	Nov. 12	Dec. 17
C	Apr. 7	July 25	July 30	Sept. 3
	Sept. 8	Dec. 27	Jan. 1	Feb. 3

The 48-day schedule allows 35 days from farrowing to weaning, 3 days for adjustment of pigs, 5 days for cleanup, and a new group of sows is moved in 5 days before farrowing for a total of 48 days. Similar 35-day schedules can be developed with weaning at 28 days, 3 days cleanup, and a new group moved in 4 days prior to farrowing.

FINISHING PROGRAM

The finishing program covers the growth and finishing of feeder pigs for slaughter (Figure 27-4). No breeding stock are kept so less management and time are required for a finishing operation. A large quantity of feed at a reasonable cost must be present for the finishing program. High-quality feeder pigs are often obtained by contract with a feeder pig producer. The growing and finishing of a pig from 45 lb to 200 to 240 lb takes 10 to 13 weeks and from 600 to 800 lb of feed. Feed conversion of the above average producer is 3.2 lb of feed per pound of gain. Thus three to four crops of pigs can be finished out in a year providing feed and feeder pigs are available. Because the growth rate

FIGURE 27-4. The finishing swine program covers growth and finishing of feeder pigs for slaughter. (Courtesy of the American Landrace Association, Inc.)

of a pig declines after 240 lb, efficiency of producing additional weight is reduced and hogs are generally marketed prior to reaching this weight.

Confinement versus Pasture Finishing

Finishing programs on good cereal grain and legume pasture reduce both feed requirements and building and equipment needs. However, with more specialization in finishing swine, confinement is used more and more in growing, finishing, farrowing, and at the nursery stage, too. This allows for mechanized handling of feeds and waste as well as better sanitation and disease control.

FARROW TO FINISH

The farrow to finish operation combines the feeder pig production and finishing operation into one. A producer finishes the feeder pigs produced so a large quantity of reasonably priced feed is required as well as a high level of management. Large investment in buildings and equipment is also required. Maximum profits are obtained because profits are not shared with other producers.

PUREBRED OPERATIONS

The purebred breeder supplies the seedstock for the commercial swineherd. High investments are required in buildings and breeding stock, and a highly skilled manager is required to produce, record, and market purebred hogs. The master purebred breeder makes maximum use of all tools available to test and improve the herds.

One of the more recent methods of evaluating purebred sows is the *Production Registry (PR)* and the *Unified Meat Type Hog Certification Program* (Chapter 25). For instance, if a sow's litter is to be a production registered (PR) litter, the sow and the boar must be registered, the litter must be nominated 7 to 10 days after birth, the litter size must be 8 or more for gilts and 10 or more for sows, and weights must be taken at birth (22 lb per litter for gilts, 28 lb per litter for sows), at 21, 35, and 56 days of age. A *certified litter* (CL) is a PR litter that has carcass evaluations of two pigs slaughtered at 242 lb or less. The standards for a certified litter adjusted to 220 lb are an age of 180 days or less, carcass length of 29.5 in. or more, backfat of 1.5 in. or less, and a loin-eye area of 4.5 in² or more. This has led to the terminology of a five-star PR sow (sow farrowing five PR litters). A pig from a certified litter has CL on its records. A *CMS (Certified Meat Sire)* is one that has sired five CL litters, and *CM (Certified Mating)* is a mating that in the past has produced a certified litter. The figures above will vary somewhat depending on the association office and its minimum standards.

STUDY QUESTIONS

1. _____ produces weaned pigs that are sold to be grown and finished on other farms.

2. Feeder pig production systems have a _____ feed requirement and _____ management level.

3. Goals of a feeder pig producer are _____ litters per year per sow with _____ pigs or more per litter marketed at a weight of at least 45 lb.

4. Selection of replacement gilts for the breeding herd should be based on visual appraisal, parental records, and predetermined _____ .

5. Swineherds help young pigs at farrowing by freeing them from their membranes and seeing that the sow _____ .

6. Management skills applied to young pigs include clipping needle teeth, evening up litters, injection of _____ to prevent anemia, and _____ for identification.

7. Male pigs are castrated at about _____ weeks of age.

8. Pigs are generally weaned at _____ to _____ weeks of age and _____ lb.

9. _____ pigs are heavier at birth and grow faster than purebred pigs.

10. Crossbred sows produce _____ than purebred sows.

11. Farrowing pigs throughout the year on a predetermined schedule is termed _____ .

12. Multiple farrowing is usually based on a _____-day or _____-day schedule.

13. _____ systems of production cover growth and finishing feeder pigs for slaughter.

14. Finishing programs have a _____ feed requirement, and a _____ management level.

15. Growing and finishing pigs from 45 to 200 or 220 lb takes _____ to _____ weeks and _____ to _____ lb of feed.

16. Above-average feed conversion of swine is _____ lb of feed or less per pound of gain.

17. _____ operations combine the feeder pig production and finishing operations into a system of production.

18. The main function of the _____ breeder is to produce the seedstock for the commercial breeder.

19. Purebred swine production has a _____ investment requirement and a _____ management skill.

20. A _____ sow is one that has produced a litter of 10 or more pigs that meet the weight requirements at 21, 35, and 56 days of age.

21. A PR litter that has a carcass evaluation of at least two pigs that meet specified standards is termed a _____ litter.

22. A boar that has sired five CL litters is a _____ .

23. CM on swine records refers to pigs that were mated to produce a _____ .

DISCUSSION QUESTIONS

1. Assuming a feeder pig producer is above average and owns 100 sows, how many feeder pigs could he produce annually? What would be the total market weight of pigs produced? At current feeder pig prices, what would be his expected gross income?

2. What standards are recommended in selecting sows for feeder pig production?

3. What breeds and in what rotation would you use them to produce the maximum hybrid vigor in your feeder pigs?
4. Develop a multiple farrowing calendar for three groups of sows on a 35-day schedule.
5. Give the details of the Unified Meat Type Hog Certification Program.

Chapter Twenty-Eight

SWINE HEALTH MANAGEMENT

The common image of a hog's dream of the perfect environment is a mud wallow filled with contented swine. Most people consider the hog an animal of filth; however, given a choice, the hog is quite sanitary, preferring clean, cool, well-kept quarters. The reason hogs are often found in mud wallows is because they have no efficient way of dissipating body heat except by conduction (lying against something cooler than body temperature).

A swine health management program should prevent disease and parasites from becoming established in the herd rather than wait to control diseases after they develop. This involves isolation of the herd from disease pathogens and parasites, proper sanitation of buildings and equipment (Figure 28-1), and a planned immunization program. Disease-free herds can remain so only if no diseases are introduced by humans or replacement stock. Visitors should be required to wear clean footwear and not allowed in the farrowing house because small pigs are very susceptible to diseases. A working routine is followed on most farms of caring for the farrowing house and small pigs first, the nursery and the older pigs (finishing and breeding herd) last. Thus diseases are not transmitted to the young pigs.

Sanitation is very important in swine managment because larger numbers of hogs are kept in close confinement as compared to other species. The usual procedure is to steam clean farrowing houses before each group enters.

FIGURE 28-1. Spraying swine quarters is an important sanitation measure in swine management. (Courtesy of Virginia Farm Agency)

The sows are washed and scrubbed thoroughly and proper disinfectants are used. It is also recommended to move sows and litters to clean pastures by hauling rather than letting them walk through areas that might be contaminated. Whenever possible, vaccination against diseases prevalent in the area of production is part of a good herd health management program.

Swine have a normal range in rectal temperature of 102 to 103.6 °F. Deviation from this norm could signal the onset of illness. Only the common problems and those of serious economic concern are covered here.

ANEMIA

Anemia is a sign of inadequate red blood cell or hemoglobin production and has been a more common occurrence since moving hog operations from outdoors to concrete or slatted floor operations. Iron deficiency is the major cause of anemia. Under natural conditions, after the first few days the piglets would be able to root around in the soil, picking up sufficient iron to enable required hemoglobin production. However, since most hog operations have now moved away from dirt, this problem has become more common. Pigs are born with minimal iron stores, and sow's milk supplies only one-seventh to one-tenth of the pigs' daily iron requirement.

Anemia is a condition primarily of rapidly growing baby pigs. The signs of anemia are rough hair coat, listlessness, and a pale color to the membrane of the eyes and mouth. Signs generally start about 10 days after farrowing and become severe by 2 to 3 weeks in unsupplemented litters. In the advanced stages, breathing becomes very labored and produces the characteristic abdominal breathing. The heart enlarges and pumps more forcefully and rapidly than normal in an attempt to compensate for the reduced blood capacity. When a severely affected pig is handled, this increased heart action may sound like "thumping" against the chest wall. Anemia, also called "thumps," lowers baby pigs' resistance to disease making them more susceptible to respiratory and intestinal disease organisms, creating major death loss.

Treatment is highly effective and usually consists of an injection of iron dextran once the condition arises. If pigs are severely affected and are comatose, a calcium gluconate or dextrose solution should be given to provide an extra boost of energy. Fresh soil for pigs to root around in and swabbing the sow's udder with iron sulfate solution may also be effective. It has become common practice to inject baby pigs at 1 to 3 days of age with an iron dextran solution in the muscle of the neck or hindquarter. This injection should be repeated at $3\frac{1}{2}$ weeks if the pigs are not on creep feed by this time. There are also iron supplemental blocks that may be offered free choice, and crystals that can be sprinkled on the floor. Most authorities agree that this is a good preventive measure, cheap, effective, and recommended as a routine practice. Thus treatment of anemia would never arise if such preventive measures are taken.

ATROPHIC RHINITIS

Atrophic rhinitis (AR) is not a disease that causes many deaths, but is does create a tremendous economic loss because of irritation to the nose, inability to eat, and a generally deteriorated physical condition. It also lowers the body's resistance to secondary lung infections brought about by the reduction of the normal filtering activity of the nasal bones. The snout is affected by decreased growth of the small, thin, bones that make up the nasal cavity (turbinate bones). The turbinates are reduced in size and are misshapen by this disease. Failure of normal turbinate growth causes a misshapen snout and severe pain.

AR develops first in young pigs about 3 weeks of age or less and becomes more chronic as the pig grows older. Initially, there is sneezing, coughing, blowing, and snorting. When severe degeneration occurs on both sides, the nose is shorter than the lower jaw. When one side is affected, the nose turns to the affected side in a "crooked nose" condition. The tear ducts may also be plugged up, creating tear discharges or dirty rings under the eyes. The nose will occasionally appear bloody on the surface.

The cause is primarily the bacterium *Bordetella bronchiseptica.* Dust and other irritants contribute to the severity of the attack. The infecting bacterium is found in most of the pig herds in the United States. It is very difficult to control because it is carried by other mammals, especially the farm cat.

Recommended treatment is to use sulfathiazole in the water and sulfamethazine in the feed continuously for 5 weeks. Sows should be treated before farrowing to reduce the organisms transferred from the sow to her newborn litter.

A rhinitis bacterin containing killed *Bordetella* organisms is now available. It is given during the first week the pig is born and again at 28 days after birth to provide protection against the disease. Sows are injected with the same preparation 4 and 2 weeks prior to farrowing, and boars are injected twice annually to provide similar protection.

Prevention is the recommended method of dealing with atrophic rhinitis. It is recommended that all animals be purchased from one source that is free of *B. bronchiseptica* and that the owner provide adequate ventilation in swine houses and good care of nursing pigs because any other disease (scours, flu, pneumonia) can make an atrophic rhinitis outbreak worse. Control of cat and rodent population and testing of each new crop of pigs for *Bordetella* is also recommended.

AUJESZKY'S DISEASE (PSEUDORABIES, FALSE RABIES, "MAD ITCH")

The signs of Aujeszky's disease, more commonly called pseudorabies, are seen in young pigs. Baby pigs that have no maternal antibodies develop high temperatures, convulsions, and paralysis, and usually die. In pigs less than 2 weeks old, death losses approach 100%. After 3 weeks, young pigs develop some resistance, and losses are considerably reduced.

In the adult swine, signs may be mild to nonexistent, or they may include abortion, stillbirths, infertility, mummified fetuses, mild respiratory signs, and lack of appetite. Death is rare for adults, but recovered hogs serve as a silent reservoir to carry the disease.

The main effect is on pregnant and nursing swine. Depending on the stage of gestation, pregnant sows may have stillborns and weak pigs (late gestation), mummification (midgestation), and embryonic death and infertility (early gestation). Abortion may occur at any stage but is most prevalent during the last one-third of pregnancy. Piglets may develop signs as early as 12 hours after farrowing, but generally the first signs are seen at 5 to 7 days of life.

The cause is a herpes virus. Since pseudorabies can occur in adult swine with few clinical signs and can persist in their system for long periods, these swine serve as a natural reservoir for the disease. The virus causes a fatal encephalomyelitis (inflammation of the nervous system) with signs of severe itching and self-mutilation—hence the name "mad itch." Other than eradication, no treatment is known to be effective. Secondary infections in adult animals can be suppressed by the use of injectable antibiotics.

The disease can be prevented by the use of a modified live virus vaccine, recently released for use only on swine. Vaccinated swine are protected for about 6 months, although they may become infected with a field strain virus and become carriers. Vaccines are restricted to use by veterinarians. Most states limit vaccination to infected or high-risk herds.

Since the major effects are on pregnant and lactating animals, vaccination of breeding animals before each breeding is recommended where applicable. This semiannual vaccination will protect both sow and litter during nursing. In herds where sows were not vaccinated, pigs can be vaccinated at 3 days of age if nursing nonimmune sows, and 3 to 8 weeks if the sows have immunity and are providing antibodies in their colostrum.

BRUCELLOSIS

Abortion is the key sign of brucellosis. It can occur at any stage of gestation but normally happens at 2 to 3 months or later. Retained placenta, temporary sterility due to inflammation of the uterus, lowered conception rate, pigs born weak, inflammation of the mammary gland, and arthritis are additional signs. Males infected with brucellosis have signs of orchitis (swelling of the testes) and lameness.

A bacterium causes the disease. Swine are most often infected with *Brucella suis*, but are also occasionally infected with *Brucella melitinisis* and the bovine variety, *Brucella abortus*. This bacterium is very resistant and hard to kill with any disinfectant unless protective organic material is removed. It remains in the soil, water, or aborted fetus for up to 60 days under natural conditions. Cresylic acid compounds and sodium orthophenylphenate will kill the organism provided the area is well scrubbed with soap and water prior to disinfection.

There is no treatment for brucellosis. Infected animals should be slaughtered. To prevent the disease, herds should be tested annually, and reactors eliminated in order to control the disease. There is no vaccine available to prevent it. Replacement animals should be from brucellosis-validated herds or animals that have had a negative brucellosis test.

CHOLERA

Although not an explosive disease, hog cholera (Figure 28-2) is a devastating problem with swine. Several stages may be present in a herd due to the often slow chronic spread, which requires 6 to 20 days to fully develop. The first warning is several hogs dying with no clinical signs observed. After the initial unexplained deaths, the major signs are lack of appetite, elevated temperatures (104 to 109 °F), involvement of central nervous system, depression, diarrhea, dehydration, weight loss, and a high death rate (Figure 28-3). Con-

FIGURE 28-2.   Listless victims of hog cholera. (Courtesy of the U.S. Department
of Agriculture)

junctivitis (severe inflammation of the eyelids) in which the eyelids may be glued together due to exudates is common. Recovered pigs often remain "poor doers" for the rest of their lives.

The cause is a virus that is specific for swine. It is a contagious disease usually transmitted directly from pig to pig, although blood-sucking insects can transmit it. It is very often passed in uncooked garbage, and for this reason it is illegal to feed raw, uncooked garbage in the United States.

There is no present treatment. The disease is under U.S. government control by means of federal quarantine and slaughter. Vaccines are no longer available for cholera. The original vaccine was found to actually cause the dis-

FIGURE 28-3.   Button ulcers in the large intestine are a common sign of hog cholera. (Courtesy of the U.S. Department of Agriculture)

ease in some cases, and the U.S. government forbade its use. Since then, outbreaks that have occurred in the United States have been largely traced back to the use of the original vaccine, which was stored in deep freezes. A good preventive precautionary practice is to cook all garbage fed to hogs. Any animals suspected of cholera should be quarantined and a 30-day isolation period maintained by a veterinarian.

DYSENTERY (BLOODY SCOURS)

In the opinion of many veterinarians, the most important intestinal disease of weaned pigs is swine dysentery. It is a contagious disease and can result in severe economic loss to pork producers.

Signs of dysentery (bloody scours) usually occur at 10 to 16 weeks of age. Unweaned animals as well as adults are susceptible. Reduced appetite is often the first noticeable sign, accompanied by soft, off-colored bowel movements, which have been described as appearing like "wet cement" or milky coffee. The stool becomes streaked with blood as the disease progresses and usually contains undigested food as well as mucus from the intestinal wall. The rear of affected pigs becomes wet and stained, and pigs take on a characteristic gaunt appearance. Death losses are not high but some sudden losses can occur early in the outbreaks.

The cause is generally accepted to be a large spirochete, *Treponema hyodysenteriae*, although other organisms may also be involved. Studies of outbreaks suggest that carrier pigs are most likely the source of infection. These pigs show no signs of the disease, having developed an immunity, but spread it to others that have not been previously exposed. Birds, dogs, man, and equipment have also been identified as mechanical carriers.

Treatment consists of using organic arsenic preparations. The preferred method of drug delivery is via the drinking water rather than in the feed. After the initial outbreak is controlled, medicated feed can be used to hold the disease in check until the pigs are marketed.

Tylosin has proved beneficial in injectable form but does not work in feed or water at currently approved levels. However, individual pigs that show extreme weakness may be injected with Tylosin. Drugs to be mixed with feed are available, such as Mecadox, Virginiamycin, and Lincomix.

Preventive methods pay big dividends. Clean, dry floors and uncontaminated feed and water will reduce losses. Steam cleaning and reduction of stress factors such as overcrowding, chilling, and poor ventilation have reduced outbreaks of dysentery.

ERYSIPELAS

Clinical signs of erysipelas vary considerably because of the three forms: acute, subacute, and chronic. In the acute form, some pigs may die very suddenly. Others may develop a high temperature and lack of appetite, appear physically sick, develop sore muscles or tender feet, have an arched back, and walk in a shuffling gait, often squealing because of pain in the feet and joints. Affected pigs may remain lying down or may protest loudly when they are forced up. About the third day of the disorder, red diamond-shaped or square patches may appear on the skin. This is considered a diagnostic clinical sign of hogs infected with erysipelas.

Subacute forms produce milder signs with fewer deaths. Pigs may show only reluctance to move and reddish skin discoloration of the ears, tail, jowl, and legs. In the chronic form (Figure 28-4), there is a persistent arthritis and swelling of joints, and the tips of the tail or ears may blacken and fall off. Although the acute form would appear to be the most dangerous type, probably more economic loss is caused from the subacute and chronic forms because of developing lameness and arthritis, a condition that is very difficult to reverse once started.

The cause is a bacterium, *Erysipelothrix insidiosa*. This organism is very difficult to control because it lives on a wide range of hosts including rats, birds, insects, in the soil, and may be carried by other swine that recover from the disease. If erysipelas is suspected, the antibiotics of choice are penicillin or chlortetracycline. If the disorder is in the later stages of development, arthritis may set in even with successful treatment. A swine erysipelas antiserum is also available that can be given at the onset of an acute infection for immediate blocking effects.

Preventive measures include control of rats, predators, flies, and cremating of dead pigs. Vaccines are available and give immunity for about 6 months following use. Vaccination can be performed at any time during the gestation period but is recommended to be given initially 3 to 4 weeks before breeding

FIGURE 28-4. Chronic swine erysipelas, showing sloughing of skin. (Courtesy of the U.S. Department of Agriculture)

gilts. The vaccines should be reinjected 3 to 4 weeks before farrowing in order to increase immunity in the baby pigs via colostrum milk. Revaccinate sows at each farrowing to maintain protection for the litter.

HYPOGLYCEMIA (BABY PIG DISEASE)

Hypoglycemia means low blood sugar. This condition is due to inadequate food intake and may develop in 24 to 36 hours after birth. Stress also plays a significant part, particularly cold stress during the first 10 days of life. Baby pigs don't have the ability to produce glucose as do adults and therefore must depend on daily intake from milk. If milk intake is inadequate or if demands are too high because of a cold barn, hypoglycemia develops. Death losses may be very high during the first few days of life. The first signs are shivering, erection of hair, squealing, weakness, a rolling of the eyeballs, and coma. Without treatment, death usually occurs within 36 hours.

The most common causes are lack of milk production in the sow and diarrhea in baby pigs. Exposure to cold and an inherited weakness may be contributing causes. Treatment involves providing a warm environment through heat lamps and supplemental feeding, and if signs are severe, an intramuscular injection of 5% glucose solution given every 4 to 6 hours. A satisfactory supplement is evaporated milk diluted with ½ teaspoon of water. Preventive measures include collection of breeding stock for high-milking qualities, providing heat lamps during colder parts of the year, and controlling litter size by redistribution of piglets.

LEPTOSPIROSIS

The main signs of leptospirosis are abortion, small weak pigs at farrowing, or a high incidence of stillbirths. The only other signs, in adult hogs, are a mild rise in temperature and a short loss of appetite. In some herds that have been exposed to the disease for the first time there is an abortion "storm." In other herds, the disease spreads slowly, causing abortions that occur over a span of a month or more. Frequency of abortions will then diminish as immunity develops. Most abortions occur 2 to 4 weeks before parturition.

Leptospirosis is caused by five strains of organisms: *Leptospira pomona, grippotyphosa, canicola, icterohaemorrhagiae,* and *hardjo.* None of these seem to produce significant clinical signs of disease in adults other than abortion, stillbirths, etc. All five strains affect the placenta, liver, and particularly the kidney of infected swine. Contamination of feed, water and pasture is through the urine of affected animals. Cattle, rats, raccoons, and other wildlife are also carriers of the organism and serve to spread the disease. Active organisms are carried in the urine of infected swine and excreted for up to 6 months after recovery. The reason for abortion is that the organism attacks the placenta (sac) and produces placental necrosis (a killing of the tissue), which kills the developing pigs.

Treatment consists of two injections given 24 hours apart of streptomycin. Other antibiotics such as oxytetracycline and chlortetracycline may be used in the feed at 400 to 500 grams per ton for 4 weeks to stop the shedding of the leptospiral organisms from infected swine.

Preventive measures can be incorporated into the management system. Vaccination should be done 2 to 3 weeks prior to breeding, but can be done at any stage, even during pregnancy. A good general recommendation would include vaccination with a five-serotype bacterin. Any new animal brought to the premises should be isolated and tested for lepto prior to admission to the herd. Routine antibiotic treatment and vaccination of new additions have also been recommended.

MYCOPLASMA PNEUMONIA (VIRUS PNEUMONIA, VPP)

Viral pig pneumonia is actually misnamed because the condition is not caused by a virus, but rather a bacterium—*Mycoplasma hypopneumoniae.* The most common sign is a dry, rasping, persistent, nonproductive cough, which is most noticeable when pigs are quiet. During the first 2 to 3 days of infection, young pigs 3 to 10 weeks of age may have mild diarrhea accompanied by this cough, but the diarrhea is very mild and short-lived and generally goes unnoticed by the producer. Slow growth and unthriftiness of litters are the major results of the disease. The original infection is quite mild, but secondary infections with other organisms, particularly *Pasteurella,* may cause severe disease losses.

There is currently no specific treatment for the initial infecting organism that causes VPP, but the secondary bacterial infection can be treated. Injectable broad-spectrum antibiotics such as tetracycline, tylosin, and lincomycin are recommended. Tiamultin, a feed additive, has been used with encouraging response.

Preventive measures recommended are to stock herds from a known VPP-free source. The organism does not penetrate the placenta; so one source of pigs free of the disease is SPF (specific pathogen free) pigs. These pigs are not immune to the attack of VPP but are free of the disease when introduced to herds.

PSEUDORABIES (See AUJESZKY'S DISEASE)

PORCINE STRESS SYNDROME

Porcine stress syndrome (PSS) is an acute, shock-like syndrome, quite common in the 1960s and 1970s. Stress was induced through routine management procedures such as handling or moving. Clinical signs often include open mouthed breathing, elevated body temperature, and (often) muscle and tail tremors.

The stress-induced condition occurs most often in heavily muscled pigs, especially the shorter legged, more compact pig. The condition is thought to be an inherited characteristic and is more common in the Poland China and some lines of the Landrace breed. Partial or total confinement of pigs appears to increase the incidence of PSS. Pigs suspected of having PSS should be provided immediate rest, and it may be advisable to tranquilize them if they are known to be PSS-susceptible.

Prevention of stress inducement through careful handling, guarding against overheating or overcrowding, and sudden changes of management are the general recommendations. Since PSS is a known inherited characteristic, proper breeding practices could, and in fact have at this writing, been most effective in reducing the incidence of PSS.

SPF (SPECIFIC PATHOGEN FREE) PIGS

Certain diseases are so costly to control and so widespread that a method was sought to raise swine that are free of the disease in an area also free of the disease. This was done successfully by strict control measures to clean up a farm—letting it lay out of swine production until the disease organisms died off or were killed. After eradication is complete, a quarantine is put into effect to keep people off unless special clothing and treated footwear are worn.

The next step is obtaining swine that are free of the diseases that are to be controlled. It is not practical to expect to produce swine that are free of all disease, but it is possible and practical to rid an area of specific pathogens, thus the name *specific pathogen free* (SPF) pigs. SPF pigs are produced by qualified veterinarians who perform Cesarean operations on thoroughly disinfected sows and take the pigs just before natural birth, rearing them on sterilized milk by artificial means in sterile confinement. The developing swine then serve as a foundation herd, being allowed to breed and farrow naturally on the clean farm.

The specific pathogens that are usually controlled by this method are atrophic rhinitis, virus pig pneumonia, and dysentery. However, the pigs are not free from the effects of all diseases.

TRANSMISSIBLE GASTROENTERITIS (TGE)

Transmissible gastroenteritis (TGE) is an infectious, transmittable disease causing a high death rate in pigs less than 10 to 14 days of age. It affects older swine also, but it is seldom fatal in adult animals. Surviving animals may shed TGE virus for up to 40 days in manure and up to 120 days from the respiratory tract following recovery. Signs of the disease are poor appetite, vomiting, scours, and weight loss. Lactating sows stop giving milk, and there are high death losses in pigs under 2 weeks of age.

In young pigs, vomiting and diarrhea are constant. Whitish, yellowish, or greenish fecal material is passed through the digestive system. Ingested milk often appears in the manure unchanged. There is rapid dehydration and weight loss, with a high mortality rate in young pigs. The disease can occur at any time of the year, and the major losses are due to poor performance and production of some runt pigs upon recovery from the initial outbreak.

The cause is a corona virus usually passed through manure and the respiratory tract of infected swine. There is no effective treatment. Avoid outside exposure during farrowing time. One form of prevention has been effective— exposing pregnant sows to the disease, creating antibodies in the milk and passing on the immunity to nursing pigs. This method of prevention is not without risk; caution and veterinary assistance are advised. A TGE vaccine is now available and is considered to be the most effective method of prevention. Prevention must rely on strict sanitation and disinfection, and possibly vaccination of pigs while they are suckling the sow. Vaccination offers the best future hope, but today, sanitation and building design are the keys to success.

VESICULAR EXANTHEMA

An incubation period of 1 to 3 days is normal before the signs of vesicular exanthema (VES) occur. These include an elevation of temperature (105 to 106 °F), followed by the development of small vesicles (blisters) on the snout, teats, udder, and feet (usually between the toes). These blisters (Figure 28-5) last 1 to 2 weeks, and recovery is usually uncomplicated; but secondary infections of the blister drains may cause swelling of the mouth and feet areas, creating temporary lameness (Figure 28-6) and reduced feed intake. There is low mortality, but occasionally sows may abort. Feeder swine may be severely stunted in growth. The cause is a virus.

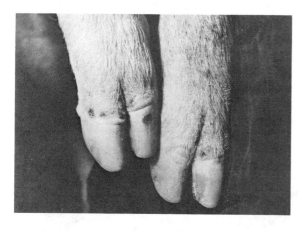

FIGURE 28-5. Feet blisters from vesicular exanthema. (Courtesy of the U.S. Department of Agriculture)

FIGURE 28-6. Hog affected with vesicular exanthema disease. Animals affected with VE may show blisters on the soft tissue just above the hoof and on the pads of the feet, causing lameness. (Courtesy of the U.S. Department of Agriculture)

The disease has been eradicated from the United States at present through quarantine and slaughter. Occasionally, outbreaks are reported in other countries. Because of its similarity to foot-and-mouth disease, the signs have been overlooked in some areas outside the United States. Continued effective control and eradication are accomplished by cooking of garbage fed (the principal source of the virus) and the use of a 2% sodium hydroxide disinfectant solution to clean premises before restocking the herd. These preventive measures are essential because there is no known treatment.

STUDY QUESTIONS

1. A swine health management program stresses _____ of diseases and parasites rather than treatment.

2. A _____ _____ _____ program involves isolation of the herd from disease pathogens, sanitation, and a planned immunization program.

3. The normal temperature range of swine is _____ to _____ °F.

Chap. 28 SWINE HEALTH MANAGEMENT

4. _____ , commonly called "thumps," results from a deficiency of iron.

5. Bacterial infection of the nasal cavity causing excessive sneezing, deformed, wrinkled snout, and painful irritation of the nose is _____ _____ .

6. Cholera in swine herds is most obviously noticed by _____ _____ .

7. Bloody diarrhea that is almost black in color in baby pigs is a sign of _____ .

8. Erysipelas resembles cholera with high temperature and _____ colored patches on the skin, but usually does not result in death.

9. Baby pig's disease, _____ , often affects entire litters; the pigs shake uncontrollably, do not nurse, and are so cold their hair stands on end.

10. Hypoglycemia is caused by low blood sugar levels and is often related to low _____ production of the sow.

11. _____ or _____ pigs at birth are symptoms of leptospirosis.

12. Transmissible gastroenteritis affects the _____ system of swine and is fatal to young pigs.

13. Vesicular exanthema in swine has the same symptoms as _____ _____ disease in cattle.

14. Mycoplasma pneumonia causes losses to commercial herds because it lowers _____ of swine.

15. SPF (specific pathogen free) pigs and herds are free from all diseases and pathogens. (True, false) _____ .

DISCUSSION QUESTIONS

1. What is atrophic rhinitis; how is it carried, treated, prevented?
2. Which two major diseases are transmitted through feeding of improperly cooked garbage?
3. Which diseases are characterized by diarrhea, and how do the signs differ?
4. What is the difference in diagnosis and treatment between hypoglycemia and anemia?
5. Name two diseases that affect the skin of hogs. What are diagnostic signs of each?
6. What is pseudorabies, and how fatal is it?
7. Name some diseases that are especially important to SPF control.
8. If SPF programs are so good, why doesn't everybody use them and eradicate all diseases?

Chapter Twenty-Nine

POULTRY—
A GENERAL VIEW

Keeping of poultry goes back in history as far as written evidence can be traced. The Bible indicates that fowls of the air were domesticated very early in human history. Ancient Egyptians kept poultry and even knew about incubation practices, using rather crude ovens to provide heat for hatching the eggs.

Chickens are probably descended from several wild species of jungle fowl originating in Southeast Asia. These birds were domesticated as early as 2000 B.C. and have been continuously subjected to extensive breeding practices to increase production of meat and eggs, two of the world's most efficient sources of human food. Estimates put the annual world production at 8 to 9 million head.

ZOOLOGICAL CLASSIFICATION

The domestic chicken, probably the world's most numerous fowl, belongs to the genus *Gallus*, species *Gallus domesticus*. The ancestors of the domestic chicken can be traced back to several species still in existence today. However, the Red Jungle Fowl, *Gallus gallus*, has the widest distribution of the wild species and is probably the chief ancestor of *Gallus domesticus*.

NUMBERS AND GEOGRAPHICAL DISTRIBUTION

For the last decade, the average number of laying hens on farms has been maintained at about 295 million, varying no more than 6.7% in any one year, and egg production has continuously increased. Broiler consumption has more than doubled in the last 20 years, mainly because chickens and other fowl are one of the few food items that are available today at prices no higher than 20 years ago.

The poultry industry has changed over the years from a few hens in the barnyard to huge, well-managed conglomerates. Approximately 82,000 commercial farms in the United States provide for 96% of the total value of poultry products sold, indicating that as farms become fewer they also become larger. It is estimated that there are almost 100 U.S. farms with sales of $1 million or more from poultry products.

One-fourth of the nation's poultry farms are located in the states of Georgia, Alabama, and North Carolina. Normally, Georgia, Arkansas, Alabama, and North Carolina account for more than one-half of the commercial broilers grown in this country, while California, Georgia, Arkansas, North Carolina, and Pennsylvania are leading egg-producing states. Georgia is the leading state in the United States in gross poultry income. However, Delaware outranks all other states when poultry population per size of state is considered. Commercial egg farms and broiler farms are highly competitive, economically managed, and scientifically supervised.

SCIENTIFIC IMPROVEMENTS

Continued selection and breeding of hens for a high rate of lay, together with scientific feeding, management, and improvements in housing, have made it possible to expect annual yields of 240 to 250 eggs per hen per year. The per capita egg consumption in the United States is approximately 285 per year. For this reason, the market will continue to be favorable for a moderate growth in both egg production and consumption.

Prior to 1920, competitive egg-laying tests were in style, and records were being made in Europe with the English Wyandotte and large-type Leghorns. These records normally reached the 200-egg per year maximum level. Then the genetics of egg production began to be better understood by American breeders. Strains of medium-sized Leghorns were developed that outperformed the larger English variety while producing more eggs at a lower level of feed. This development led to still further breeding techniques, such as crossing of inbred lines, producing strains of chickens that lay uniformly high and are efficient feed converters.

More recently, the trend has been toward reducing the size of the hen through use of sex-linked genes for body size. This may result in a smaller bird without excessive reduction in egg size. Smaller birds eat less feed and can be caged in smaller areas, thus consuming substantially less feed and lowering the cost of chicks produced for the industry.

Progress has been made in the broiler industry by application of many principles discovered in layer research. Production of commercial broilers has grown from 34 million in 1934 to eight times that figure today. This rapid expansion has been possible largely because of the previously discussed nationwide increase in egg production, allowing for selection from a larger poultry population.

In addition, the hatchery business, a year-round operation in the United States, has evolved to keep pace with the layer and broiler chick demand. The change has been spectacular, literally taking the hatching and incubating processes off the farm.

Because of the advancements in hatcheries and layer and broiler production, poultry exports from the United States have been very gratifying, now amounting to 84 million pounds of broilers, 13 million dozen hatching eggs, and 25 million baby chicks annually.

BREEDS OF CHICKENS

In the United States, it is evident that there is less concern over maintaining purebred strains in chickens than there is in large animals. The American Standard of Perfection lists nearly 200 varieties of chickens, but only six appear to be of much commercial importance. From a purebred standpoint, these are the White Leghorn (Figure 29-1), the White Plymouth Rock (Figure 29-2), the Barred Plymouth Rock (Figure 29-3), the Rhode Island Red (Figure 29-4), the New Hampshire (Figure 29-5), and the Dark Cornish (Figure 29-6).

The National Poultry Improvement Plan (NPIP) has been responsible for supervision of breeding practices and techniques in the United States to develop the high-quality, high-producing form of poultry in existence today. The NPIP was started in 1935, administered jointly by the U.S. Department of Agriculture and an official of each state agency. Coordination of the plan is the responsibility of the Department of Agriculture. The objectives are to improve production and market qualities of chickens and to reduce losses from diseases. Divisions of the plan are set forth in the Department of Agriculture publication, "The National Poultry Turkey Improvement Plans," available from the Animal Husbandry Research Division, ARS, Beltsville, Maryland 20705.

The vast majority of chickens in the United States are hybrids of one kind or another. Strain crosses, breed crosses, or crosses between inbred lines make up the bulk of modern poultry. An interesting fact is that Single-Comb White Leghorns are considered to be as efficient layers as any of the inbred lines. This is one example of a purebred strain maintaining its superiority over a crossbred bird in at least one trait.

FIGURE 29-1. White Leghorns.

FIGURE 29-2. White Plymouth Rocks.

FIGURE 29-3. Barred Plymouth Rocks.

FIGURE 29-4. Rhode Island Reds.

FIGURE 29-5.　New Hampshires.

FIGURE 29-6.　Dark Cornishs.

ANATOMY OF THE CHICKEN

Skeleton

The skeleton of the chicken is shown in Figure 29-7. The bones of almost all fowl are pneumatic (hollow). The hollow spaces connect with the respiratory system, making it possible for a bird with a broken wing to actually breathe through the wing, a fact long noted in wounded birds by hunters. Twelve percent of the bone structure in chickens is of the unique medullary type of bone. This is a network of tiny spines that lace the hollow structure, along with marrow, which serves the wild bird as a substance for egg formation when dietary calcium levels are low.

Muscles

Although there are smooth muscles of the intestine and cardiac muscles of the heart, skeletal muscle is the most important for poultry producers. The breast is the largest skeletal muscle because of the need for flight in the wild bird. This muscle has been genetically improved upon by breeders of domestic species. Birds have red muscle and white muscle, corresponding to dark meat and light meat. This differentiation in color is due to the myoglobin contained in red muscle. Myoglobin is the oxygen-carrying red pigment in the muscle of chicken.

Lungs and Air Sacs

The respiratory system consists of the lungs, which are relatively immobile, and four pairs of air sacs and one median air sac ranging in position from the neck to the abdomen. Air movement into and out of the lungs occurs primarily due to pressure changes in the air sacs rather than action of the lungs themselves. These four pairs of air sacs connect to a single median sac in the thorax. The entire system connects the lungs, passages, sacs, and hollows of the bones to create an interrelated respiratory system. The respiratory apparatus is important not only for breathing but also for losing heat through evaporation. Because of the lack of sweat glands, when the temperature exceeds 80 °F, chickens have to rely on evaporative cooling, which takes place through the air sacs and lungs.

The Skin and Feathers

The skin of poultry is without glands except for the uropygial (oil gland, preen gland), which secretes oil used by the chicken to "dress" the feathers with a

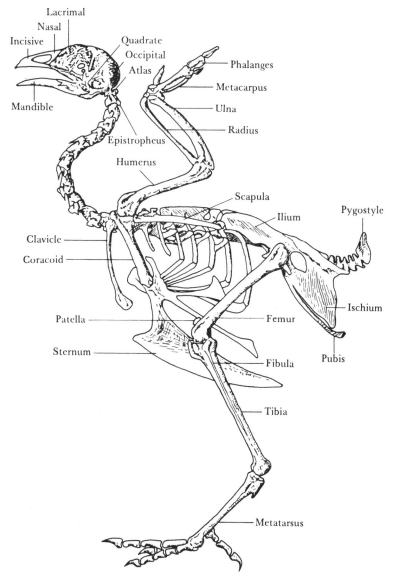

FIGURE 29-7. The skeleton of a fowl. [Reprinted by permission from L. E. Card and M. C. Nesheim, *Poultry Production*, 11th edition (Philadelphia: Lea & Febiger, 1972)]

protective coating through the act of preening (combing the feathers with the beak). Poultry have many types of feathers, all originally used to aid in flight, to protect, and to provide warmth. Noting the types of feathers is not relevant to our discussion except in the case of the annual renewal (molting), which is an expense to the bird's net physical production. Molting is discussed in Chapter 30.

Endocrine Glands

Endocrine glands act through chemical stimulation in the bloodstream to regulate every phase of poultry body processes. The product of these glands—hormones—are produced by such organs as the testes, ovaries, thyroid, pituitary, adrenal, pancreas, gastrointestinal tract, and brain. Some of these hormones can also be released by nervous reactions to stress, noise, or light, providing proof of an endocrine–nervous system interaction. One such interaction with light is covered in Chapter 30 as a positive form of manipulation by man.

STUDY QUESTIONS

1. Chickens probably originated in Southeast _____.
2. Ancestors of the chickens were probably several wild species of _____ or _____.
3. The species name for the most common domestic chicken is _____ _____.
4. One-fourth of the nation's poultry farms are found on the east coast of the United States. Of the leading poultry-producing states, _____ ranks tops in gross poultry income.
5. The most notable exception to geographical concentration of layers is the state of _____.
6. High-producing layers should lay an average of _____ eggs in a year.
7. Per capita consumption of eggs is about _____.
8. The trend today is toward a _____ (smaller, larger, giant) laying hen.
9. Improvements made in layer research led to the rapid expansion of the _____ industry.
10. The most common type of laying hen today is the _____.
11. Of the six breeds most used in commercial production in the United States, the _____ and _____ breeds are red in color.
12. The hollow bones of chickens connect with the _____ system.
13. The respiratory apparatus serves a secondary function as a method of losing _____.

DISCUSSION QUESTIONS

1. Where are most commercial laying hens concentrated? Why?
2. Why are almost all commercial laying hens crossbreeds? Are there exceptions? Give examples.

3. How do chickens "sweat"?

4. How do bones of chickens differ from those of domestic animals previously covered in this book?

5. Of the 200 varieties of chickens available to poultrymen in the United States, how many are of any real commercial value? Name them.

6. Assuming an average population of 5 million people per state in the United States, how many dozens of eggs would be consumed annually by the average state?

Chapter Thirty

REPRODUCTION IN THE FOWL

Breeding practices have radically changed over the last few decades from a few hens and an occasional incubator on the farm to a giant cooperative effort between the foundation breeder and the hatcheryman. The breeder develops the right kind of bird and supplies the foundation stock to the hatchery operation for the production of commercial chicks. Hatcheries handle marketing and distribution. This chapter will briefly cover the principles of physiology of reproduction that the breeder must know.

THE MALE REPRODUCTIVE SYSTEM

Figure 30-1 illustrates the male reproductive organs, which are greatly simplified in comparison to the larger domestic species such as cattle, swine, sheep, and horses. The male reproductive organs are the *testes, ductus deferens,* and a rudimentary copulatory organ located in the *cloaca.* The male fowl differs from other domestic animals in that the testes do not descend into a scrotum but remain in the abdominal cavity along the backbone close to the anterior portions of the kidneys. Testes produce live sperm, to fertilize the egg of the

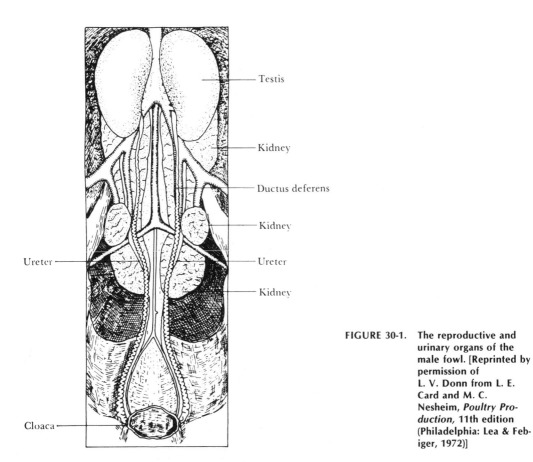

Ureter

Cloaca

Testis

Kidney

Ductus deferens

Kidney

Ureter

Kidney

FIGURE 30-1. The reproductive and urinary organs of the male fowl. [Reprinted by permission of L. V. Donn from L. E. Card and M. C. Nesheim, *Poultry Production*, 11th edition (Philadelphia: Lea & Febiger, 1972)]

female, and the male hormone, androgen, which is responsible for secondary sex characteristics in males, such as bright red combs, plumage, and the crowing response.

As in other species, sperm is produced in the *seminiferous tubules* deep within the testes. Sperm released from the tubules enter the *ductus deferens,* small tubes that conduct the sperm along a pathway to the *cloaca.* The ductus deferens does not open into a copulatory organ, as in other species, but rather into small *papilla* (finger-like projections). These projections are located on the dorsal wall of the cloaca and serve as semen transporting organs. The male fowl also has a rudimentary copulatory organ that has no connection to the ductus deferens and is located on the ventral part of the cloaca. Mating with the female is mostly a matter of joining cloacas long enough for semen injection.

The male responds to light just as the female does (see light response, discussed later in this chapter). Males that are to be used for natural matings should receive the same form of light stimulation as females in order to produce the most viable semen in the largest quantities.

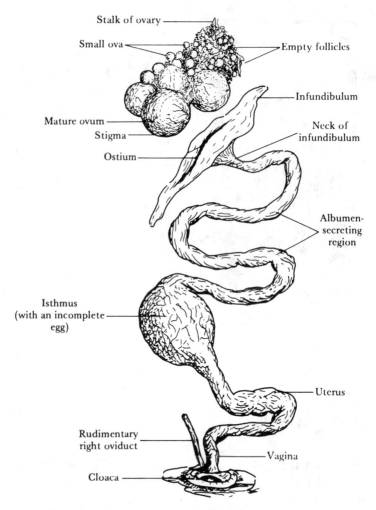

Stalk of ovary

Small ova

Empty follicles

Infundibulum

Mature ovum

Stigma

Ostium

Neck of infundibulum

Albumen-secreting region

Isthmus (with an incomplete egg)

Uterus

Rudimentary right oviduct

Vagina

Cloaca

FIGURE 30-2. The ovary and oviduct of a chicken. [Reprinted by permission of A. L. Romanoff and J. Wiley & Sons, Inc. from L. E. Card and M. C. Nesheim, *Poultry Production*, 11th edition (Philadelphia: Lea & Febiger, 1972)]

THE FEMALE REPRODUCTIVE SYSTEM

Figure 30-2 illustrates the female reproductive system. It should be noted that female fowl normally have only the left *ovary* and *oviduct* fully developed. During incubation, the right side fails to develop and by hatching time has degenerated to only a rudiment. There are five clearly defined regions of the oviduct: (1) the funnel or *infundibulum* that is responsible for picking up the yolk after it has been ovulated; (2) the *magnum* that secretes albumen or the white of the egg; (3) the *isthmus* that secretes the shell membranes; (4) the *uterus* (shell gland) that secretes the shell; and (5) the *vagina* where the egg is temporarily stored and expelled when fully formed.

The female reproductive system functions initially through stimulation by the hormone FSH (follicle stimulating hormone) from the anterior pituitary, which causes development of the mature follicles (yolks). This FSH production is normally stimulated by increasing periods of light brought about in wild birds by the lengthening days of spring. FSH production may be artificially induced by the lighting systems of man. The *ovary* is influenced by this stimulation to begin production of its own hormones, *estrogen* and *progesterone*. Estrogen causes an increase in blood calcium, protein, fats, vitamins, and other substances necessary for egg formation. Estrogen also stimulates separation of the pubic bones and an enlargement of the vent in preparing the hen for laying. Progesterone acts on the hypothalamus gland to produce LH *(luteinizing hormone)* from the anterior pituitary, which causes release of the mature yolk from the ovary to the funnel or infundibulum. If sperm are present at this point, a fertile egg will be laid. Otherwise, production will continue, but infertile eggs are produced.

BREEDING PHYSIOLOGY

Puberty

Sexual maturity occurs in chickens about 22 to 26 weeks of age. The stocking rate for male to females, under natural mating systems, is 1 to 12 for chickens. Ten to fourteen days should be allowed to be relatively certain of maximum fertilization in mass-mated flocks. This permits males to mate with all females at least two to three times.

Egg Formation

The ovary of the laying hen contains from 1000 to 3000 follicles, varying in size from microscopic to the size of a yolk. The smaller yolks begin to grow rapidly approximately 10 days before they are released into the infundibulum. The yolk is enclosed by a *follicular membrane,* which attaches it to the ovary. This membrane has one section that appears as a broad line, a readily visible streak lacking in blood vessels, called the *stigma.* This is the point at which the yolk ruptures and releases the ovum at the time of ovulation. Because nutrients are transferred across the follicular membrane from the bloodstream to the *ovum,* some blood is occasionally released along with the yolk because it doesn't break right at the stigma, creating a blood spot.

The yolk is then received by the infundibulum. Occasionally, a yolk will drop into the body cavity, where it is reabsorbed. The yolk that enters the infundibulum goes directly to the magnum, the longest portion of the oviduct.

This is where the albumen (egg white) is secreted to surround the yolk. The yolk continues in a rotating motion downward to the lower portions of the oviduct. The shell membranes are added in the isthmus. Two membranes, an inner and outer shell membrane, are formed at this point. Normally adhering to each other except at one point, these membranes separate at the large end of the egg to form the air cell.

The egg remains in the uterus (shell gland) for the longest period of time. The eggshell is formed here, a process requiring approximately 20 hours. The shell is composed almost entirely of calcium carbonate deposited on a matrix of protein and mucopolysaccharide. The final layer or covering on the shell is known as the *cuticle*, an organic material protecting the egg from invasion by harmful bacteria and serving as a seal to reduce moisture loss. It is interesting to note that the major source of the calcium carbonate that goes into shell formation is the bicarbonate ions of the blood. Bicarbonate is created through a mixture of carbon dioxide and water in the presence of the enzyme carbonic anhydrase. When a hen pants in hot weather, she increases water loss by evaporation from the respiratory tract, causing a reduction of carbon dioxide and bicarbonate ions in the blood. This is thought to be the reason for thin-shelled eggs being laid in very hot weather.

Fertilization and Development

Sperm from fowl retain their fertilizing capacity for a considerably longer time than mammals. Chicken sperm have been known to be viable for as long as 32 days after insemination, but weekly insemination *is* needed to assure highest fertility. The sperm, after mating, are stored in natural folds in the oviduct of the female. These folds are sometimes referred to as sperm nests. As the yolk enters the infundibulum, the walls of the oviduct are stretched, liberating the sperm to fertilize the egg. This fertilization takes place at the *germinal disc* on the yolk.

In the event of fertilization, the embryo begins development around this well-defined germinal disc. The area is clearly visible to the naked eye when broken out of the egg. Within 48 hours, a chick embryo has established an intricate type of blood circulation between itself and the life-sustaining yolk. Since there is no placenta as in the other species discussed in this book, the fowl embryo has to depend on this intricate blood vascular network to carry out the necessary functions of bringing nutrients in and ushering out waste products.

By the end of the third day, the embryo has a full complement of membranes, known as the *allantois, chorion,* and the *amnion* (see Figure 30-3). The allantois, which at first serves to store excretory wastes, later merges with the chorion to form the *chorio-allantois.* The major part of this combined membrane is closely associated with the shell. This membrane serves as the respiratory organ for the developing embryo until the pulmonary type takes over about 24 hours after hatching. By the end of the first one-third of the incubation period, the general outline of the embryo is fully recognizable. Also by

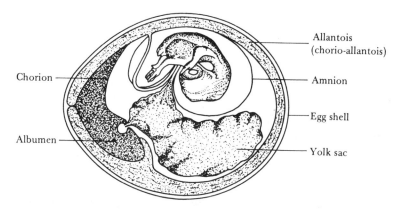

FIGURE 30-3. Diagrammatic illustration of a 10-day-old chicken embryo, showing some important membranes.

that time, most of the essential internal systems, such as the lungs, nervous, muscular, and sensory systems, are developed. The sex of the chick embryo can be determined as early as the fifth day of incubation. By the middle of the incubation period, embryos of most domestic poultry species are fully covered with down (the first feathers).

As with other species covered in this text, the embryo partly floats within fluid in the amnionic cavity. It is important for this to happen in order to protect the developing embryo and allow it free movement. This freedom of movement is especially important in the chick embryo and must prevail to within the last 3 or 4 days of actual hatching or malformations can occur that endanger the life of the newly hatched chick. The egg must be turned several times a day in the incubator to prevent the embryo from adhering to the chorio-allantois membrane. Under natural conditions, the hen shifts the egg several times a day out of instinct. The shell and the membranes also serve to further protect the developing embryo from harmful microorganisms or molds. Additional protection is further added by the mildly bacteriostatic action of the albumen.

Malpositions in fowl embryos may not be given much thought, but there is a natural position for poultry. About midway through the incubation period, the embryo assumes the normal position, which is to lie on its left side along the longest axis of the egg. The head should be tucked under the right wing facing the large end of the egg. Any other position is considered a malpresentation.

Oviposition (Laying of the Egg)

Eggs are normally formed with the small end first as they move through the oviduct. Curiously enough, if the hen is not disturbed while laying, most eggs are laid large end first. It is not known why this is so, but it is known that the egg just prior to being laid is turned horizontally (not end over end) 180 degrees just before the hen lays. Ovulation in chickens normally occurs about 30 minutes after a previous egg is laid. A time interval for this ovulation may vary from 7 to 74 minutes. An overall average interval between successive eggs laid in the same clutch is 27 hours.

The Clutch

The *clutch* is the number of eggs laid by a hen on consecutive days. The *clutch cycle* is terminated on the day an egg is not laid. Poor layers may have cycles of one or two eggs, whereas good layers may range up to 200 or more. It is likely that clutches are under the influence of hormone secretions, which accounts for this variation. The position of an egg in the clutch affects its weight. Normally, the first egg in a clutch will be the heaviest. The difference may be only very slight in the case of hens with very long clutches, or there may be a wide range of egg weights in hens with a shorter cycle.

As mentioned before, ovulation normally occurs 30 minutes after the laying of the previous egg. However, if an egg is laid after 2 P.M., the next ovulation usually will not occur for 16 to 18 hours, resulting in no egg being laid. This produces a termination of the cycle. The exact manner by which this occurs is not clear. It is thought to be related in some way to approaching darkness, which delays ovulation, bringing about termination of a clutch. Darkness may have some effect on the release of the luteinizing hormone, which induces ovulation.

The Completed Egg

The parts of an egg are shown diagrammatically in Figure 30-4. The chief components of an egg are the germ spot *(blastoderm)*, the yolk, the white, the shell membranes, and the shell. The germ spot is associated with the yolk and represents the nonfertile living part of the egg. Note the light and dark layers

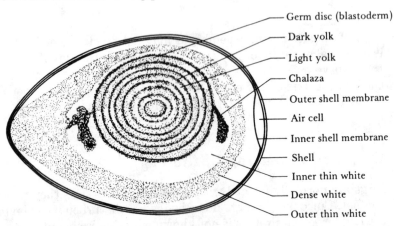

Germ disc (blastoderm)
Dark yolk
Light yolk
Chalaza
Outer shell membrane
Air cell
Inner shell membrane
Shell
Inner thin white
Dense white
Outer thin white

FIGURE 30-4. Vertical section showing the parts of a fresh egg.

Chap. 30 REPRODUCTION IN THE FOWL

of yolk rings and the germinal disc that floats to the top. If the egg is fertilized, this area is referred to as the blastoderm (which becomes somewhat larger than the germinal disc) and represents embryonic development. The white of the egg has two twisted cords composed of albumen extending from each end of the yolk. These cords are the *chalazae*, formed as a result of rotation of albumen around the yolk during the laying process. The inner and outer shell membranes protect the egg mechanically by providing a durable, strong container and join together, except at the large end where they separate to form the air cell. The shell encapsulates the contents to provide a thin but surprisingly strong outer surface.

The completed egg is also very stable, being resistant to spoilage because of the shell's resistance to penetration by microorganisms. The cuticle surrounding the eggshell adds another dimension to protection by filling the pores of the shell and retarding moisture loss and bacterial penetration. Internally, the egg is also protected, particularly with an antibacterial protein, *lysozyme*, which is contained in the egg white and helps retard spoilage. The egg is one of nature's great miracles. It is one of the most nearly perfect foods because it is a complete self-contained survival kit for the avian embryo.

Broodiness

Under the influence of the hormone *prolactin* from the anterior pituitary, some chickens spend unnecessary amounts of time sitting on a nest and hatching and rearing their young. When this maternal instinct is so strong that the hens are continuous nest sitters, it is considered an undesirable characteristic. A broody hen is one that ruffles her feathers, clucks continuously, sits on the nest continuously, and has a very mean disposition. Broodiness is considered a disadvantage in poultry production because while the female is in the broody disposition eggs are not normally laid. It is an inherited characteristic. A practical way of interrupting broody hens, on a small-scale basis, is to enclose the hen in a well-ventilated, wire-bottom pen to discourage nesting.

Molting

The end of the egg-laying season is usually marked by *molting*, the annual loss of feathers and growing of new ones. A general molt in hens will involve tail and flight feathers. Molting is related to reproduction in that the ovary remains active until the time of molting. In the female, this feather replacement signals the end of egg production for the season. Molting in the male does not take place all at one time, as it does in the female. The male continuously molts a few feathers. His sexual activity is not diminished as it is in the female.

Light Response

All species of poultry respond to light-dark reactions. Both natural and artificial light stimulate the fowl's reproductive processes. The ratio of light-dark periods and the rate of change of light influences the time of oviposition. Wild birds normally begin nest building, mating, and oviposition during the early spring because of lengthening light periods. Egg laying normally ceases during the fall season with decreasing light. This is nature's signal to the mating system. Domestic birds respond to light in the same fashion. Through human control of the environment, the bird's reproductive system can be manipulated to cause egg laying at any season of the year. Light stimulates the optic nerve to act on the hypothalmic–pituitary complex to release and increase supplies of FSH. This hormone in turn initiates ovarian activity, resulting in ovulation and oviposition.

If egg laying begins at too early an age, through increasing the light to very high intensity levels or too long a day length, the eggs may be small. If egg laying is delayed until a later age, the eggs are generally much larger. Since pullets are reared throughout the year, it is important to manage light so that they start egg laying at the proper period of their development and with a size of egg that is compatible with market requirements. The role of light in poultry production has lead to management schemes that completely house poultry so that light, temperature, humidity, and other factors can be controlled. Complete confinement of laying chickens is the rule, rather than the exception, in today's modern poultry science. This gives the operator full control over light exposure, which is extremely important.

The following examples will illustrate some of the systems now in use:

The Step-Down, Step-Up System. A light-tight housing system is not necessary in this case. Pullets maturing in days that are increasing in length are provided with just enough artificial light so that the light can be reduced by 1 hour or slightly less every 2 weeks until pullets reach 22 weeks of age, at which time the light is increased in a similar fashion until a total of 14 to 16 hours is attained.

A Modified Step-Down, Step-Up System. A light-tight house is required in this case. Twenty-one hours of light is the starting point for chicks. The light is decreased 45 minutes each week until the chicks reach 12 weeks of age, then the light is reduced to 7 hours per day until 21 to 22 weeks of age. At that time, the light is maintained 12 to 13 hours per day and increased by 1 hour each month until 16 hours of light are reached.

A Constant Light System. A light-tight house is required in this case, starting with 14 hours of light and kept at this level until the chicks are 12 to 14 weeks of age. The light is lowered to 9 hours per day and maintained until 20 weeks of age. At 20 weeks, the light is increased to 14 to 16 hours and held constant.

The Auburn System. This is also a constant light system that requires a light-tight house. This system was developed by Auburn University. The chicks are kept at 6 to 8 hours of light from hatching until 20 weeks of age.

Then the light is gradually increased by 18 minutes per week until 16 hours per day are reached. This level is maintained.

Although these are not all of the light management systems in use today, these four examples will give the reader an idea of man's manipulation of light to force production at the time when eggs are required for the year-round market.

Hens that are laying should never be subjected to decreasing amounts of light. This is simple to control in a light-tight house. If hens are laying under a natural lighting system in a non-light-tight house, lights can be increased to maintain a steady period of 14 to 16 hours per day, even though the outside light is decreasing far below that. The reason for maintaining this light is that decreasing light levels will lower egg production.

Chickens do not respond to all wavelengths of light. Normal incandescent bulbs are quite commonly used and work sufficiently well. However, orange and red lights are generally considered most effective. The level of light measured to yield 0.5 to 0.9 footcandle of light power is considered to be a most efficient effective level. This level should be maintained at the darkest points of exposure for the hen, usually the corners of the house. Excessive light or very bright light is unnecessary and economically unsound.

Flashes of intense light every hour, rather than a continuous regimen of 13 to 14 hours of continuous light, have also proven to be beneficial. Flashes of intense light every hour throughout a 24-hour period may approach the routine system in effectiveness but they have not proven to be more beneficial than the normal 14 to 16 hours of light, followed by a dark rest period.

As an example of the physical arrangements to produce the actual lighting, for hens on clean litter, use one clean 60-watt bulb with a shallow dome reflector 7 ft above the floor for each 200 square feet of floor area. This will place the light bulbs 14 ft apart over the entire width and length of the house. In cages, 40-watt bulbs, 8 ft apart, 4 ft above the birds will provide sufficient light.

HATCHABILITY

Incubation in birds corresponds to gestation in domestic farm animals. Several factors are necessary for adequate levels of hatchability in eggs.

Temperature

The shell of the egg serves not only as a mechanical protection to the developing embryo but also as a main source of calcium. Exposure of eggs to low temperatures, such as freezing, before the eggs are incubated will destroy the viability for hatching purposes. However, eggs kept at a temperature of 20 to

35 °C (68 to 95 °F) may allow for limited development, but the future ability to survive will be greatly lowered. Even under optimum conditions, the egg rapidly loses its ability to have a high hatchability when the preincubation period is extended beyond 7 days. In order to have the highest percent of survival in chicken embryos, it is normally necessary to store the eggs for no longer than 7 days and to store hatching eggs at 12.5 °C (54.5 °F). This allows for developmental processes to continue without adverse effect once they are placed in the incubator. After the egg is placed in the incubator, the machine is the controlling factor in hatchability. Disregarding some minor differences in heat requirements for incubation between species, a range of 37.5 to 40 °C (98 to 104 °F) is considered optimum within the incubator, depending on the type of incubator.

Humidity

The egg is approximately 70% water. For that reason, it is important to maintain a humidity level that would prevent water loss from inside the egg. Preincubation storage of hatching eggs should be at 85% relative humidity, and 60 to 65% during incubation. This water is necessary inside the egg environment to allow for excretion of the embryo's metabolic wastes and to serve as a heat regulator, much like a car radiator transfers heat through water.

Velocity of Air

Although rate of air moving through an incubator does not appear to have much influence upon hatchability, it is necessary for the developing embryo to receive a constant supply of oxygen through fresh air. This movement should not be extreme because carbon dioxide (CO_2) is also needed for the movement of calcium from the shell into the developing embryo. If the air velocity is too high, no carbon dioxide is allowed to develop, negating this effect.

Energy for Survival

The total energy stored in an egg is extremely important for the developing embryo, not only during incubation but also for short periods thereafter. Twenty-five percent of the energy in the egg is used for development of the embryo. The rest becomes incorporated into the developing chick. All but a minute part of the fertilized avian egg is utilized by the chick. This capsule storage represents a highly concentrated form of energy provided by the hen. The avian embryo has to depend upon this maternal organism during both incubation and postincubation periods.

BROODING

The handling, rearing, and growth of the chick after hatching is referred to as brooding. With the exception of nutritional needs, there are few undisputed recommendations for the brooding period. Yet in many ways this is the most critical period in the life of the young bird. Successful brooding is best described as an art rather than a science.

Housing

Brooding requires adequate housing, to protect chicks from extreme changes in the environment; feed and water; and control of parasites and disease. Manure removal or absorption by the litter is important. The house may have slotted floors, solid floors, or some other combination, but its basic purpose is to provide comfortable quarters for the chicks.

Temperature

The brooders that provide heat for chicks in the house are extremely important for overall ability to survive and for growth. The key to temperature is not providing a constant level throughout the house but to provide for a range of temperature from 27 to 31 °C (80 to 95 °F) during the first 3 weeks of the chicks' life at the level of brooder. This gives chicks the opportunity to find a temperature that is most comfortable to them. They quickly learn to seek their comfort level. Most trouble in brooding comes not so much from temporary changes in variable temperature but from continuous exposure to temperatures that are too high or too low, with no opportunity for the birds to move to a more comfortable zone.

Brooding Equipment

Portable brooders are made in many styles and sizes. The gas-heated types are most common in commercial operations. Oil-heated brooders, as well as hot water or hot air types, are used in some areas. On a small scale involving just a few baby chicks in a "backyard operation," the electric brooder is common.

STUDY QUESTIONS

1. The rudimentary copulatory organ of the male chicken is found in the _____.

2. The testes of the male bird do not _____ as in domestic mammals.

3. The right ovary and oviduct of the female normally do not _____.

4. A follicle, or _____, is developed under the influence of the hormone _____.

5. The shell portion of the developing egg is secreted in the _____.

6. The earliest a chicken should be expected to be reproductively active is _____ weeks.

7. Under natural mating systems one rooster is mated to about _____ hens for _____ weeks.

8. When a yolk doesn't break right at the stigma, a _____ _____ is often the result.

9. The most important function of the cuticle is to prevent contamination of the egg by _____.

10. Panting by hens in hot weather could conceivably cause _____ eggshells.

11. Chicken sperm inseminated in the oviduct may remain viable for as long as _____ days.

12. The embryo of a chick compared to most mammals has no _____.

13. Malformations can occur in chicks if the eggs are not _____ regularly.

14. The number of eggs laid by a hen on consecutive days is known as a _____.

15. It is thought that luteinizing hormone, which induces ovulation, may be influenced to be released by _____.

16. Broodiness causes a hen to _____ (start, stop) egg laying.

17. The annual loss of feathers is called _____.

18. Egg laying can be delayed by a reduction of available _____.

19. Hens that are laying should never be subjected to _____ levels of light.

20. For best results in hatching, eggs should be stored no longer than _____ days at _____ °F.

21. The optimum temperature for hatching eggs is a range from _____ to _____ °F.

22. Humidity during incubation should be _____ to _____%.

23. Brooders should provide chicks with a wide _____ of temperatures.

DISCUSSION QUESTIONS

1. Discuss the frequency cycle for layers and potential annual production.
2. Give a detailed explanation of how, when, and why the various fractions of an egg are formed.
3. Discuss the unique features of both male and female avian reproductive tracts, compared to those of most domestic mammals.
4. Discuss the key features necessary for successful incubation.
5. What effect does the hormone prolactin have on hens?
6. What effect does the annual growing of new feathers have on female and male reproduction?
7. Give a detailed explanation of how an egg is fertilized, how it must be handled to hatch, and what is the normal position of the embyro.

Chapter Thirty-One

FEEDING POULTRY

As a result of the tremendous increase in agricultural production throughout the world, new varieties of grain and protein-type crops have been made available in abundance. Because of a reoccurring grain surplus, man in many parts of the world has seen fit to transform these crops into more desirable and edible forms of food for human consumption. Two of the most prized forms are eggs and poultry meat.

FEED INGREDIENTS

Corn and soybean meal are generally produced in excess throughout the world and form the backbone of modern poultry rations. In general, the major sources of other feeds are meat packing by-products, fish meals, wheat, rice, and corn milling by-products. Others that are often used include yeast, distillery by-products, milk by-products, alfalfa leaf meals, and synthetic ingredients.

Corn is the most important grain used in poultry rations, sorghum grains (milo) rank second, and wheat ranks third. Both animal and vegetable protein supplements are used for protein in poultry rations. Common calcium supplements include ground limestone, crushed oyster shells, and bonemeal. Phosphorus is supplied in the form of bonemeal and bicalcium phosphate.

Salt (sodium chloride) is a mineral supplement that should not exceed 0.5% of a ration. Too much salt causes an increase in water consumption in poultry and creates the undesirable management problem of very wet feces.

Some nonnutritive additives in poultry rations include antibiotics, arsenicals, nitrofurans, drugs, and antioxidants. Many of these, however, are under surveillance and threat of removal from the market by the Federal Drug Administration. Grit is an unusual nonnutritive additive for poultry, but it has a very definite use and purpose, especially if birds are consuming coarse, fibrous feeds. The purpose of grit is to help the gizzard grind food materials since the chicken has no teeth and few other facilities to break down feedstuffs mechanically.

ESSENTIAL NUTRIENTS

Chickens are extremely susceptible to deficiencies of every sort under the modern system of keeping them caged or housed away from natural pasture or barnyard where many essential elements are consumed by accident. In general, the essential nutrients for poultry are the same as for other classes of livestock previously discussed (water, carbohydrates, fats, proteins, minerals, and vitamins). Poultry require fresh, clean water at all times because of the high water content of eggs and meat. Carbohydrates are needed in poultry rations and are generally present in a highly concentrated form, such as grain. Fats are used more readily for poultry rations than for other domestic animals to increase the energy content of the feed to promote faster growth or higher energy levels for broiler production. Proteins are used in both the animal and vegetable form in order to maintain a balanced diet of essential amino acids. Minerals are required at adequate levels. Vitamins are needed for adequate health, growth, and reproduction. Table 31-1 gives the basic feed and protein requirements for laying hens.

Recommended vitamin and mineral levels for chickens are given in Table 31-2. Since poultry rations are highly concentrated, the carbohydrate content is normally sufficient for adequate growth and, if not, it can be easily modified by adding fat to increase the energy level. However, feeding requirements other than those given in Tables 31-1 and 31-2 are generally beyond the scope or needs of an introductory text. Perhaps more meaningful aids to feeding can be found in Tables 31-3, 31-4, and 31-5 later in this chapter.

DIGESTIVE SYSTEM OF THE FOWL

For illustrative purposes, the monogastric (simple stomach) digestive tract of the fowl is divided into the various parts shown in Figure 31-1. The brief dis-

TABLE 31-1. NUTRIENT REQUIREMENTS FOR LAYERS

	Feed[a] consumed (lb) by 100:					
	4-lb Hens		5-lb Hens		6-lb Hens	
Eggs/100 hens/day	Per Day	Per Dozen Eggs	Per Day	Per Dozen Eggs	Per Day	Per Dozen Eggs
0	15.8	—	18.6	—	21.2	—
10	16.7	20.1	19.6	23.6	22.1	26.6
20	17.6	10.5	20.4	12.2	22.9	13.7
30	18.4	7.4	21.3	8.5	23.9	9.6
40	19.3	5.8	22.2	6.7	24.7	7.4
50	20.2	4.8	23.1	5.5	25.6	6.1
60	21.1	4.2	24.0	4.8	26.5	5.3
70	22.0	3.8	24.9	4.3	27.5	4.7
80	22.9	3.4	25.8	3.9	28.1	4.2
90	23.8	3.2	26.7	3.6	29.2	3.9
100	24.7	3.0	27.6	3.3	30.1	3.6

		Protein (Percent of Diet)		
		Stage of Egg Production		
Feed per Hen per Day (g)	Feed per 100 Hens per Day (lb)	1	2	3
85	18.7	21.0	18.8	17.7
90	19.8	20.0	17.8	16.7
95	20.9	19.0	16.9	15.8
100	22.0	18.0	16.0	15.0
105	23.1	17.1	15.3	14.5
110	24.2	16.3	14.5	
115	25.3	15.7		
120	26.4	15.0		
Protein intake required (g/hen/day)		18	16	15

Source: L. E. Card and M. C. Nesheim, *Poultry Production,* 11th edition (Philadelphia: Lea & Febiger, 1972). Reprinted by permission.

[a]Feed assumed to contain 1350 kilocalories of metabolizable energy per pound.

cussion of the function of each part will provide the basic knowledge necessary to understand introductory poultry nutrition.

Mouth

Chickens do not have teeth or serrated edges on their beaks, so chewing cannot take place. The tongue provides some assistance in eating because of the forked section at the back which forces food down into the gullet. Little saliva is secreted in the mouth to aid in swallowing.

TABLE 31-2. RECOMMENDED PRACTICAL LEVELS OF NUTRIENTS IN FEEDS FOR CHICKENS (IN PERCENT OR UNITS PER POUND)[a]

	Starting Chicks and Broilers	Growing Chicks and Broilers	Laying Hens	Breeding Hens
Vitamins				
Vitamin A (IU)	5000	3000	4000	5000
Vitamin D$_3$ (IU)	500	300	500	500
Vitamin E (IU)	5	4	—	7.5
Vitamin K$_1$ (mg)	1	1	1	1
Thiamin (mg)	1	1	1	1
Riboflavin (mg)	2	2	2	2.5
Pantothenic acid (mg)	6.5	6	2.5	7.5
Nicotinic acid (mg)	15	15	12	15
Pyridoxine (mg)	2	1.5	1.5	2
Biotin (mg)	0.06	0.05	0.05	0.08
Folic acid (mg)	0.6	0.18	0.18	0.4
Choline (mg)	600	450	500	500
Vitamin B$_{12}$ (mg)	0.005	0.003	0.003	0.005
Linoleic acid (%)	1.2	0.8	1.4	1.4
Inorganic Elements				
Calcium (%)	1.0	0.8	3.7	3.7
Phosphorus (available) (%)	0.5	0.5	0.55	0.55
Sodium (%)	0.15	0.15	0.15	0.15
Potassium (%)	0.40	0.40	0.40	0.40
Chlorine (%)	0.15	0.15	0.15	0.15
Manganese (mg)	25	25	15	15
Magnesium (mg)	250	250	250	250
Iron (mg)	40	40	20	20
Copper (mg)	5	5	5	5
Zinc (mg)	20	15	10	10
Selenium (mg)	0.07	0.07	0.07	0.07
Iodine (mg)	0.17	0.17	0.15	0.15

Source: L. E. Card and M. C. Nesheim, *Poultry Production,* 11th edition (Philadelphia: Lea & Febiger, 1972).

[a]These recommended levels may be converted to amounts per kilogram of diet by multiplying by 2.2 the requirements stated in IU or mg.

Esophagus (Gullet) and Crop

The gullet is a canal that leads to the crop and continues on to the proventriculus. It has great ability to expand. Food is stored in the crop temporarily. Some softening and predigestion takes place here due to the action of enzymes.

Glandular Stomach (Proventriculus)

The stomach secretes gastric juice, principally hydrochloric acid, and the enzyme pepsin, which acts to break down proteins to amino acids.

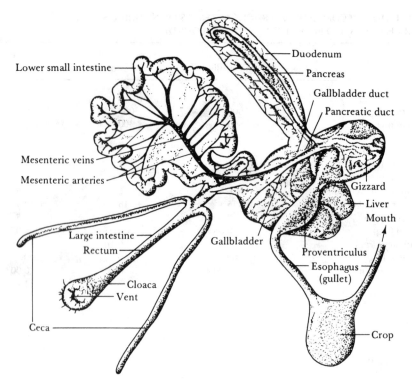

FIGURE 31-1. The digestive system of a fowl. [Reprinted by permission from L. E. Card and M. C. Nesheim, *Poultry Production,* 11th edition (Philadelphia: Lea & Febiger, 1972)]

Gizzard (Ventriculus)

The gizzard is composed of a horny, lined structure which is heavily muscled. The involuntary grinding action of the muscular gizzard has a tendency to break down food products much as teeth would do. Feeding of grit is controversial but often practiced in order to assist the gizzard. It is especially beneficial when birds consume whole grains. Crushed granite, oyster shell, or other hard, insoluble material may be used as a free choice additive (made available, not mixed in the ration).

Small Intestine

Most of the digestion takes place in the small intestine. Intestinal juices are enzymes secreted in order to split sugars and other nutrients into more simple forms, where they are transported to the bloodstream. Peristaltic action (contraction of the smooth muscle) also takes place here to move materials through the system to the ceca and rectum.

Ruminants have the ability to digest cellulose, but because the passage of food through the digestive system of the chicken is very rapid, and the only

Chap. 31 FEEDING POULTRY

location where bacteria occur to any extent is in the ceca, little digestion of fibrous feed can occur.

Ceca

Corresponding to the appendix in humans, the ceca has no known function at all. Although some bacteria are found in these two blind pouches, so little digestion of fiber occurs that, for all practical purposes, it can be disregarded.

Large Intestine and Cloaca

The large intestine is a continuation of the digestive tract leading from the junction of the ceca to the cloaca. The cloaca is a common junction for the outlets of the digestive, urinary, and genital systems through the vent.

TABLE 31-3. EXAMPLES OF PULLET STARTING AND REARING RATIONS

	Starter 0-6 Weeks (lb/ton)	Grower 6-12 Weeks (lb/ton)	Developer 12 Weeks—Maturity (lb/ton)
Ingredient			
Barley	400	—	—
Corn, yellow	925	1305	1360
Oats	—	—	300
Standard middlings	—	200	—
Soybean meal, 50% protein	400	280	150
Meat and bone scrap, 50% protein	160	100	100
Corn distillers' solubles	50	25	—
Alfalfa meal	50	50	50
Dicalcium phosphate	—	15	15
Limestone	—	10	10
Salt, iodized	5	5	5
DL-Methionine or methionine hydroxy/analog equivalent	1.5	1.25	0.80
Vitamin and trace mineral mix	10	10	10
Analysis			
Protein (%)	21.5	17.7	14.4
Metabolizable energy (kcal/lb)	1325	1340	1390
(kcal/kg)	2915	2950	3060
Calcium (%)	0.97	0.91	0.90
Available phosphorus (%)	0.58	0.54	0.50
Fat (%)	3.5	3.9	4.1
Fiber (%)	3.5	3.3	3.9
Linoleic acid (%)	1.3	1.6	1.6
Methionine (% of protein)	1.95	2.0	2.0
Methionine + cystine (% of protein)	3.60	3.7	3.8
Lysine (% of protein)	5.0	4.9	4.4

Source: L. E. Card and M. C. Nesheim, *Poultry Production,* 11th edition (Philadelphia: Lea & Febiger, 1972). Reprinted by permission.

TABLE 31-4. EXAMPLES OF LAYER AND BREEDER RATIONS

	Layer		Breeder
	Early Stages of Production (lb/ton)	After Peak Production (lb/ton)	For Young Pullets in Early Stages of Production (lb/ton)
Ingredient			
Corn, yellow	1330	1285	1265
Wheat shorts	—	100	—
Stabilized grease	20	20	30
Soybean meal, 50% protein	250	300	260
Fish meal, 60% protein	25	—	80
Meat and bone scrap, 50% protein	130	50	50
Corn distillers' solubles	50	—	100
Alfalfa meal	30	40	50
Dicalcium phosphate	10	30	20
Limestone	130	160	130
Salt, iodized	5	5	5
DL-Methionine or methionine hydroxy/analog equivalent	0.80	1.0	0.60
Vitamin and trace mineral premix	10	10	10
Analysis			
Protein (%)	17.0	15.6	17.4
Metabolizable energy (kcal/lb)	1330	1310	1346
(kcal/kg)	2925	2880	2960
Calcium (%)	3.3	3.45	3.1
Available phosphorus (%)	0.58	0.52	0.58
Fat (%)	5.7	4.2	5.2
Fiber (%)	2.3	2.7	2.6
Linoleic acid (%)	1.5	1.5	1.7
Xanthophyll (mg/lb)	9.0	9.3	9.7
(mg/kg)	19.8	20.5	21.3
Methionine (% of protein)	2.0	2.0	2.0
Methionine + cystine (% of protein)	3.7	3.7	3.6
Lysine (% of protein)	5.1	4.8	4.9

Source: L. E. Card and M. C. Nesheim, *Poultry Production,* 11th edition (Philadelphia: Lea & Febiger, 1972). Reprinted by permission.

Accessory Organs

The liver and pancreas contribute to digestion, although food does not pass through them. In addition to many other functions, the liver secretes bile, which is stored in the gallbladder and used by the body to emulsify fat in preparation for digestion. The liver also stores readily available energy (glycogen) and breaks down protein waste by-products into uric acid for excretion by the kidneys. The white portion of chicken feces is mostly uric acid and represents the fowl's system of urinary excretion. The pancreas secretes en-

Table 31-5. EXAMPLES OF BROILER STARTER AND FINISHER RATIONS

	Starter Ration (lb/ton)	Finisher Ration (lb/ton)
Ingredient		
Corn, yellow	793	1300
Milo	300	—
Stabilized grease	70	50
Soybean meal, 50% protein	630	260
Corn gluten meal, 60% protein	50	100
Fish solubles, dried	10	—
Fish meal, 60% protein	—	50
Poultry by-product meal	—	100
Corn distillers solubles	50	50
Alfalfa meal, 17% protein	25	40
Dicalcium phosphate	40	25
Limestone	15	10
Salt, iodized	5	5
DL-Methionine or methionine hydroxy/analog equivalent	1.6	—
Vitamin and trace mineral premix	10	10
Analysis		
Protein (%)	23.5	20.5
Metabolizable energy (kcal/lb)	1410	1480
(kcal/kg)	3100	3260
Calcium (%)	0.93	0.81
Available phosphorus (%)	0.55	0.52
Fat (%)	6.1	6.1
Fiber (%)	2.6	2.6
Linoleic acid (%)	1.6	1.9
Xanthophyll (mg/lb)	9.5	16.8
(mg/kg)	20.9	37.0
Methionine (% of protein)	2.0	2.0
Methionine + cystine (% of protein)	3.6	3.7
Lysine (% of protein)	5.2	5.0

Source: L. E. Card and M. C. Nesheim, *Poultry Production,* 11th edition (Philadelphia: Lea & Febiger, 1972). Reprinted by permission.

zymes (amylase, trypsin, lipase) to aid in the digestion of carbohydrates, proteins, and fats. Sugar metabolism is also regulated by production of the hormone insulin.

FEEDING WITH A PURPOSE

Although many people in the world still feed relatively small numbers of fowl on the farm by hand, for the most part, poultry feeding has become a very large, highly specialized, production of eggs and meat. The science of feeding has become so technical that almost all of the commercial growers purchase their feed rather than mix their own. However, for purposes of illustration, sample rations will be given for those who wish to experiment on their own or to raise small batches of poultry. These rations may serve as a guideline.

Table 31-3 illustrates rations that are to be used for replacement chicks (pullets). The main purpose of rations for a replacement chick is to produce growth without bringing the chick into an early lay. Table 31-4 illustrates layer and breeder rations. Breeder rations are usually more complex than layer diets to make sure that all necessary nutrients are put into the egg from which the chick will hatch. Table 31-5 gives a broiler starter and finisher ration. The main purpose of these rations is to produce the most rapid growth possible on the least amount of feed, with satisfactory finish in the shortest economical period.

STUDY QUESTIONS

1. Too much salt in poultry rations can cause the undesirable management of _____ manure.
2. A desirable, nonnutritive product commonly used in rations to assist poultry in utilizing coarse feeds is _____ .
3. Some assistance is provided chickens in swallowing because of the unusual shape of their _____ .
4. Some softening and predigestion of feed takes place in the _____ .
5. The ventriculus, or _____ , serves as the chicken's _____ .
6. The small intestine is where most _____ takes place.
7. Very little _____ can be digested by chickens.
8. The digestive system shares a common junction with the _____ and _____ system.
9. _____ _____ causes the white portion of chicken manure.
10. The digestion of carbohydrates and proteins is regulated by enzymes secreted by the _____ .

DISCUSSION QUESTIONS

1. Assuming you have a flock of 510 caged layers, averaging 5 lbs in weight, laying 348 eggs per day on the average, how much total feed will be required per month? How much for each dozen eggs produced? What percent protein in the feed would you require from your feed supplier?

2. Define the following:
   a. crop
   b. gizzard
   c. proventriculus
   d. grit
   e. ceca

   f. uric acid
   g. bile
   h. amylase
   i. oyster shell
   j. trypsin

3. What are the recommended levels to feed growing broilers for the following?
   a. vitamin A
   b. salt
   c. riboflavin

   d. calcium
   e. phosphorus
   f. iron

4. Give the ingredients and amounts needed to mix a 100-lb batch of feed for a starter broiler ration. How will this ration differ from a finishing ration to be used later?

Chapter Thirty-two

DISEASES AND PARASITES OF POULTRY

Chickens will not grow nor lay eggs up to potential if they are parasitized or diseased. The chance of a disease outbreak is great under modern confinement procedures, so prevention of disease outbreak is the key to health. Treatment of diseases, as opposed to preventive medicine, is costly and often unsuccessful.

Disease in this text is defined as any departure from normal health. This can be caused by microorganisms, nutritional deficiencies, or unusual environmental stress. The discussion of diseases in this text will be limited to those problems caused by microorganisms, environmental conditions, external and internal parasites, and behavioral problems. The strategy of disease control will include stimulating the body's natural defense mechanisms (immunization), sanitation, quarantine, and good environmental management.

It is important for the reader to understand that the terms *horizontal* and *vertical transmission* are used repeatedly in this chapter in dealing with the transmission of diseases. Horizontal transmission refers to the spread of disease by contact with litter, inhaling dust particles containing organisms, and infestation by vectors such as insects, wild birds, or parasites. Horizontal transmission is the most common form of disease transfer. However, some diseases are also transmitted through vertical means, that is, the passage of disease-producing agents from the dam to the offspring through the egg. A number of diseases are spread by both horizontal and vertical transmission (for example, pullorum, lymphoid leukosis, and mycoplasmosis).

The poultryman should be aware of the deviation from normal health that signals the onset of a disease. Often the first sign of a problem is a decrease in feed and water consumption. Other specific signs may include diarrhea, paralysis, respiratory difficulty in breathing, skin conditions, bloody or unusually wet droppings, or other signs that distinguish diseased fowl from normal birds. Many states have diagnostic laboratories that confirm the suspicion of a disease outbreak by the posting (autopsy) of a dead bird or the sacrifice of a bird for analysis. Birds that die, apart from the normal expected losses, should be preserved through refrigeration or packing in ice until a qualified veterinarian or diagnostic laboratory personnel can conduct a postmortem.

Some 80 separate diseases or parasitic problems are recognized by diagnostic laboratories in regions throughout the United States. Millions of pounds of poultry are rejected for human consumption by inspectors annually due to diseased birds that do not succumb to an outbreak but are a loss to the industry nonetheless. Leukosis, airsacculitis, and septicemia are some of the most important causes of rejection by poultry inspectors during slaughtering procedures. The term *septicemia* is a non-specific condition referring to rejection of the carcass because of such things as anemia, edema, dehydration, or other indications of disease. Septicemia, however, does not specify what the disease might be.

In a successful laying flock, a 10 to 12% death rate is considered normal in an average production year. With broiler flocks, the normal maximum annual death loss is about 4%. Any death losses in excess of these two figures should be taken as a serious condition that should receive prompt attention by the poultryman.

The diseases or problems discussed in this chapter are grouped to include diseases of the respiratory tract, tumor diseases, salmonelloses, miscellaneous diseases, external parasites, internal parasites, and behavioral problems.

DISEASES OF THE RESPIRATORY TRACT

Respiratory diseases are among the most troublesome in the poultry industry. A generalized term, airsacculitis, is often used to characterize any respiratory infection. This term refers to an infection of the air sacs and is not descriptive of a specific disease. A flock with respiratory problems will show signs of difficult breathing. Gasping, wheezing, coughing, and nasal discharge are among signs that indicate respiratory distress.

Mycoplasmosis (CRD or Chronic Respiratory Disease)

Mycoplasmosis is a disease of the air sacs in which the sacs become filled with exudates (mucous fluid). The lungs harden and breathing is obviously impaired. The death rate may be quite high. A reduced growth rate is evident during and shortly after recovery. Broilers that do recover may be condemned at carcass inspection. The principal cause is the organism *Mycoplasma gallisepticum*, a primitive organism, in association with other viruses and bacteria. Both vertical and horizontal spread of the disease are common. Treatment of hatching eggs with antibiotics has proven successful in breaking the vertical transmission. Isolation and proper sanitation thereafter to establish mycoplasma-free breeding flocks, similar to SPF hog operations, have been established to prevent horizontal transmission.

Infectious Bronchitis

This disease causes an infection of the trachea, the air sacs, and the bronchi leading to the lungs. Coughing, wheezing, and difficult breathing are characteristic signs. Death is common in young birds, and lower egg production is a consequence of the disease in adults. The cause is a microorganism that is spread horizontally, and there is no treatment for the disease once an outbreak occurs. Control is handled through vaccination using a vaccine that can be administered on a mass basis by inclusion in the drinking water.

Newcastle Disease

Newcastle disease is characterized by difficult breathing, rattling, coughing, and sneezing. In a small portion of the affected birds, central nervous signs are also observed, including muscular incoordination and partial paralysis, particularly of the head and neck muscles (Figure 32-1). Death rates can be high in broilers. In laying hens, it is not uncommon to observe a complete failure of the flock to produce eggs. Although egg production may recover to normal levels following an outbreak, the eggs are often of poor quality for some time thereafter.

Transmission is by horizontal means. The disease was first encountered in England near Newcastle-on-Tyne and was given the name Newcastle because of this. Control of Newcastle disease is primarily by inoculation. Vaccines may be used in the drinking water, as a spray or dust to treat the entire flock, in wingweb inoculation, or through other direct methods.

Laryngotracheitis

The characteristic signs of this disease are gasping, coughing, rattling, wheezing, and an occasional attempt to dislodge accumulations of mucus by loud outcries. The birds often extend the head and neck upwards with the mouth wide open when inhaling. While exhaling, the head is drawn back and lowered with the mouth closed. The causes of laryngotracheitis is a virus that infects the larynx and trachea of the windpipe.

FIGURE 32-1. Newcastle disease is characterized by muscular incoordination and a particular paralysis of the neck muscles, as shown. (Courtesy of the U.S. Department of Agriculture)

Transmission of the disease is by horizontal means. Infected adults have lowered egg production initially, but after the disease has run its course, affected birds appear normal. A portion of the recovered birds will remain carriers that can spread the disease to healthy birds even though they themselves do not appear to have any remaining problems. For this reason, it is recommended that recovered birds be removed and new stock vaccinated when entering the flock for production. Old birds that have suffered from laryngotracheitis should be discarded even though they appear normal.

Although the condition can be quite severe, outbreaks normally occur only occasionally. Therefore, routine vaccination for laryngotracheitis is not recommended unless there has been a problem in the house before. Once an outbreak occurs, flocks in the nearby houses should be protected by prompt vaccination.

Infectious Coryza

Signs of this form of respiratory disorder include discharges from the nostrils and eyes. The spread of coryza is rather slow but can be speeded up by contact with birds using the same drinking water. This condition is similar to the common cold in humans. Infectious coryza is caused by the bacterium *Hemophilus gallinarum*.

Treatment with drugs such as sulfathiazole in the feed or injections of streptomycin have been effective. However, medication is not the answer to control. Separation of affected birds, disposal of old hens at the end of the year, and isolation rearing in a controlled, clean environment are considered keys to prevention of infectious coryza.

Aspergillosis

A respiratory disease characterized by gasping and rapid breathing. Often there is a loss of appetite and an increase in thirst. Birds lose weight and exhibit some central nervous system (CNS) signs. A fungus causes the condition. Outbreaks usually occur only when moist conditions in litter will support the growth of mold. Spores from the fungi enter the air and are breathed in by the birds, resulting in aspergillosis. There is no treatment. Prevention is accomplished by keeping feed and litter low in moisture content to prevent mold.

AVIAN TUMOR DISEASES

Two tumor diseases, Marek's and lymphoid leukosis, represent a most critical problem in poultry production today. Although they are separate diseases, they are frequently grouped together and referred to as the avian leukosis complex. A virus is the cause of tumors, but there has been no indication that man or other species are similarly affected.

Marek's Disease (Range Paralysis)

Although adults are not totally immune, Marek's disease is primarily a disease of young, growing chickens, 3 to 5 months old. The key signs are paralysis of the legs, wings, and neck. Blindness may also occur, and a high death rate is not unusual. Birds that do recover may have a condemned carcass (broilers) or suffer a lower egg production (layers). A horizontally transferred virus causes the disease. Vaccines are available to develop immunity in poultry, and some strains have also been selected for natural resistance to Marek's disease.

Lymphoid Leukosis (Big Liver Disease)

Younger chickens are not completely exempt from lymphoid leukosis, but it is primarily a disease of chickens over 18 weeks of age. Laying hens are most affected. The initial signs are loss of weight and condition. Autopsy reveals tumors which may affect any part of the body except the nerves. The liver is frequently enlarged and pale, and shows lesions.

The cause of lymphoid leukosis is a group of viruses that can be spread both horizontally and vertically, although the horizontal spread is a very slow form of transmission. There is no treatment or preventive vaccine, but since the faster form of transmission is vertical the recommended practice is to eliminate infected breeder hens so that hatching eggs are free of the disease. Infected hens can be identified by a blood test.

COCCIDIOSIS

Coccidiosis is normally a controllable disease, but it is one of the most expensive to the poultry industry. Affected birds are obviously ill. Signs are weak, droopy, anemic birds, often showing blood in the droppings. Death rates in the severe outbreaks can be very high. The cause is any number of protozoans that not only attack the intestinal tract but also have a stage of their life cycle outside the body as well. The name *coccidiosis* comes from the fact that coccidia are initially picked up from the litter by foraging birds. The coccidia cause ulceration of the cells lining the intestine. Bacteria can then invade the damaged areas to cause infection. Intestinal bleeding shows up in the droppings. The coccidia spend 4 to 7 days in the bird before passing out in the feces. The external stage then starts. If the litter is sufficiently warm and moist, a process called sporulation takes place within 24 to 48 hours, and the infection can start anew.

Treatment and prevention involve reducing bird concentration, keeping litter dry, and the use of drugs called anticoccidials. Caging layers can eliminate the disease by breaking the cycle. Broilers or breeders can be protected by dry litter and the use of anticoccidials in the feed to prevent the infection. Virtually all commercial rations for broilers now routinely use an anticoccidial compound.

SALMONELLOSES

Pullorum and fowl typhoid, two diseases caused by organisms of the genus *Salmonella*, will be briefly discussed in this text. Salmonella organisms attack several animal species, including man, and constitute serious health hazards. Symptoms in humans include severe intestinal disturbances and sometimes death.

Pullorum

Baby chicks show the only clinical signs of pullorum. A few days after hatching, affected chicks huddle together, lose appetite, may show labored breathing, and often develop a whitish diarrhea (Figure 32-2). Death rates can be very high. Posted birds reveal lesions in many organs including the heart,

FIGURE 32-2. Pullorum disease can be deadly to baby chicks. A whitish diarrhea, as shown in this chick, is a characteristic sign. (Courtesy of the U.S. Department of Agriculture)

spleen, kidney, lungs, and digestive tract. Adult birds that have recovered from pullorum carry the disease but show no signs of it, although they are responsible for the vertical transmission of the infection. The cause is a bacterium, *Salmonella pullorum*. Both vertical transmission through infected eggs and horizontal spread are possible while chicks are still in the incubator.

Prevention is the key to control. Where pullorum has been a problem or is suspected, it is common to fumigate the incubator and eggs with formaldehyde gas followed by strict sanitation measures between hatches. The national poultry improvement plan in the United States set in motion the annual blood testing of breeding flocks to eliminate infected breeder birds. By eliminating the possibility of vertical transmission in this way, pullorum has almost been eradicated in America.

Fowl Typhoid

Signs of fowl typhoid are quite similar to pullorum. The cause is another of the salmonella organisms, *Salmonella gallinarum*. Although the drug furazolidone has been effective in reducing losses to this organism, good hatchery sanitation and elimination of infected adults are keys to control.

Paratyphoid Infections

Salmonella organisms other than those that cause pullorum and fowl typhoid are collectively referred to as paratyphoid infections. Clinical signs are the same as for the two previously discussed diseases. The drug furazolidone has been effective in controlling losses and horizontal spread of organisms.

OTHER DISEASES

Although there are numerous other poultry diseases that may cause significant losses in other parts of the world, only those of current economic concern in the United States will be covered here. They are fowl pox, infectious synovitis, and infectious bursal disease.

Fowl Pox

Black raised scabs on the comb, wattles, face, earlobes, shanks, and feet make this disease easy to recognize. Birds become very ill, egg production is reduced, growth is retarded, and hatchability and fertility are reduced. The cause is a virus. Although the signs are similar to chicken pox in man, the virus is not the same one despite the name and will not infect man. The disease is spread horizontally by other chickens and by mosquitoes.

Treatment of noninfected birds with a live virus vaccine will protect birds against infection. However, routine vaccination is not recommended except in areas where it is known to be a threat. The more practical treatment is to remove and isolate infected birds at the first sign of fowl pox and vaccinate the remainder of the flock.

Infectious Synovitis

The hocks and joints of growing broilers swell with an arthritic condition in this disease. Broilers 4 to 12 weeks old are most commonly affected. The hock joints and foot pads swell, making many birds lame and reducing their growth rate. The cause is a microorganism that invades the joints, producing excess fluids in the joint area and surrounding tissue. The organism is spread horizontally and vertically.

Treatment of hatching eggs has not been effective in controlling the organism. Therefore, the generally accepted theory is that control is possible only through production of flocks selected free of the organism. Prevention currently consists of using a low level of antibiotics administered in the feed to control the infection and keep horizontal spread to a minimum.

Infectious Bursal Disease (Gumboro Disease)

The bursa of Fabricus is a sac in the cloaca of the bird. In infectious bursal disease, chickens 3 to 6 weeks old develop a whitish diarrhea, become dehydrated and may show darkening of the muscles or muscle hemorrhage. These signs may persist 5 to 7 days, and death losses may reach one-third of the flock, although it is not usually so severe. Recovered birds develop immunity.

The cause is a virus that is difficult to eradicate once an outbreak occurs. Commercial vaccines are available for control in areas where the disease is known to be a problem. One such area is Gumboro, Delaware, where it was first discovered in 1962—hence the name "Gumboro disease."

EXTERNAL POULTRY PARASITES

Because of irritation to the host and possible transmission of diseases through horizontal means, external parasites should be controlled. Elimination of pests also improves the overall health and production of birds. External parasites mainly include lice and mites and to a lesser extent chiggers, ticks, fleas, bedbugs, and flies.

Lice

About 40 species of this whitish, wingless, flattened insect are known to infest poultry. Adult lice lay their eggs around the vent at the base of the feathers. Young lice appear transparent and reach full size about two weeks after hatching. They spend their entire life cycle on the host and generally spread to other chickens by direct contact. Damage to the host is caused by the biting and chewing of the insects on feathers, scales, and skin. Growth rate and egg production are lowered due to irritations.

Control of lice by the use of approved pesticides is quite effective, although several applications may be necessary to kill the insects that may have hatched since the time of first application. Treatment should include the body of the bird as well as roosts and house. Young chicks placed in a clean house with no older birds should, therefore, remain free of infestation.

Mites

These insects are smaller than lice and are barely visible to the unaided eye. The two most common species are the red mite and the northern fowl mite.

Red mites spend part of their life cycle living and breeding in cracks and crevices of poultry houses, feeding on poultry only during the night. Northern fowl mites are found on the bird and in the house at all times, being the most common vector of caged hens in the United States.

Damage to chickens is caused by the sucking of the mite, which lives primarily off the blood and lymph. Loss of blood, anemia, and carcass blemishes or condemnation are cause of losses. Control of mites requires that approved pesticides be used that are applied to both house and bird in order to eliminate the pest from habitat and host.

Other External Parasites

Problems are sometimes caused by less frequent infestations of chiggers (actually a larval stage of mites), ticks, fleas, bedbugs, and flies. In general, control of these pests is similar to the methods discussed for lice and mites since the insects live both on and off the host.

INTERNAL POULTRY PARASITES

The major internal parasites of chickens are worms, of which there are several types. Roundworms or tapeworms constitute the most commonly encountered forms. Large roundworms are unsegmented, white, round internal parasites that may measure 2 to 5 in. in length. Infestation begins when foraging chickens eat the worm eggs, which hatch in the digestive system. Larvae enter the intestinal wall where they live to maturity. Eggs pass out in the droppings and start the cycle again.

Other roundworms that spend all their lives (larval and adult stages) in the host may inhabit the intestinal tract. The cecal worm is a small, white roundworm that inhabits the cecum of the chicken. These worms have been particularly troublesome because they are carriers of disease-producing protozoans.

Several pieces of tapeworms may inhabit the intestinal tract of chickens. It is important to know that tapeworms require an intermediate host such as snails, slugs, beetles, crayfish, or flies so that control must include external intermediate hosts as well as internal eradiction. Signs of heavy internal parasite infestation in the chicken may include lowered production, unhealthy appearance, anemia, inactivity, ruffled feathers, and drooping wings.

Control is largely a matter of sanitation. Clean litter and clean houses are the keys to preventing infestations. If infestations are verified, there are commercial drugs called antihelmintics that are available for deworming parasitized birds. Various antibiotics may also be used at low levels in the feed to protect against some forms of worms.

BEHAVIORAL PROBLEMS

Cannibalism and hysteria are the two most common social problems of rearing chickens in high concentrations of confinement.

Cannibalism

The habit of birds pecking one individual, or each other, to death is called cannibalism. The cause is not well understood, but it occurs frequently under confinement conditions. Some speculative causes have been suggested, such as nutrient deficiencies, overcrowding, not enough feeding or water space, or too much light in the house. Whatever the cause, the only correction and standard prevention is debeaking, the removal of about two-thirds of the beak. This is usually done at 4 to 6 weeks of age to produce the minimum stress. An electric debeaking machine is commercially available that cuts the beak and cauterizes the wound at the same time to aid clotting and speed healing.

Hysteria

Under some conditions, such as excessive, unusual noise, rapid change in light, or quick movement, broilers in open houses fly into one corner of the house, causing many deaths due to suffocation. In caged layers, birds may try to fly, which results in broken wings or necks.

The causes of hysteria are unknown, but care should be taken not to frighten birds. Some poultry managers use radio broadcasts piped into houses to accustom birds to the sound of human voices and noise. Another technique used by some is to knock several times before opening a door to the poultry house. This supposedly draws the birds' attention to the door, lessening the surprise of door movement.

STUDY QUESTIONS

1. A disease that is spread through contact with contaminated litter would be an example of _____ transmissions.
2. Vertical transmission of disease occurs through the _____.
3. A very broad term for disease of a nonspecific nature such as an anemia, edema, and so on, is _____.
4. Airsacculitis is a generalized term referring to a disease of the _____ tract.
5. A characteristic sign of paralysis of the neck muscles in which the chicken curls its head toward the breastbone is a _____ disease called _____.

Chap. 32   DISEASES AND PARASITES OF POULTRY

6. A respiratory disease similar to the human cold is called _____ _____ in chickens.

7. Marek's disease or lymphoid leukosis refers to a rather common _____ condition.

8. Coccidiosis in layers can be prevented by simply _____ the birds.

9. Blood in the droppings should be taken as a possible sign of _____.

10. A high death rate in baby chicks, labored breathing, and whitish diarrhea are characteristic signs of _____.

11. Infectious synovitis is a swelling of the _____ and is best prevented by low level _____ or selection from _____ _____ flocks.

12. The most common external parasites of chickens are _____ and _____.

13. The major internal parasites are several types of _____.

14. Cannibalism is most commonly controlled by _____.

15. Thunder and sudden movement sometimes causes damage or losses due to a behavioral problem known as _____.

DISCUSSION QUESTIONS

1. Which disease of poultry can be prevented by vaccination?

2. Name the diseases known to be vertically transmitted in poultry.

3. Describe the signs of internal parasite infestation in poultry and general recommendation for treatment.

4. Define the following:
   a. horizontal transmission
   b. septicemia
   c. airsacculitiss
   d. hysteria
   e. coccidiosis

5. Name the causes, other than microorganisms, that can cause disease. Name the measures to control diseases.

6. What are the very first signs of an impending problem and what immediate steps should be taken once it is recognized that a problem exists?

7. Numerous birds begin coughing and wheezing. Which diseases, if any, would you suspect and how would you determine the possible cause to explain your suspicions over the phone to a veterinarian?

Chapter Thirty-Three

THE BROILER AND LAYER INDUSTRY

Although the management of chickens either for meat or eggs is highly complex and scientific, a few basic management suggestions may be helpful in orienting the beginning poultryman.

BROILERS

The proportion of chicken meat provided by commercial broilers has grown from about 5% in 1935 to well over 90% currently. Most all commercial broilers are grown today on a contractual arrangement or reared by large companies that have their own market outlets.

Chicks are normally bought from companies that specialize in producing special strains of chickens for either meat or egg production. Straight-run (unsexed) chicks are usually ordered for broiler production. White-feathered broilers are also always in demand, making a straight-run, white-feathered chicken the favorite. Deliveries direct from the hatchery to a clean house streamline the modern, efficient system and relieve the producer from details of a breeding program.

Size of Operation

With mechanized feeders and waterers, one full-time operator can care for 36,000 to 45,000 broilers at one time. Of course, some part-time producers may have only a fraction of that amount, and some large establishments house 100,000 or more at one time. About 18 minutes per 1000 birds per day is required to feed using automatic feeding equipment; 30 minutes without automatic equipment.

Housing

Generally speaking, broiler houses are 30 to 40 ft wide and as long as necessary to accommodate the number of birds desired. Six-tenths of a foot per bird is considered the minimum space requirement in completely environmentally controlled houses; 0.75 ft^2 per bird in conventional houses. The house should be constructed so that birds are protected from heat, cold, or other inclement weather, thus providing a comfortable environment for maximum growth potential. Most houses today are well ventilated, insulated, and have controlled temperature and automatic lights, feeders, and waterers. Litter (sawdust, shavings, ground corncobs, etc.) should be 4 in. deep.

Production Expectations

An average of 3.5 lb liveweight at 50 days of age should be the goal if the poultryman is to compete and profit in today's market. Feed conversion should average 2.00 lb of feed per pound of gain or less. Some instances have been recorded of conversions of 1.8, although this is unusual. Death losses should not be more than 1% to be considered normal.

Grades of Broilers (Fryers), Other Classes

The grades established by the USDA are: A or No. 1 Quality; B or No. 2 Quality; and C or No. 3 Quality for live birds. Dressed birds are graded U.S. Grade A, U.S. Grade B, and U.S. Grade C. Criteria for the grades are based on conformation, fleshing, fat covering, and defects.

A special class of broiler chicken that should be mentioned is the Rock Cornish game hen or Cornish game hen. This is a 5- to 7-week-old immature female weighing no more than 2 lb ready-to-cook weight, selected from Cornish or Cornish cross matings. Cornish hens are very popular at convention banquet affairs.

Other classes of chicken meat are: roaster (a young chicken more matured than the acceptable criteria for broilers); capon (a castrated male, under 8 months of age); stag (a male chicken under 10 months of age but showing developing sex characteristics); cock or rooster (mature male); and hen or stewing chicken (mature hen, usually older than 10 months of age, often culled from laying operations).

Normal and Abnormal Signs

If chicks or pullets are spread out evenly over the house, this is a good sign that health and social adjustments are normal. Birds crowded against the walls are signs that they are too hot; huddling closely and crowding around the heat source (brooder) or at corners of the house means they are too cold. Gasping or panting could mean a disease problem or too much heat. If litter is packed, wet, or giving off a strong ammonia odor, there could be a ventilation problem that should be corrected before a more serious problem such as respiratory disease develops.

Watch for telltale signs such as blood in the droppings (disease), rough feathers (parasites), pale combs (anemia), and any other abnormalities that may signal the beginnings of health problems. A few minutes of careful observation each day will soon train the eye to distinguish between normal and abnormal signs.

LAYERS

Except for a few areas where premium markets exist for brown eggs, it is recommended that layers of white eggs (white leghorns or leghorn crosses) be used for commercial egg production. The key to success in the competitive egg business is to use high-producing layers regardless of breed, cross, or strain. Most emphasis should be placed on performance records of selected type in laying tests.

Housing

The construction of layer houses will depend on size of operation, environmental conditions, and individual preferences. Houses may even be multiple-story arrangements. The more common housing designs are the floor type, slat or wire floor type, and the cage.

The floor or litter-type house is an open house covered with 6 to 8 in. of litter (as opposed to 4 in. for broilers) to absorb the moisture from the droppings over an entire laying year. Nests and roosts are provided in addition to feeders and waterers. The slat or wire floor house uses either wooden slats turned edgewise, wire floors, or a combination of the two. This allows for easier manure removal and requires no litter. Nests, roosts, automatic feeders, and waterers are used.

Some variation of one or more hens in wire cages is probably the most popular design in the United States. The more common type is one hen per cage (8 by 16 in.) or two hens per cage (also 8 by 16 in.) with hardly more than room enough to turn around. A few colony cages (about 25 hens per large cage) have gained acceptance. The advantages of the cage system are that droppings are well placed, more birds can be housed in a given area, eggs roll on the slanted cage bottom to an easy collection point, and internal parasites are eliminated. Disadvantages, such as difficult manure removal and outbreaks of flies, have been noted.

Production Expectations

A good average production rate is 20 eggs per month for commercial high-producing layers. Death losses of 1% per month are to be expected as normal; unhealthy or diseased birds should be culled. Layers should not be kept in the flock over 19 months because declining production is inevitable by this age. Many birds are force-molted at this time and brought back into production for another 6 to 8 months. A feed conversion goal of less than 4½ lb of feed per dozen eggs is considered attainable and in keeping with good management practices. Eggs should be collected five times daily, cleaned immediately, and stored under refrigeration at 13 °C (55 °F), 75 to 80% relative humidity.

Grades of Eggs

The *Egg Grading Manual*, Agricultural Handbook No. 75, USDA, Agricultural Marketing Service, gives a guide to all egg grades. Basically, eggs are graded according to similar characteristics of weight and quality for three marketing outlets: (1) consumer grades—Grade AA or Fresh Fancy, Grade A, Grade B, and a size classification of Jumbo, Extra Large, Large, Medium, Small, and Peewee; (2) wholesale grades—U.S. Specials, U.S. Extras, U.S. Standards, U.S. Trades, U.S. Dirties, U.S. Checks (used in wholesale channels of trade and may be resorted to conform to consumer grades); (3) U.S. procurement grades—special designations for institutions and the armed forces.

Eggs for the consumer trade are candled (held to a special lighted device similar to an X-ray, so that the inside contents may be seen to determine quality based on size and condition of the white, yolk, and air cell. Soundness of the shell is also estimated and weight (or size) is expressed in ounces per dozen. Weight, however, is distinctly different from quality and has no bearing on grade. Color does not influence grades but sometimes enters into descriptive marketing channels as "whites" or "browns."

Normal and Abnormal Signs

Alert birds with red color to comb and wattles, well-bleached beaks and shanks (indicating good laying production), roughened feathers (old feathers, lack of molt), good appetite, and condition are signs of healthy producing flocks. Abnormal conditions include early molt, shrunken head parts, yellow shanks, excessively soiled feathers, poor egg shell thickness, reduction in feed consumption, and lowered egg production. These signs can be early indications of disease, parasites, environmental problems such as excessive heat or cold, improper ration, poor management of light, and so on.

STUDY QUESTIONS

1. The term *straight-run* refers to _____ chicks.
2. The color of almost all broilers today is _____ .
3. A completely environmentally controlled broiler house for 1000 birds should contain _____ ft^2.
4. A typical broiler should weigh at least _____ pounds live weight at the end of _____ weeks of feeding.
5. A broiler that gains 3 lb should be expected to consume about _____ lb of feed.
6. The immature hen so popular at banquets is the _____ breed.
7. Birds that appear to be crowding the walls of a house, avoiding the center, may be too _____ .
8. Poor ventilation combined with wet litter could lead to _____ problems.
9. Rough-appearing feathers could be a sign of _____ .
10. A laying flock of 10,000 birds may have normal death losses of _____ birds per year.
11. Laying hens are seldom kept in the flock beyond _____ months of age.
12. Fifty dozen eggs can be produced on about _____ lb of feed.
13. The largest and smallest size classification eggs are _____ and _____ .
14. The highest consumer egg grade is _____ .

DISCUSSION QUESTIONS

1. Define the following:
   a. candling   d. straight-run
   b. capon   e. checks
   c. stag
2. Name the consumer grades and size classification for eggs.
3. Suppose that you won first prize at an agricultural fair—1000 baby chicks, straight-run, broiler type. The prize included a buy-back arrange-

ment at double the market price. When could you guarantee delivery, how many could you reasonably expect to deliver live, what would be the total live weight, how much total feed would be consumed, and what would be your gross income on today's prices?

4. What size building would be needed to house 100,000 broilers under one roof?

5. A city dweller has just inherited a laying hen operation of 40,000 hens. He asks you to take over operations for him in the next few days and give him a report on what to expect in annual egg production, total feed required per dozen eggs per month, replacement schedule, and total annual death losses. What are reasonable expectations?

Chapter Thirty-Four

TURKEYS, DUCKS, GEESE, PIGEONS, AND GUINEA FOWL

TURKEYS

At one time, most turkeys were reared in small numbers on farms, but commercial production today has replaced the small producer. Highly specialized producers grow turkeys (native to America) for the Thanksgiving and Christmas markets in the United States and Europe. In recent years, turkeys have been making strides toward increasing their share of the fowl meat market by year-round sales, especially appealing to the diet fad in the United States. Turkey is not only nutritious, but devoid of excess fat, producing a low-cholesterol meat.

By production of eggs all year round, specialists using several flocks and artificial lighting can produce large numbers of day-old poults, which are fattened by other specialists. The integration of breeding, rearing, and marketing has led to turkey production groups and through proper advertising has created an increasing demand for oven-ready birds.

Definition of Common Terms

Poult. Young bird under 8 weeks of age.
Turkey grower. Bird from 8 to 26 weeks of age.

Turkey hen. Female, over 26 weeks of age.

Turkey stag, tom, or cock. Male adult over 26 weeks of age.

Breeds

Although many breeds exist in the United States, there are five principal breeds in serious production today.

1. *Broad Breasted White.* White feathered breed, 14 to 22 lb (6.35 to 10 kg), males, 25 to 40 lb (11.3 to 18.2 kg). Hens are poor layers (50 to 60 eggs per season).
2. *Broad Breasted Bronze.* Similar to Broad Breasted White. Feathers have terminal edging of white, copper bronzing on tail and wings.
3. *American Mammoth Bronze.* Similar to above, but less developed breast muscles.
4. *White Beltsville.* Small, white, females 10 lb (4.54 kg), males 14 lb (6.35 kg), egg production good (100 to 120 eggs per season).
5. *Hybrids.* Inbred lines from various breeds are crossed to produce faster growing, fleshier, more efficient poults than the parent stock.

Breeding

Breeding birds are normally only kept for one laying period because of the cost of carrying them over to the next season. Hens of the small type may be bred at 30 weeks of age, the larger type at 36 weeks.

Stags can be used for breeding at 34 weeks of age for small types, 40 weeks for larger varieties. One stag per 10 to 15 hens is recommended for pen mating; flock-mating is 15 to 20 hens per stag. Birds should be mated at least 3 weeks before eggs are needed for hatching. Because the spurs of stags may lacerate hens during mating, canvas-back saddles with a loop on either side to go over each wing are used to protect the hens and provide a surface for marking to keep track of mating records.

Fifty to 120 eggs per breeding season may be expected in the first clutch. If hens are held over for the second season, this may be reduced to 40 to 75 eggs. Without artificial lighting and under natural conditions, most eggs are produced during the spring and summer.

In extremely broad breasted varieties, stags may not be able to mate with females because they lose their balance due to this extreme conformation, which has been bred into them by man's selection principles. In this case, semen is collected from stags and hens are inseminated every 3 or 4 weeks at the start of the breeding season and every 2 or 3 weeks toward the end of the breeding season. An alternative method is to inseminate once weekly during the breeding season.

Incubation.   Incubators are kept at 99.5 °F (37.5 °C), 62% relative humidity for 24 days, and 75% relative humidity for 4 days; 28 days total are needed for hatching. At 24 days, eggs are moved from the setting section to the hatching section of the incubator.

Management.   After hatching, young turkeys are kept 4 weeks in specially designed, draft-free brooders. Temperatures are kept similar to those required for baby chicks, but poults are recommended to have twice the floor space. At 4 to 8 weeks of age, poults are often moved to hay box brooders (although they may be left in the tier brooders). Hay brooders have a section between the ceiling and the roof covered with hay on a wire frame to trap heat produced by the birds. Wire floors are usually fitted above the ground to prevent parasite infestation. In this way, no artificial heat or light is required.

At 8 weeks of age, poults become stronger and may be moved to relatively inexpensive pole barns. These are cheap structures, often made of corrugated iron for a roof and wire-netting for fronts and sides. Earth floors are customary. Turkeys require 3 to 5 ft^2 (0.28 to 0.47 m^2) of floor space. Pole barns are frequently constructed face to face with a service track between them. More elaborate structures may be used as alternative housing and, in fact, may be essential in colder climates.

Legend indicates that wild turkeys are crafty, intelligent, and smart, but commercial varieties must be protected from every imaginable predator and disease. Turkeys also have strange behavior patterns. Numerous instances have been recorded in very wet climates of poults standing under the drain area of a roof with mouth wide open where some of them promptly drown. Why they do this is unknown, but speculation is that the turkey is not the brightest of birds. However, given proper protection from the elements, predators, and with the right kind of management, turkeys can be a rewarding enterprise.

Nest boxes are usually supplied in breeding operations, at least one box per three hens. Each nest is recommended to be 1.5 ft (0.46 m) wide, 1.5 ft (0.46 m) deep, and 2 ft (0.62 m) high at the front, sloping to 1.5 ft (0.46 m) at the rear. A suitably designed trapdoor on the front can allow for accurate egg laying records and is commonly called "trap nesting." The bird is released after the egg is collected for later incubation.

Feeding

Grit is needed by turkeys in the same way that other fowls need it. However, because of peculiar behavior by poults, granite grit must be used with caution, sprinkling it on the food every day for the first two weeks and once a week for the next four weeks to prevent young poults from gorging on it. At 6 weeks of age, hoppers of grit may be utilized using adult-size grit, free choice.

Poults have very poor eyesight and, for the first week or so, may not find food and water troughs and die of starvation, unless bright strip lights are fitted over troughs; colored glass marbles are often used in the feed to catch the light and attract the bird's attention. Pecking at the marbles encourages them to feed. Some growers sprinkle feed on corrugated cardboard floors, causing poults to notice and peck at it.

A turkey starter diet (Table 34-1) or a commercial game bird feed, high in energy (and about 28% protein), is fed for the first 8 weeks. The crumb form of feed is preferred over mash because it may cause a clogging on the beak and curling of the tongue. From 8 weeks of age on, a slightly lower energy feed (see Rearer, Table 34-1) is recommended. Birds are usually switched to a Finisher (Table 34-1), at 14 weeks, and harvested at about 16 weeks of age. Grass or green chop may be used at 12 weeks of age and thus reduce concentrated feed requirements by 10%.

Most large producers separate sexes at 8 weeks of age because stags, which grow more quickly than hens, can be fed a higher-energy ration. Also, hens do better when they do not have to compete with stags for food.

TABLE 34-1. TYPICAL TURKEY RATIONS (%)

Foodstuff	Starter	Rearer	Finisher	Prebreeder	Breeder
Maize	25	30	25	25	25
Wheat	25	37.5	31.2	—	19
Barley	—	—	25	20	15
Wheat middlings	10	5	5	40	20
Soybean meal	22.5	10	5	7.5	10
Fish meal	12.5	7.5	2.5	2.5	5
Meat and bonemeal	2.5	5	5	—	—
Limestone	—	—	—	3.7	4.7
Fat	—	2.5	—	—	—
Mineral/vitamin supplement	2.5	2.5	1.3	1.3	1.3

Birds being reared for breeding rather than slaughter can follow the same regimen previously discussed up to 16 weeks of age. The ration is then changed to a 15% protein mixture (see Prebreeder, Table 34-1) until 4 weeks prior to laying, when a 16 or 17% protein ration (see Breeder, Table 34-1) is fed.

Show Feeding

Some readers may want to raise turkeys as only a small project or may fit only a few birds for shows and fairs. It is recommended that turkey raisers work closely with their county agent and follow specific recommendations that utilize local feeds and take into account specific management as dictated by environment. However, in the absence of any formal guidance, the following guidelines might serve as a starting point for further experimentation.

Wet Mash Recipe. Use a typical finisher ration and mix it with yellow cracked corn (80% finisher, 20% corn). Add water or milk until a paste is formed. Place the mash in a separate container. Let the birds eat as much as they can in 30 minutes. Be sure to clean out the container and make a fresh mixture each day. Do not start feeding wet mash earlier than 3 weeks before the show.

The percentage of corn may gradually be increased but more than 50% corn is not recommended. Feed wet mash once daily, but keep dry finisher out. Be careful not to overfeed.

Cooked Ration

1. Mix 1 to 2 quarts of powdered milk (or 2 oz of milk for every ounce of corn).
2. Place in large pan. Do not use a galvanized utensil.
3. Add 16 oz of yellow corn chops (1 oz per bird).
4. Add ¼ lb (1 stick) margarine.
5. Bring to boil and cook until margarine completely melts.
6. Continue to cook for an additional 3 to 4 minutes. Stir mixture thoroughly during this time.
7. Set aside to cool. The corn will absorb a large quantity of the milk during cooling period. The longer the mixture cooks, the greater amount of milk absorbed. In some cases, additional milk will need to be added to produce a watery paste.
8. Cook fresh daily. To be fed once a day. If cooked corn finisher diet is used, keep dry finisher out and do not feed the regular wet mash.

Make sure that fresh dry feed is fed twice or three times daily. Do not overfill feeders because they should be almost empty at each feeding. Wash water containers thoroughly each day.

Raise feeders and waterers each week about 1 in.

Use wet mash *or* cooked ration, *not both*.

Feed amount of cooked feed that birds will eat in 30 minutes.

Health

Turkeys are subject to many of the same diseases that affect other poultry, covered under diseases of chickens. In general, a turkey should be observed to see that it does not sneeze or gasp and has no discharges from the eyes or nostrils. These are early signs of disease and should be investigated by growers at the onset. Other early signs of disease are reductions in consumption of food and water, a body temperature that varies considerably from the normal 107 °F (41.7 °C), and normal pulse rate of 150 beats per minute. Turkeys normally receive a deworming agent and aureomycin, or similar antibiotic (preventive program), according to manufacturer's directions to maintain reasonable health and prevent losses.

DUCKS

Commercial duck production has been practiced for thousands of years in China and was mentioned as early as 2000 years ago by the Romans. In the United States, which produces about 10 million ducks annually, commercial production is concentrated around Long Island, New York. Ducks are excellent producers of meat and eggs and are much less susceptible to diseases than chickens. Some breeds of duck will reach market weight (7 lb) in 8 weeks and lay 150 or more eggs per year.

Breeds

With the exception of the Muscovy, it is thought that most ducks originated from the wild Mallard. Although no one is sure of its ancestry, the Muscovy, which orginated in South America, is thought to have descended from a different line entirely. The major distinguishing characteristic of the Muscovy is that it has very sharp claws in addition to its web feet, roosts in trees, and has a very mean disposition. There are many breeds of ducks, both ornamental and domestic, but we shall concentrate on the more common and utilitarian ducks.

Khaki Campbell.   One of the best breeds for egg production, originally bred by a Mrs. Campbell in England at the beginning of the twentieth century. It has the capabilities of laying up to 300 eggs per year. For this reason, it has great popularity worldwide.

Indian Runner.   This breed originated in Malaya, about 1870. Varieties include the White, the Buff, and the Fawn and White. The breed has a most peculiar upright stance from which the name Runner is derived. The breed lays about 180 eggs per year, on the average.

Welsh Harlequin.   A very popular breed in Great Britain, derived from Khaki Campbell stock. It is similar in egg production to the Khaki but superior as a table bird.

Aylesbury.   A snow white, heavy bird, producing 100 eggs per year and weighing up to 10 lb.

Pekin.   Introduced from China about 1870, the most desirable table duck in Australia and North America. Average annual egg production is about 130 per year.

Rouen.   A heavy French breed with a poor laying record of about 90 greenish eggs per year.

American Buff Duck (Buff Orpington).   Orginated in Great Britain, averages 240 eggs per year, and is considered a good dual-purpose duck (eggs and meat).

Cayuga and Black East Indie.   Ornamental breeds, but in high demand as table ducks by gourmets because of their game bird flavor. They are poor layers but have an excellent record as good hatchers and mothers.

Crested Duck.   Named for its characteristic top knot. It is of British origin, medium weight. Unfortunately, breeding stock is not readily available.

Muscovy.   The Muscovy is a breed of duck worth having around if for no other reason than to hatch other duck eggs. Muscovys are very broody and thus serve the breeder well who has a few eggs to hatch from other breeds. Muscovys are wild and high flying, so their wings need to be clipped in captivity. Numerous instances occur where Muscovys escape from pens and return to the wild. They are unique in that they have sharp claws on their web feet and roost in trees. Some Muscovys are also raised for the commercial market because of their rather gamey flavor. Breeders quickly learn that they are vicious birds and must be handled with precaution to prevent scratches and bites.

Duck-Keeping Systems

Ducks are waterfowl and, as such, must have access to a certain amount of water. However, a pond is not a necessity, provided ducks are allowed enough water to completely submerge their head in water and occasionally ruffle their feathers in it. The head submersion is important in order to keep the eyes and nostrils clean. Otherwise, incrustations may occur that produce a variety of disorders and diseases of the optical and respiratory system.

Basically, there are five methods of keeping ducks: the free-range system, the Dutch system, straw yards, the verandah system, and the small house and run.

The Free-Range System.   The stocking rate on 1 acre (0.5 ha) is 100 ducks year-round or half a dozen ducks per 50 ft^2 (4.6 m^2). Although ducks are very hardy, they usually will need shelter in severe weather and at night, particularly in colder climates. An old poultry house can serve well for this purpose, but ducks will need a small ramp to walk up to the floor because their legs are weak and easily damaged. If there are no existing structures, a couple of bales of hay with a sheet of tin across the top can serve the purpose just as well.

The Dutch System.   A series of long, narrow pens fronting a stream or shallow trough with free flowing water is used to provide ducks with a suitable environment. A 5-ft fence is usually used to enclose the area, the ground is often covered with gravel, a duck house with ramp is provided, and the ground slopes toward the running water for a self-cleaning arrangement.

Straw Yards.   A fenced-off area is covered with fresh straw, which is continually added as old straw becomes soiled. This system naturally relies on a source of cheap and dependable straw.

The Verandah System.   The verandah system often utilizes an old poultry house with an extended galvanized wire floor attached to the poultry house. The walls and roof are also covered with galvanized wire, creating a large cage, preferably with the floor high enough over a concrete apron that droppings fall through the verandah and allow for easy cleaning. The ducks will stay in the verandah area on the wire floor for the major part of the day but have access to the poultry house at night and in very severe weather.

The Small House and Run.   A small enclosure attached to a duck house constitutes the run. Half of this area may be covered with smooth cement and the grassy area kept closed off during wet weather or when ducks eat up existing vegetation. Ducks do well on cement provided it is smoothly finished. Grassy areas can be given a rest to promote regrowth, and ducks turned back onto the full run when vegetation and weather permits.

Breeding

The normal system for breeding ducks is flock mating. The male/female ratio is usually 1:7. Incubation of duck eggs is 28 days, the same as for chickens. The only exception is the Muscovy duck, which requires 33 to 35 days. Duck eggs hatched in an incubator are handled in the same way as chicken eggs except that more moisture is required at hatching time. Follow the directions for machine incubators. If eggs are being hatched under broody ducks, it is important that the ducks be allowed access to water during this period because nature has endowed the broody duck with the desire to ruffle her feathers and immerse her total body in water at this time—which provides the extra moisture necessary for proper hatching.

Small operators may simply leave the rearing of ducks to the mothers. Commercial breeders, however, use commercial brooders. About 100 to 150 ducks are placed under each brooder. Temperatures will be about 90 °F the first week, 85 °F the second week, 75 °F the third week, and 65 °F the fourth and fifth weeks. At 6 weeks of age, ducks are moved to fattening pens where no heat is usually required.

Feeding

If special duck rations are not available, the operator may use, from the first day to approximately 1 month of age, broiler starter crumbs, which is a good, convenient feed for ducklings. After 4 weeks of age, when ducklings are growing rapidly, a broiler finisher ration will provide nutritional needs. Water and feed should be kept some distance apart to prevent ducks from wasting feed by alternating between feeder and waterer. From 16 weeks onward, ducks come into adulthood and can be fed accordingly. Layers, pellets, or mash can be fed either wet or dry. The ducks will come into lay at about 16 weeks to continue the cycle of life.

Uses

Ducks may be used for a variety of purposes. Commercial producers use them for meat, down, and eggs. Table ducks are in demand in hotels and restaurants and are considered a delicacy in many parts of the United States with high Chinese populations, such as San Francisco. The down feathers are in demand for quilts, bedcovers, pillows, and winter clothing of various sorts.

Several misconceptions about duck eggs are that they "taste strong," "have to be boiled before you can eat them," and that "you get salmonella food poisoning from them." None of these rumors can be substantiated except for the transmission of salmonella, but even this is not the case unless eggs are laid on damp, dirty straw, or in a muddy environment at the edge of a pond. Clean, fresh duck eggs should be regarded with no more suspicion than hen eggs. It should also be noted that salmonella organisms only multiply at temperatures exceeding 41 °F; thus refrigeration becomes an added preventative measure.

GEESE

Definition of Common Terms

Geese. Plural, referring to males, females, adults, and young.

Gander. The mature male adult of geese.

Goose. The mature female of geese.

Gosling. A young goose or gander prior to adulthood.

History

The goose is said to be one of the oldest domesticated animals. Geese were among the few animals discovered in drawings on the walls of King Tut's tomb. Looking back to customs in ancient China, 4000 B.C., the favorite parental gift to the bride and groom was a pair of live geese, symbolizing a long, faithful marriage. Geese in the wild are monogomous and mate for life. As the years rolled on, the goose came to be known as the "wedding feast bird," cherished not only for faithfulness and fidelity, but for good luck.

Not much else is known of the history of the goose. Even during colonial times, the goose was classified as "a silly bird, too much for one and too little for two." However, today's marketed goose is far more meaty and tender; most are packaged and fresh frozen for greater convenience and wider availability. Today's commercial goose is "too much for two" and actually enough for a whole family to enjoy.

Our colonial forefathers made great use of goose and goose products. Early New England settlers had to struggle to keep warm during the grim winter. They quickly learned to rely on the soft feathers and down of the goose to keep them comfortable in bed. Goose grease cleaned and polished the black boots of our forefathers and, when the Declaration of Independence was signed, a goose quill declared a nation on July 4, 1776.

As strange as it may seem, some animal authorities equate the goose with cattle in usefulness and compatibility. If the world should have to divert the usage of grain from animals, there would be only three surviving species adaptable to animal agriculture: the ruminant (those species that chew a cud), the rabbit, and the goose. All are forage consumers.

Unlike other forms of poultry, the goose is basically a "roughage burner." This fact was well known in the pioneer days because nearly every small farm had its flock of geese to provide down for pillows, eggs and a delicate meal for special occasions. Geese were so practical and useful during the revolutionary days that Benjamin Franklin made a strong appeal to the founding fathers to appoint the goose as the symbol of the United States. Little is known of the history of geese, but we do know that they were extremely important in the development of the country. A record of the first Thanksgiving meal by our Pilgrim Fathers listed some venison, two turkeys, and 70 geese.

The goose is capable of taking care of itself on very rough land. Although geese like water, it is not necessary to have lakes, ponds, or running streams for them to be of benefit in making use of grasses and weeds. In the wild, one gander mates with one female for life. However, under domestication, a gander will "consent" to taking care of four or five families.

A good example of how geese are managed on a large scale today can be found in the state of Minnesota. Low lying areas that cattle, sheep, or goats would find unattractive can serve the web footed goose perfectly. One grower ran several thousand goslings on nothing but Reed's canary grass (a rather undesirable roughage). The goslings were turned on to the grass at 4 weeks of age, left for 14 weeks, and marketed without any form of supplementary feed. In another case, a small acreage of cattails was cleared for other uses by 5500 goslings (Figure 34-1).

The ability of the so-called "weeder geese" has long been recognized for its cultivation of certain crops. All geese have some dislikes when it comes to forage consumption. They do not care for mint, strawberries, asparagus, or cotton. Consequently, before the advent of herbicides, "weeder geese" kept the rows spotless in areas where these crops were grown. Because of herbicides and residues, "weeder geese" are not used much today with the exception of mint and strawberry fields.

FIGURE 34-1.

Another interesting use of geese involves their unusual habit of "honking" when strangers encroach on their territory. A large distillery in Scotland was having difficulty with theft of their products stored in open yards where valuable kegs of liquor were aged. Although the area was fenced and watch dogs were used, thieves were able to outwit the distillery by dropping female dogs in estrus over the fence. Losses continued to mount until the distillery got rid of the dogs and put in a flock of geese. The geese quickly learned the people on duty and were silent, except when a stranger neared the storage area. Geese then let out a clatter of protest calling attention to the potential thieves. Thereafter, theft from breaking and entering was completely eliminated.

Commercial goose production supplies birds for special occasions in many parts of the world. In the United States, geese are served as a substitute for turkey during major holidays such as Thanksgiving and Christmas. In Europe, especially England, the most festive occasion is St. Michael's Day, September 29. Legend has it that if you eat goose on Michaelmas Day, you will never want for money all year. In some parts of the United States, especially Pennsylvania, St. Michael's Day is also celebrated in the same way.

The most unusual method of producing a goose product has to be claimed by the French. For many years, the French have practiced the art of "stuffing" of geese, force feeding by hand with a funnel more than the goose cares to consume. This causes an enlarged liver. Normally the liver will easily fit in one's hand, but "stuffing" increases the size of the liver to 1.5 to 2.5 lb in weight—and a price approximately 10 times the price of an average goose carcass. These giant livers are then flavored with truffles, a fungus that grows primarily in southern France, to produce the gourmet delight known as gooseliver paté. It is world famous and very expensive. An interesting side story to the finding of truffles is the way in which French farmers discover this fungus, which grows about the depth of potatoes underground. Truffles have defied domestication and so must be hunted in the wild. Hogs have a keen taste and nose for them, so some French farmers train a few hogs to work on a leash like a dog and scour the countryside in search of truffles. When the hog makes a find, the farmer pulls him off the area and ties him to a tree while he digs the truffles (which normally wholesale for about $150 per pound) for later inclusion with the gooseliver and numerous other foods.

The undisputed leader in goose production in the United States is Pietrus Foods, Inc., 112 Pine Street, Sleepy Eye, Minnesota 56085. Geese are hatched and, when only a few days old, are shipped to farmers. They buy them back for processing in their slaughter plant after they have weeded a field or cleared an area of grass, weeds, and so on. Approximately 100,000 geese are hatched, sold, and bought back in this manner.

Legends of the goose abound, but often are based on fact. Take the case of the long lifespans attributed to geese. In the nineteenth century, it was traditional, especially in European countries, for a mother to give her daughter a prize goose so she would have a start on eggs, down for pillows, and meat for the table when offspring were numerous enough. Many daughters held a sentimental attachment to retaining the same goose in the family and would give her daughter the family goose. Reports indicate that some geese were handed down over several generations so that records of geese living to be 100 years or longer were not unusual.

Breeds of Geese

Geese are raised mostly in the north central portion of the United States and make up only about 0.5% of poultry raised in America. There are only 10 recognized breeds of commercial importance. They are Toulouse, Emden, Pilgrim, Roman, Buff, Sebastopol, African, Chinese, Canada, and Egyptian.

The *Toulouse* is the largest and most popular breed of geese in the United States, weighing from 20 to 26 lb. Plumage is dark gray on the back, fading to light gray, turning to white on the breast and abdomen. The Toulouse will average 15 to 35 eggs per year.

The *Emden* is pure white, closely feathered, adults weighing 18 to 20 lb. The Emden lays fewer eggs than the Toulouse but is a better sitter. In Australia, the Emden is now the most popular breed.

The *Pilgrim, Roman, Buff, Sebastopol, African, Chinese* (excellent foragers and lay more eggs than any other breed), *Canada*, and *Egyptian* breeds are of less economic importance but might be expected to perform similar to the Toulouse and Emden.

Breeding

As mentioned previously, geese in the wild mate for life and are monogomous (one gander mating with a goose for life). However, under commercial production, one gander will accept three geese in his "set" or family. Some of the lighter breeds such as Roman, Chinese, or Buff may accept four or five.

A goose begins laying in the winter. She selects her own spot and continues until she has a clutch of about 20 eggs. Even though the spot selected by the goose may seem impractical, it is almost useless to try to move them. The best recommendation is to build a crude shelter around the goose. This can be done with a few bales of hay and a tin roof.

Broody chicken hens have been used to hatch goose eggs, but they are too heavy for the hens to turn them, so this will have to be done by hand. Also, two or three eggs are perhaps maximum for an average size broody hen to handle.

Artificial incubation of goose eggs can be handled in the same way as chicken eggs, except that goose eggs need a slightly higher humidity. A daily fine spray from a garden hose mist usually provides the desired results. Incubators also usually have specific instructions on temperature and humidity setting for various eggs and instructions should be followed closely.

Feeding

As previously mentioned, geese can derive all their feed from only pasture. The stocking rate is approximately eight geese per acre for year-round grazing, heavier if clearing is the objective. Some reports indicate that geese do not digest forage any better than some other large birds, but they consume five times as much and, because of their web feet, are able to utilize wet marshy lands that would be undesirable for most other livestock.

For geese confined to a pen, a good all-round feed is the following mixture, fed twice daily; 1 part oats, 3 parts green chop, 2 parts wheat. The greens can be fresh grass, leafy vegetables, such as lettuce, cabbage, even dandelion leaves. When green chop becomes unavailable and/or fattening ration is desired, the following may be used: 1 part wheat, 1 part barley, 1 part oats.

Water must be provided to geese in quantities large enough that their heads can be immersed completely under the water. Otherwise, corrosions of the beak, nostrils, and eyes may become a serious problem. If this requirement is met, lakes, ponds, or other bodies of water are not necessary for successful rearing.

Additional Information

To get started in the geese business, to purchase weeder geese, and/or to obtain more detailed recommendations on feeding, breeding, and care for geese, the reader is referred to Pietrus Foods, Inc., 112 Pine Street, Sleepy Eye, Minnesota 56085. This modern hatchery handles geese and other poultry throughout the world.

PIGEONS

Pigeons, since antiquity, have been used for food, sport, racing, entertainment, even for communication (the carrier pigeon). For meat production, the most desirable form of pigeon is the squab. Curiously, few people have tasted this delicacy, and even fewer have raised them. Yet, the meat is dark and rich, with a full-bodied flavor. Tender and moist, squab ranks along with veal, lobster, or suckling kid (young goat). Squabs were consumed by nobility since the time of the ancient Greeks, but today are mainly found in exclusive hotels, restaurants, and, because of their ease of digestibility and low fat content, in a few hospitals.

An immature pigeon, 25 to 30 days old, the squab loses its maximum degree of plumpness, tenderness, and taste soon afterwards. Squabs, produced from special strains, varieties, or breeds, can be easily raised, at minimal expense, and make a fine project for the backyard enthusiast or serious poultryman.

Breeds of Pigeons

Pigeons may be divided into three major categories: (1) exhibition, (2) meat production, and (3) performance—tumbling and homing. Regardless of the use, serious breeders need to be very selective to produce the most desirable traits for the intended function. Tumblers (aerial pigeon acrobats) are selected on the basis of vigorous and controlled performances in the air, show pigeons are selected for color patterns and plumage, and squab producers select on the basis of raising the largest number of plump, healthy birds over the maximum length of time.

The beginner must be aware that not all pigeons are useful for all purposes. For instance, show birds may be large, beautiful, and well-balanced, but are often slow breeders, not suitable for commercial or even serious backyard squab production.

Squabbing Breeds

Hundreds of breeds and varieties of pigeons are available throughout the world, but only a few seem suitable for use by serious enthusiasts. The White King is the principal breed used for squabbing in the United States. If stock can be located, the Red Carneaux is an ideal breed for the beginning producer. This breed, used extensively earlier in the century for squab production, lost favor among commercial producers when the larger white varieties came on the scene. Perhaps more readily available in Europe, the Red Carneaux is still used (mainly for exhibition) and many prolific strains may still be found. The selection of a breed is not as important as a good strain within the breed. Whichever breed is selected, the beginning squabber is advised to check the laying record, feed efficiency, and longevity of potential birds within a strain. Selecting a strain within one of the following breeds is recommended as most suitable for the beginner (Table 34-2).

Characteristics of High-Quality Squabbers

Several characteristics mark a good pair of breeding birds for the production of high-quality squabs. Healthy, vigorous, disease-resistant birds are a necessity in any program. In addition, squabbers should be docile. Aggressive birds steal nests, attack other pigeons, and create expensive disturbances.

Good squabbers are attentive parents, feeding squabs generously. Males should be "good drivers" (mating behavior in which the male coaxes the female to her nest). Dominant male mating behavior occurs just before the female starts laying, continuing throughout the egg-laying period.

Squabbers should average 14 to 15 squabs per year for at least 5 years. Select medium-sized strains. Oversized birds often crush eggs and, in general, are less productive.

Pairs should be selected to produce squabs with white skin as most demanded by the popular consumer market. This can be disregarded for home producers. Selection for birds that have tight feathers is recommended because these birds also have fewer pin feathers, making them easier to clean.

In general, undiluted colored birds are recommended. Diluted birds (dun, yellow, and silver) have been observed to produce weak squabs. Good squabbers produce broad, thick, solid squabs weighing approximately 1 lb in weight.

Considering all the characteristics for good squabbers mentioned previously, a condensed, simplified starting point for selecting a strain is to look at the production records of a pair being considered for purchase. If the pair is young (2 to 3 years old), 16 to 18 high-quality squabs should have been produced in the past 12 months. If the pair is old (5 to 6 years old), 12 or more high-quality squabs should have been produced in the same time period.

Reproduction

Although there is some variation within each breed, a general conclusion is that a pair of pigeons used for squabbing should parallel the following statistics:

1. Mating starts between 5 and 8 months of age.
2. Peak production occurs between 12 and 18 months of age, continuing through 2 or sometimes 3 years.

TABLE 34-2. RECOMMENDED BREEDS FOR SQUABBERS

Breed	Floor Space per pair	Approximate Feed Consumption	Dressed Squab Weight	Comments
Large Breeds (26 to 32 oz, Mature)				
American Swiss Mondane[a]	$3\frac{1}{2}$–4 ft^2	115 lb/yr (pair and squabs)	1 lb	Very large and docile. Limited use for commercial production and perhaps too large for the backyard producer.
White King	$3\frac{1}{2}$–4 ft^2	115 lb/yr (pair and squabs)	1 lb	Highly recommended, most popular commercial breed available.
Silver King	$3\frac{1}{2}$–4 ft^2	115 lb/yr (pair and squabs)	1 lb or more; slightly larger than the White King	Very large breed for commercial producer but not as popular as the White King. Stock may be less available.
Auto-sexing King[a]	$3\frac{1}{2}$–4 ft^2	115 lb/yr (pair and squabs)	1 lb	Becoming popular but more expensive and less available than the King breeds.
Auto-sexing Texan Pioneer[a]	$3\frac{1}{2}$–4 ft^2	115 lb/yr (pair and squabs)	1 lb	More expensive and less available than the regular Kings, but like Auto-Sexing King, recommended for home production.
Medium-Sized Breeds (23 to 27 oz, Mature)				
Red or White Carneaux	$2\frac{1}{2}$–3 ft^2	100 lb/yr (pair and squabs)	$\frac{3}{4}$–$\frac{7}{8}$ lb	Used extensively long ago, still highly recommended, but stock may be hard to locate. The White Carneaux may be more available than the Red.
American Giant Homer	$2\frac{1}{2}$–3 ft^2	100 lb/yr (pair and squabs)	$\frac{3}{4}$–$\frac{7}{8}$ lb	Popular, new, upcoming breed. Beautiful color patterns. Stock may be difficult to locate. Some males and females have different colors or patterns by which they can be differentiated at birth (auto-sexed pairs).
Small Breeds (15 to 25 oz, Mature)				
Hungarian (Blue, Black, or Red)	$2\frac{1}{2}$–3 ft^2	100 lb/yr (pair and squabs)	$\frac{7}{8}$ lb	Color patterns most beautiful of the squabbing breeds, used mainly for exhibition, but recommended if fast-producing pairs can be located.
Squabbing Homer (Working Homer)	$2\frac{1}{4}$–$2\frac{1}{2}$ ft^2	90 lb/yr (pair and squabs)	$\frac{3}{4}$ lb	Prolific, hardy, but small for commercial production. Red, white, or blue strains are recommended. Ideal size for backyard production and home consumption.

[a]Recommended for the more experienced squabber.

3. Profitable commercial production usually does not exceed the fifth or sixth year. (Less than 12 squabs in 12 months is considered the minimum level of profitable production.)

It is recommended that to avoid disappointment the beginning squabber buy only from sellers who have production records on their birds. Buying older birds from a recognized operator with production records is a great way to get started. Replacements from these older birds can then lead to a satisfying level of production. If more money is available, younger pairs can be purchased from reputable breeders, but never buy birds solely on the basis of how they look or the glowing recommendations of the breeder unless records are available to back up production claims.

The mating behavior of pigeons is quite different from other fowl. The natural urge to mate is very strong, and the male shares some of the chores of building the nest, sitting on the eggs, and rearing the young.

Pigeons mate for life, but if one of the partners dies or is separated by man where they cannot see or hear one another, the birds will select new partners within a matter of days. However, if the birds are separated by man and placed back into a common pen, the original partners will ordinarily return; so strict management is needed to produce desired matings unless the birds are allowed to select their own partners and are left alone.

Mating Behavior. The male starts the courtship procedures and amid much strutting, the male ballons his crop, ruffles his feathers, drops the wings slightly, and produces quite a show. If the female accepts the male, then the pairing will be for life.

Soon after mating, the male will select and coax the female to the nesting spot. The behavior to force, coax, and keep the female in the nesting site is called "driving." The male brings nesting materials to the female, and she constructs the nest in the box.

Egg Laying. After the nest is complete, or nearly so, the female will lay the first of two eggs. The second egg is normally laid within 24 hours of the first. Two squabs per mating should be expected. Brooding of the eggs (sitting) begins immediately, shared by both partners. The female does most of this, but the male does relieve her for short periods about midmorning and midafternoon. The first egg should hatch in 17 to 18 days followed by the second, 48 hours later.

Feeding

Nutrition of the pigeon is quite similar to other fowl with one major exception. Pigeon grit is absolutely essential for successful grinding and digestion of grains fed to the birds. Ordinary poultry grit is not suitable and should never be used. A suitable pigeon grit formula is 40% medium-crushed oyster shell, 35% limestone or granite grit, 10% medium-sized hardwood charcoal, 5% ground bone, 5% ground limestone, 4% iodized salt, and 1% venetian red.

Although pelleted pigeon feeds may be used, most successful growers prefer a whole grain form of feed. It is recommended that only seasoned (aged) grains be used. For some peculiar reason, freshly harvested grain used in pigeon rations can cause severe diarrhea and even death in squabs and adults. A typical ration of seasoned grains might consist of whole corn (39.5%), cowpeas (22.7%), hard red wheat (19.8%), and milo (18%).

The New Jersey Agriculture Experiment Station recommends the following:

1. Whole yellow corn   40%
2. Kaffir corn   20%
3. Durham wheat   15%
4. Canada peas   15%
5. Hemp seed   2.5%
6. Millet   2.5%
7. Hulled oats   2.5%
8. Buckwheat   2.5%
9. Crude protein minimum content should be 14%.

If growers are away from home or have irregular hours, self-feeders can be used to provide feed to the birds. If the hand feeding method is practiced, it should be done twice daily, always at approximately the same hour. The normal recommendation is between sunrise and 9:00 A.M. and between 4:00 P.M. and sunset. Consumption of grain per day is from one-fourth pound per pair for smaller breeds to about one-third pound for medium and larger breeds. With self-feeders, the birds regulate their own intake, but can be expected to consume about the same amount.

Feeding Ritual. Feeding behavior for pigeons is quite different from other fowls. Both parents normally share equally in this chore. Unlike other birds that place food, such as worms, into the mouth of the young, squabs place their beaks into the mouths of the parents to consume a cottage-cheese-like substance called pigeon "milk." This secretion from the lining of the crop is unique in the animal world, a phenomenon shared only by pigeons and their cousins, the doves. As the production of pigeon's milk subsides, the squabs' developing digestive systems utilize more and more grain.

Immature birds are harvested for the squab market at 25 to 30 days of age. Squabs, kept for breeders, at 4 to 5 weeks of age have feathers filled out, and although they will still be unable to fly, will flutter to the floor, learning location of feeders, waterers, and grit hoppers. Sexual maturity occurs at about 4 months for the male, 6 months for the female. The new young should be removed from the pens at this time, otherwise, disturbances will be created by their seeking of mates within the pen.

Housing

Wild pigeons seek nesting places that are high above the ground, protected from wind, rain, and predators. Human modification utilizes similar principles through housing. Basically, pigeon shelters are of two types: (1) single-pair housing and (2) multiple-pair housing. Rabbit hutches, old chicken coops, and wooden and wire cages of all sorts can be used. The basic requirement is to prevent exposure to the elements, protect against predators, and maintain a dry, sanitary environment. The open front or entrance should always be faced in the direction of greatest exposure to the sun with the possible exception of tropical areas.

If birds are kept in a rural environment, they may be allowed to fly free with a pigeon box simply placed on top of a pole. Serious producers, however, have much more control, usually using a wire cage flyway that attaches to the nesting boxes. If more than eight pairs are used, multiple-pair housing is recommended. Conversion of an old garage or existing farm building is usually the preferred method of getting into the business or hobby. Consult your county agent for USDA information on specific sizes and plans for pigeon houses. Basically, all that is needed for the hobbyist is an enclosed area with nesting boxes (see Table 34-2 for space requirements).

Equipment

Equipment for feeding and watering pigeons need not be expensive and often can be made out of scrap material. The main consideration to remember is that birds must be allowed space to place their head in the feeder or waterer, but not be such a wide opening that the body can pass through. Otherwise, contamination with feces, feathers, and trash can create a problem with disease and efficiency.

Feeders are of two types: self-feeders and trough types. A similar device is used to provide grit. One other possibility that may be an advantage in raising squabs is the use of nesting bowls, made of ceramic, plastic, or even paper. These are devices that provide a concave surface when placed inside the nesting box to simulate the right shape, often preventing young squabs from falling out of the nest or developing abnormal leg and hip problems.

In addition, solidly placed perches (4 to 6 in. wide) should be provided in the fly pen and directly in front of the nest boxes. The approximate perch height in the fly pen is 4 ft, 4 to 6 in. away from the recommended one-inch poultry wire netting used to enclose the flyway.

Management

Although some observation of the nest and eggs is necessary, too much inspection may cause abandonment by the parents. Observe the nest from a distance but never disturb the pigeons at night because they may leave the nest until morning, losing the warmth necessary for developing eggs. This could cause the embryos to die.

If one or both eggs appear to be abnormal, broken, undersized, or if only one egg remains intact, one or both eggs can be removed and discarded. The pair simply starts over again with a new clutch.

About 20 days after the eggs have been laid, the nest should be checked for two normal squabs with closed eyes, nude bodies, and healthy appetites. If only one egg has been hatched, wait a day or so and break open the other to determine if the embryo has died, in which case it should be removed. The single squab can be raised by the parents, or moved to foster parents that similarly have only one young. The original pair then can start over for most efficient production.

About 10 days of age, the squabs should be checked again, at which time eyes should be open and pin feathers begun to develop. At this point, they should be taking small amounts of grain along with the pigeon milk.

At 25 days of age, the young squabs should be checked for those that will be harvested for meat. This normally occurs between the 26th and 28th days. The key to harvesting is when pin feathers under the wings and along the body are fairly well filled out. If the pin feathers appear to be immature, then the harvest may be postponed for one or two days.

If replacements are to be kept rather than squabs produced for harvest, a pair of young are placed in a pen together until they mate and begin nesting. Although inbreeding is not as critical in pigeons as it is in other animals, cross-mating is generally recommended—taking the male from one set of parents and placing him with a female from another set of parents.

Sex Differentiation. Popular belief among the uninformed is that a clutch (pair of eggs or squabs) is always composed of a male and a female. Although this is generally the case, it is far from being true. Numerous instances of clutches producing members of the same sex occur.

It is almost impossible to differentiate the sex of immature birds and for the less than expert, the safest and simplest form of sex differentiation is simple observation. Females will be smaller, do less strutting during the mating period, and do not make a complete circle when strutting. The female will also place her beak inside the male's beak in the process of "cooing and billing."

The male will appear larger, more rugged, generally has a thicker neck, struts in a complete circle, and, when doing so, will often spread his tail feathers, dragging the ground, and dropping his wings in similar fashion. As stated previously, mature birds that have already selected a partner, may have to be separated where they cannot see or hear one another in order to produce the desired cross with another mate.

Positive identification and accurate record keeping can be easily employed through the use of leg bands. The common procedure is to use colored plastic rings that are expandable and removable to temporarily identify members of a pair. More serious breeders use serial-numbered aluminum leg bands that are placed on the squabs legs at 6 to 10 days of age. After 10 days of age, these bands cannot be removed and therefore provide positive identification.

Many feed companies provide appropriate record forms that may be available through local feed dealers. The customary essential records to keep are number of eggs laid, number of eggs hatched, number of squabs raised, weight of dressed squabs, and comments about peculiarities of the pair from which the squabs came. Such records become indispensable in making future matings, and serious breeders rely heavily on this to improve production and efficiency.

Additional Sources of Information

When searching for information on breeders of pigeons, several possible local sources exist, such as the yellow pages of the telephone book (look under poultry or poultry dealers), county agricultural agents, and feed stores. For a wide variety of information, the *American Pigeon Journal* (Box 278, Warrenton, Missouri 63383) contains classified sections listing breeders throughout the country. Similar journals are available in England, Australia, and other countries.

The National Pigeon Association (P.O. Box 3488, Orange, California 92665) and the International Federation of Homing Pigeons (107 Jefferson, Belmont Hills, Pennsylvania 19004) represent breeders and growers of most every pigeon breed that would be of interest to squabbers and others. Contact with these clubs or associations can put the breeder in touch with additional specialty clubs that represent, for example, only one breed, such as the American King Club. Finally, county and state fairs throughout the world have pigeon exhibits included with the poultry. Virtually every available breed of pigeon of importance will be exhibited here and contacts can be made for information on purchase.

GUINEA FOWL

Between 1840 and 1950, guinea fowl were most numerous in general farms throughout the United States. Usually referred to as "guineas," they were brought to America by settlers from Italy, Spain, Portugal, France, England, and Holland. guinea were popular during the colonial days and until shortly after World War II because they ranged the land for most of their feed, raised their own young without assistance, were great insect eaters, always came home to roost, were good to eat on special occasions, and were the first burglar alarms. When predators or strangers approach the farm, they make a blood-curdling, clattering noise and from their perch high atop trees, cannot easily be silenced. Years ago, Elbert Hubbard wrote:

> Many farmers keep guineas as sentinels or watchers for the farm. When hawks, coons, stray cats or dogs, or strange humans come around, they set up a terrific shriek or clatter to drive the intruder away and alert other animals or men for

help. The guinea usually flies up on a board fence or other perch, and shouts forth a torrent of Billingsgate defiance. No bird has a vocabulary equal to a guinea. It is so profane that it is unprintable. Epithet, ridicule, sarcasm, and cuss words are sent forth in rapid fire. guinea take after dogs, cats, humans, hawks, skunks, foxes, and snakes, but dogs are held in special abomination.

History

One of the earliest written records of the guinea fowl is found in the Egyptian dynasty of 3000 years ago, when the Pharoah's expeditions brought them back from the Sahara Desert. Early settlers of America brought them to the eastern and southern states, where they became a food for the poorer class of people. The birds' ability to reproduce themselves and forage for most of their food in the south made them a popular addition to our early farms and plantations.

In 1920, the U.S. Census reported 2,400,000 guineas on American farms, but as small farms disappeared, their numbers dwindled to less than 10% today in the United States. However, guineas are extremely popular in France, where it is considered the Sunday bird of that nation, replacing the chicken and rabbit for Sunday dinner. Estimated consumption of the gamey, slightly robust flavored guinea is 50 million birds a year in France.

In the United States, a few commercial producers are producing guinea hens in lots of 5000 and up to produce uniform market guineas that will cook the same each time one is roasted or fried. Dressed guineas are attractively plump, with a darker skin than other upland birds. Many chefs consider the guinea notch above partridge, pheasants, quail, and grouse.

Characteristics of guinea

Guineas are terrific insect eaters, eat few plants other than grass, and do not scratch, as does the chicken. However, they are less docile and more easily agitated compared to other forms of poultry and remain rather distant to man, always walking a thin line between a domestic and a wild bird. If allowed to roam free, guineas like to roost in the treetops at night. When the weather drops below 10 °F, their feet will freeze; so if they do not seek shelter, they must be driven inside an old chicken house or other shelter with straw or other type of insulating ground cover.

If guinea hens are given the run of the farmstead, they will hide their nests in the grass, brush, or briar patch until about 30 eggs are laid; then set on them for 26 days to hatch. The wild hens lay their eggs on the ground or on the dry vegetation, making no attempt to build a nest, but they will fight any intruder that tries to take their eggs, and the male will join the female in the fight, whether they are wild or not.

In France, the breeder hens and cocks are kept in cages—four hens per cage—and all are artificially inseminated. The houses are windowless with light, heat, humidity, and ventilation completely controlled. Lights are dimmed off, then gradually turn on, so as not to shock or startle the layers.

Guineas are very flighty and easily frightened. To get maximum egg production, the birds must be kept as calm as possible. Breeders in cages lay 36 to 40 weeks, and during this period, produce an average of 170 eggs per hen.

Market guineas, after 4 weeks of age, are raised on the floor in dark houses to keep them quiet. The baby keets are started on a 24% protein starter with full light for 4 weeks. Then light is gradually reduced until houses are dark most of the time. Lights are dimmed on for feeding watering several times daily, then gradually returned to darkness. This method of growing market guineas to 10 to 12 weeks produces an extremely plump, tender bird that is well finished.

The guinea hen is not considered a good mother, as she tends to leave the nest before all eggs hatch and will take the baby keets through grass before the dew has dried off. It is best to take the eggs away from the guinea hen and either set them under a chicken hen for raising, or set them in commercial incubators and raise them in a brooder house. The fertile eggs may hatch from 95 to 100%—they often appear to just jump out of the shell. Keets are active shortly after hatching.

Feeding a Guinea Keet to Market at 12 Weeks

Day-old keets raised in a commercial brooder house should be fed a 24% protein ration for 4 weeks because the keets grow extremely fast during this time. Then drop to 22% protein between 4 and 8 weeks, and finish on an 18% protein finishing ration.

High protein starter and grower ration closely follows the guinea's natural eating habits in the wild. Along with grass and weed seeds, the guinea is a meat eater—the meat on range is insects—and in commercial feed, protein comes from meat scraps, soybean meal, and alfalfa meal.

Guinea are used by many fruit and vegetable farmers to help combat fruit flies, beetles, aphids, and gnats where sprays are not desirable. Young guineas do need a good supply of small grit for grinding grain and insects in their gizzards. Coarse hard seeds aid in quick digestion. Grit assists guineas to more quickly digest a balanced pellet, mash, or grain ration.

Guinea fowl make a sensible addition to the small farm, provided one can stand the shrieks of "nature's burglar alarm." Four guineas per acre in potato patches have been reported to completely control potato beetles, attesting to their effectiveness as a controller of insects. Some farmers and small acreage owners consider them an important part of their organic gardening program for this feature alone. A final interesting observation is that most poultry writers have never heard of any disease attacking this rough, tough, noisy bird.

For more information, breeding stock, or dressed birds, contact Pietrus Foods, Inc., 112 Pine Street, Sleepy Eye, Minnesota 56085.

STUDY QUESTIONS

1. A turkey poult is a young bird under _____ weeks of age.
2. A male turkey adult over 26 weeks of age may be called a _____, _____, or _____.
3. Large variety stags should not be used for breeding purposes until they are over _____ weeks of age.
4. The recommended tom-hen ratio for flock mating is one tom to _____ to _____ hens.
5. Incubators for turkey eggs are kept at _____ °F.
6. Turkey eggs should take _____ days to hatch.
7. One hundred turkeys would require _____ to _____ square feet of floor space.
8. Special care must be exercised with grit because poults may _____ themselves on it.
9. Normal turkey body temperature is _____ °F.
10. It is thought that all ducks originated from the wild Mallard except for the _____.
11. The Khaki Campbell breed of duck may lay up to _____ eggs per year.
12. The most vicious breed of duck that must be handled with care is the _____, but it has a great reputation for _____ other duck eggs.
13. Ducks need have only enough water to submerge their _____ completely.
14. Incubation of duck eggs is _____ days, except for the _____ which may require as much as _____ days.
15. Commercial producers use ducks for _____, _____, and _____.
16. The goose primarily eats _____.
17. Geese used in mint, strawberry, or asparagus fields are called _____ geese.
18. Legend has it that some geese may live to be _____ years of age or older.
19. Gooseliver _____ is made from geese that have been force fed.
20. The most popular and largest breed of geese in the United States is the _____.
21. The most desirable form of pigeon meat is the _____.
22. The principal breed used for squabbing in the United States is the _____ _____.

23. _____ to _____ squabs per year can be expected from a good pair of pigeons for at least _____ years.

24. Pigeons may be mated as early as _____ to _____ months of age.

25. Egg laying in pigeons is unique because they always lay in a clutch of _____ eggs.

26. _____ grain should not be used to feed pigeons.

27. A secretion from the lining of the crop used to feed squabs is called _____.

28. Squabs are usually harvested on the _____ to _____ day of age.

29. The three basic categories for raising pigeons are _____, _____; _____ _____; and _____.

30. A bird that is often used as a watchdog is the _____.

31. A baby guinea fowl is called a _____.

32. Probably the most outstanding characteristic of the guinea fowl is its ability to resist virtually any form of _____.

DISCUSSION QUESTIONS

1. Outline a management program for establishing a breeding program for turkeys. Include breeding season, age to be bred, tom-hen ratio, mating records, precautions to be taken, incubation, and any other management procedures necessary to progress from one generation to the next.

2. What are the peculiarities of turkey feeding habits, and what practices are recommended to counteract them?

3. If you had no incubator, but wished to produce the maximum number of duck eggs and ducks, which two breeds would you select and why?

4. Of the five management systems of keeping ducks, which would you use and why?

5. Discuss stocking ratio and temperature management for typical commercial duck brooders.

6. Someone confronts you with the rumor that duck eggs are not good or safe to eat. What reassurance can you give them?

7. Geese are primarily roughage burners and selective feeders. Discuss every practical use for a large flock of tame commercial geese utilizing these principles.

8. Discuss the characteristics needed for good squabbers and outline an annual management program. Include housing, type and quality of feed, egg-laying peculiarities, level of production, and expected longevity.

9. Name at least three reasons why guinea fowl are kept on farms and ranches. What is about the only precaution that must be taken to ensure their good health?

Chapter Thirty-Five

THE HORSE INDUSTRY

ORIGIN AND HISTORY OF THE HORSE

There are two theories of the origin of the horse, both backed by scientific evidence, both subject to a great deal of faith and speculation. Until recently only the theory of evolution proposed by Charles Darwin in 1859 was given credit for close, scientific scrutiny. More recently, the theory of special creation has also received wide attention and scientific compilation of evidence.

THE THEORY OF EVOLUTION

Basing their speculations on fossil remains found of a horselike animal with four functional toes and one splint on the front feet, and three functional toes and one splint on the hind feet, some scientists believe there is a five-toed, pre-horse fossil, remains of which have yet to be found. The theory is that a curious little four-legged mammal (Figure 35-1), standing 7 or 8 inches high at the shoulder, developed around the Rocky Mountain area of the United States. Its weight might have been that of our common house cat of today; its body covering was part fur, part hair.

FIGURE 35-1. Artist's conception of the hypothetical pre-horse, fossil remains of which have not been found. Speculation has it that this earliest ancestor of the horse had five toes with claws, was covered by part fur and part hair, and stood 6 to 7 in. high at the shoulder.

It lived in a marsh-like environment, browsing on leaves and marsh plants, and used the toes to distribute weight. Its teeth were not suitable for grazing on tough, coarse grass and few predators then existed to stimulate modification by the "survival of the fittest" process of natural selection. Although no fossil remains have yet been found of this hypothetical creature, some think that it will one day be found. It could possibly be given the name Paleohippus (from the Paleocene epoch plus hippus, Greek for horse).

First Known Horse—Eohippus

Approximately 13 million years after the pre-horse existed, the first "known" ancestors of the horse evolved (Figure 35-2). It was a small creature, no bigger than a fox, ranging from 10 to 18 in. tall at the shoulder and had four toes on its foreleg plus a splint (theory has it that this was a remnant of a toe from Paleohippus) and three toes plus a splint (Figure 35-7) on its hind legs. Small, alert ears; dog-like, furry coat; swishing tail, it needed a rather long face to accommodate its 44 teeth (today's horse has 36 to 42 teeth, depending on age and sex).

The distinguishing characteristic of Eohippus as compared to its theoretical predecessor is the direction of growth of the feet. Front feet, adapted from the original five toes, had been reduced to four. One of the toes on the front feet had retracted to become a splint. The rear feet had been modified also, now containing three toes with one split indicating the withering away of toes not needed. Both the front and back toes had been replaced, however, by tiny hooves instead of the original claws.

FIGURE 35-2. (Left) Eohippus or the "Dawn Horse," originally called Hyracotherium when first discovered in Europe, is the first ancestor of which fossil remains have been found. It was no larger than a fox. Note the four toes on the front feet and three toes on the rear feet. (Right) The same fossil remains were the basis for this reconstruction of Hyracotherium, which scientists opposed to evolutionary theories claim was not an ancestor of the horse.

Chap. 35 THE HORSE INDUSTRY

493

Modification—Mesohippus

About 40 million years ago, the evolutionary process began to change the horse even more. At this time, during the Oligocene Epoch, appeared Mesohippus (Figure 35-3). Mesohippus developed larger than its ancestors, standing 24 in. high at the shoulders, about the size of a Collie dog. Its feet were still clinging to pads but tiny hooves had developed on three toes on the front and three toes on the back feet.

Totally New Adaptation—Merychippus

During the Miocene Epoch, about 25 million years ago, Merychippus (Figure 35-4) appeared as a totally new, adapted type of horse. This was a sleek, three-toed animal with erect mane; it was about 40 in. high and had three toes on both front and back feet.

FIGURE 35-3. Mesohippus, stage three in evolutionary development of the horse. The face and legs had begun to lengthen, feet still clung to pads, but tiny hooves had developed on each of the three toes on front and back. Size had increased to about 24 in. at the withers.

FIGURE 35-4. Merychippus, with its erect mane, sleek three-toed arrangement, and greater size (40 in. at withers), appeared as a totally new adaptation.

"One Toe" — Pliohippus

About 10 million years ago, during the Pliocene Epoch, the horse evolved into Pliohippus (Figure 35-5). This ancestor had only one toe, or hoof, on each foot. Pliohippus was the first true monodactyl (one-toed animal) of evolutionary history.

Final Model — Equus

Finally, about 2 million years ago, the horse as we know it today, *Equus caballus,* emerged as a rather large, magnificent creature (Figure 35-6). About 8000 years ago, *Equus* became extinct in North America and was not to return until the Spanish brought horses to the New World in the fifteenth century.

Starting with the mysterious "undiscovered" five-toed ancestor, the horse showed great persistence and adaptability to changes on the earth, ad-

FIGURE 35-5. Pliohippus, the first truly one-toed horse, about 46 in. at the withers, was the nearest approach to modern-day *Equus*.

hering to those mutations that proved to have the greatest chance for future survival (Figure 35-7).

The horse deserted its ancestral home in North America, or died out, or was driven out, for unknown reasons, perhaps because of climatic changes, famine, or a rampant disease. By the year 6000 B.C., it was extinct in the western hemisphere. Prior to that time, horses were magnetically pulled by some mysterious force westward in their subconscious quest for survival. We do not know why horses migrated into Asia but evolutionists think that this movement was the reason for survival of the species. Although they became extinct in North America under mysterious circumstances, the speculation is that they wandered into Mongolia to develop into wild horses that became ancestors of the modern animal. The Arabian, Percheron, Clydesdale, Appaloosa, Zebra, Ass, and all equines had their beginning in North America.

THE THEORY OF SPECIAL CREATION

The only other theory of origin that logically deals with the known laws of science, natural phenomena, and the geological record is the concept of special creation. This theory is the belief that all life forms were created by God and

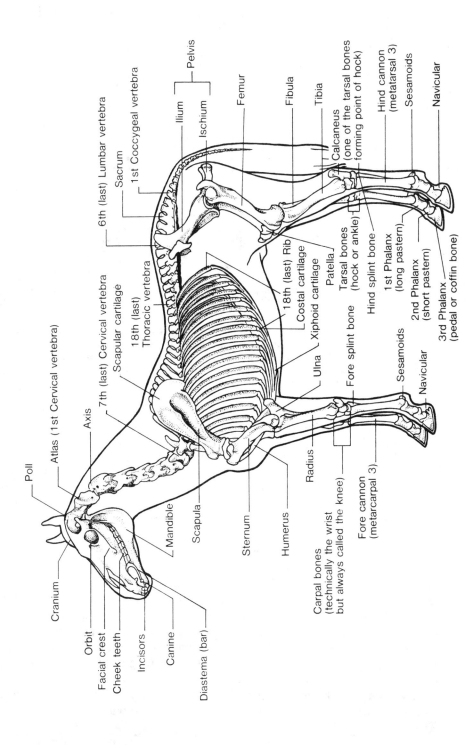

FIGURE 35-6. The skeleton of *Equus* is made up of some 210 bones.

Poll

Cranium

Orbit

Facial crest

Cheek teeth

Incisors

Canine

Diastema (bar)

Mandible

Scapula

Sternum

Humerus

Radius

Carpal bones
(technically the wrist
but always called the knee)

Fore cannon
(metarcarpal 3)

Sesamoids

Navicular

Atlas (1st Cervical vertebra)

Axis

7th (last) Cervical vertebra

Scapular cartilage

18th (last)
Thoracic vertebra

Ulna

Fore splint bone

6th (last) Lumbar vertebra

Sacrum

1st Coccygeal vertebra

Ilium

Ischium

Femur

Fibula

Tibia

Calcaneus
(one of the tarsal bones
forming point of hock)

Hind cannon
(metatarsal 3)

Sesamoids

Navicular

18th (last) Rib

Costal cartilage

Xiphoid cartilage

Patella

Tarsal bones
(hock or ankle)

Hind splint bone

1st Phalanx
(long pastern)

2nd Phalanx
(short pastern)

3rd Phalanx
(pedal or coffin bone)

Pelvis

497

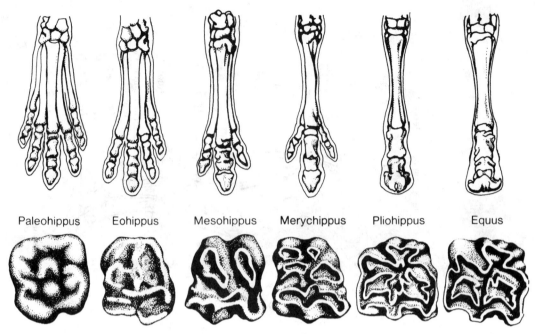

Paleohippus Eohippus Mesohippus Merychippus Pliohippus Equus

FIGURE 35-7. Feet and teeth adaptability was likely the key to evolutionary survival. As the horse changed from hypothetical Paleohippus, a forest browser, to a plains animal needing speed to escape predators, the number of toes was reduced. The change of diet from soft forest browse to abrasive grasses that needed grinding resulted in ridges of enamel on the teeth.

designed to reproduce according to their "kind." If two animals could breed and produce an offspring, they were of the same "kind." These creative "kinds" are not subject to unlimited change as required by Darwin's theory of evolution. That is, a fish cannot evolve into an amphibian, later a reptile, still later a mammal. Special creation does not allow for the possibility of a small, multitoed, rodent-like creature evolving into a horse. The fact of diversification, however, is recognized. Diversification means allowing for considerable variation in the outward characteristics of the offspring (horses, zebras, asses, etc.).

Special creation views the origin of the horse as follows. In the beginning, God created the original equid, referred to in this text as *Proequus*. As he reproduced according to his "kind," the principle of diversification began to work producing variations, as provided for and limited by his genetic boundaries, in each generation. Because of the horse's tremendous capacity for variation within the fixed "kind," early horse was able to adapt quickly to various climates and conditions. This capacity for rapid adaptation provided the animal with a tremendous survival advantage, and only superficial change in appearance.

Proequus lived before the great global catastrophe popularized through the story of Noah and the Ark. Evidence exists world-wide to indicate that a great flood did occur.

After the waters receded, the only horses to survive and repopulate the earth were those carried by Noah on the ark. They began their exodus into the post-flood world from the mountains of Ararat in Turkey where the ark came to rest. It is at this point that the evolutionary theory and special creation theory can merge without serious conflict.

Despite the rich diversity of superficial changes between the ass, horse, and zebra, creationists insist that they are all of the same biblical kind, each can cross with the other, and they are not evolving into other species. Thus, creation scientists hold the Darwinian concept of evolution has not taken place.

ZOOLOGICAL CLASSIFICATION

The horse belongs to the animal kingdom of the phylum *Chordata*, which means they have a backbone, class *Mammalia*, which means they suckle their young, order *Perissodactyla*, which means they are nonruminant, hooved animals, family *Equidae*, and species *Equus caballus*. The donkey, which is used to cross on the horse to produce a mule, is of a different species, *Equus asinus*.

As we might suspect, the horse's first use was not as a beast of burden or for pleasure. A campsite in southern France was unearthed during this century, and archeologists estimated that it contained the bones of 100,000 horses or more. This indicates that primitive man hunted horses and consumed them as a source of food. Later, horses were milked and still are in some parts of the world. Mares may give up to 4½ to 5 gallons of milk per day. In some parts of the world, this milk is prized and preferred over cow's milk. Table 35-1 gives the latest distribution and population data in the United States. Worldwide equine population is approximately 71 million.

TABLE 35-1. AMERICAN HORSE COUNCIL EQUINE POPULATION DATA

State	1984 AHC Estimate	1971 USDA Estimate
Alabama	173,000	129,000
Alaska	11,000	3,000
Arizona	158,000	100,000
Arkansas	171,000	128,000
California	843,000	406,000
Colorado	170,000	125,000
Connecticut	43,000	30,000
Delaware	11,000	8,000
Florida	178,000	145,000
Georgia	165,000	107,000

TABLE 35-1. CONTINUED

State	1984 AHC Estimate	1971 USDA Estimate
Hawaii	13,000	10,000
Idaho	171,000	125,000
Illinois	308,000	300,000
Indiana	130,000	75,000
Iowa	63,000	120,000
Kansas	199,000	150,000
Kentucky	213,000	170,000
Louisiana	190,000	150,000
Maine	37,000	25,000
Maryland	68,000	61,000
Massachusetts	39,000	21,000
Michigan	219,000	169,000
Minnesota	210,000	157,000
Mississippi	189,000	149,000
Missouri	228,000	188,000
Montana	218,000	250,000
Nebraska	108,000	78,000
Nevada	78,000	50,000
New Hampshire	30,000	20,000
New Jersey	44,000	31,000
New Mexico	131,000	90,000
New York	202,000	250,000
North Carolina	152,000	140,000
North Dakota	52,000	35,000
Ohio	237,000	205,000
Oklahoma	294,000	230,000
Oregon	160,000	125,000
Pennsylvania	135,000	98,000
Rhode Island	11,000	9,000
South Carolina	66,000	70,000
South Dakota	116,000	85,000
Tennessee	226,000	180,000
Texas	792,000	625,000
Utah	141,000	95,000
Vermont	17,000	12,000
Virginia	152,000	118,000
Washington	172,000	125,000
West Virginia	53,000	40,000
Wisconsin	148,000	110,000
Wyoming	67,000	48,000
D.C. and Territories[a]	32,000	27,000
Totals	8,197,000	6,197,000

[a]Includes Puerto Rico, Virgin Islands, Guam, and American Samoa.

COMMON HORSE TERMS

Several terms are used in describing the sex, age, or condition of a horse, and these should be kept in mind by the informed horse enthusiast. A *stallion* is the term given to a sexually intact male horse over 3 years old, while the term *stud* refers to a stallion used for breeding or to the breeding establishment or the farm at which the stallion is in service. This has led to the old saying, "King Lightening the 3rd standing at stud." A *mare* is a mature female, and a *filly* is a young female up to 3 years of age. A *gelding* is a castrated male of any age. A *colt* is an intact male up to 3 years of age, a *foal* is a young male or female under 1 year of age, a *weanling* is a young horse of either sex just weaned.

BREEDS OF HORSES

No place in the world are there more varieties and breeds of horses than in the United States. These breeds came from the other parts of the world or were developed in the United States for a specific purpose. Some of them exist very much in the same form in which they originated. Others have changed considerably to meet modern demands.

SADDLE AND HARNESS HORSES

American Saddle Horse

The American Saddle Horse (Figure 35-8) was developed in the plantation states of the south. Almost all American Saddle Horses can be traced back to the great 4-mile race stallion Denmark, foaled in 1839. This breed's outstanding characteristic is that it is comfortable to ride for long distances and is able to work both in harness and under saddle.

Appaloosa

The Appaloosa (Figure 35-9) is a beautiful horse with a distinct color pattern. It was developed by the Nez Perce Indians in what is now Washington state. The name Appaloosa came from the word *palouse,* a French word meaning grassy plain. The Nez Perce developed this horse from Spanish spotted horses and instilled great quality in it. It has been said that the Nez Perce were never defeated on horseback by the U.S. Cavalry during the Indian wars. It was not until the Nez Perce signed a peace treaty with the United States that they were subdued. Legend has it that after the Nez Perce signed the peace treaty and moved to the reservation, the U.S. Cavalry herded all the Appaloosas they

FIGURE 35-8. The American Saddle Horse is easy to ride for long distances and can work in a harness. (Courtesy of the U.S. Department of Agriculture)

could find into a canyon and killed them, thus ending the threat of a fast-moving, proud nation. If this legend is true, it could account for a lack of Appaloosa blood today. The outstanding characteristic of this breed is its color pattern, spots over the hip and loin preferably, although some Appaloosas have no spots at all.

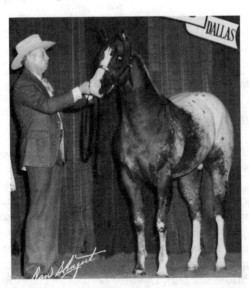

FIGURE 35-9. The Appaloosa has a distinct color pattern. Although some are nearly solid in color, spots over the hip are preferred. (Courtesy of Don Shugart Photography, Grapevine, Texas)

Chap. 35   THE HORSE INDUSTRY

FIGURE 35-10. The Arabian is noted for speed, stamina, and beauty. Shown is Baskette, AHR 36751, 1971 U.S. Top Ten Arabian Mare. (Courtesy of the International Arabian Horse Association)

Arabian

The Arabian (Figure 35-10) probably came from Egypt but was developed in its current form in Arabia and has been bred in the United States since the colonial period. For instance, George Washington rode a gray Arabian charger. The Arabian's outstanding characteristics are speed, stamina, and beauty. It should also be noted that the Arabian has a gentle disposition, seeming almost to prefer companionship with humans, a quality that endears it to its owner. Known as an "easy keeper," the Arabian is able to maintain good condition on low levels of grain or pasture, which makes it an economical horse to own.

Hackney

The Hackney is descended from a Thoroughbred stallion that was crossed on native English mares from Norfolk County. This breed was originally a saddle horse, but it was also used as a harness horse and is considered today to be the leading and one of the few heavy harness horses of the world. The Hackney can be either full sized or a pony in measurement. Used today primarily for

FIGURE 35-11. All Morgan horses today are the descendants of the outstanding stallion Justin Morgan. (Courtesy of the American Morgan Horse Association, Inc.)

showing in the United States, the Hackney once recorded speeds of 17 miles per hour sustained travel over long periods on poor roads.

Morgan

The story of the Morgan (Figure 35-11) is a most unique one. All Morgans today are the descendants of one outstanding stallion named Justin Morgan. As was the custom in New England in the early nineteenth century, an outstanding horse took on the name of its owner. Justin Morgan was a man who developed a rather small colt into a horse that was reputed to be able to outrun, outjump, outpull, or outdo any horse in any contest in which he was pitted. History records that Justin Morgan was truly a remarkable stallion that never lost any race or contest. Even more remarkable was his ability to pass on his characteristics to his descendants. The Morgan today is mostly used as a saddle horse and is considered to be a quiet, reliable all-purpose animal.

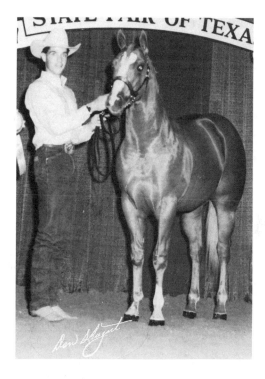

FIGURE 35-12. A grand champion Palomino. (Courtesy of Don Shugart Photography, Grapevine, Texas)

Palomino

The Palomino (Figure 35-12) came by way of California from Spanish stock. It was once known as the horse of queens, in Spain, and commoners were forbidden to ride or own a Palomino because of the royal connections of the breed. The Palomino is not a true breed, but rather a type whose color can always be produced by crossing a chestnut with an albino. A Palomino bred to a Palomino may produce a chestnut, an albino, or another Palomino. Its most outstanding characteristic is its golden color, varying from light to dark; the Palomino's conformation, quality, and performance vary with the particular bloodlines involved.

Pinto

The Pinto color is characteristically brown and white, but the unique black on white color is referred to as *Piebald*. Although not a breed in the conventional sense, the Pinto and Piebald are represented by the American Paint Horse Association (Figure 35-13). An interesting characteristic of the Paint (Pinto or Piebald) is its "glass eyes," so called because they are white or milky in appearance. Glass eyes are not discriminated against in the association and do not indicate poor vision or lack of intelligence as some elder horsemen used to believe.

FIGURE 35-13. Grand Champion Paint Stallion. (Courtesy of Don Shugart Photography, Grapevine, Texas)

American Quarter Horse

The American Quarter Horse (Figure 35-14) is a truly American innovation. At the beginning of America's settlement, towns were hacked out of the wilderness; and for entertainment, people naturally migrated to these centers of

FIGURE 35-14. An excellent Quarter Horse. Close coupled topline and long underline are characteristic of the breed.

supply distribution. Racing between neighbors became a familiar pastime. Most of the open areas led down the main street of town and the distance was usually no longer than one-quarter of a mile. The farmers began to develop a breed of horse that would compete favorably in the local town races, and thus the Quarter Horse had its origin under these circumstances. There are now two types of Quarter Horse: The racing-type Quarter Horse, which developed through the infusion of Thoroughbred blood to improve speed over race track conditions, and the working-type Quarter Horse, which is used for roping and working cattle and pleasure riding. Therefore, there are two distinct types in the judging or selection of Quarter Horses also.

Shetland Pony

The Shetland Pony was originally developed in the Shetland Islands about 100 miles north of Scotland and only 400 miles from the Arctic circle. It was developed originally from native stock for use in the coal mines to pull heavy loads in very small spaces. The native pony was ideally suited for this type of work, and selection procedures enhanced its qualities even more. It is a small, rugged breed that has found favor all over the United States as a child's mount. Aside from its many positive points is a tendency to be short tempered and stubborn, as the Shetland owner has found, but nonetheless, this pony breed has endeared itself to the equine public.

Standardbred

A Thoroughbred stallion named Messenger shaped the Standardbred breed (Figure 35-15). This stallion was crossed on native mares that were natural trotters or pacers. Standardbred blood contains the blood of Arabians, Barbs, Hackneys, and Morgans. The name came from the necessity for the horse to make a standard mile time of 2 minutes, 30 seconds for a trot and 2 minutes, 35 seconds for a pace before it could be registered. Its outstanding characteristic is its strict use as a harness race horse. The Standardbred is to the harness racing industry what the Thoroughbred is to flat racing.

Tennessee Walking Horse

The Tennessee Walking Horse (Figure 35-16) was developed in the state whose name it bears. Some Standardbred, Thoroughbred, American Saddle Horse, and Morgan blood went into its development. The Tennessee Walker is called the gentleman of the equines because of its gentle disposition and way of going. Perhaps its most endearing quality is the gliding sensation felt by the rider. The Tennessee Walker's gaits include a very easygoing flat foot walk, a running walk that covers a great deal of territory but is very comfortable for the rider (the Tennessee Walker does not trot), and the canter.

FIGURE 35-15. The Standardbred horse is used strictly as a harness race horse. (Courtesy of the United States Trotting Association)

FIGURE 35-16. The Tennessee Walking Horse is known for the gliding sensation noted by riders. Shown is the 1974 World's Grand Champion Tennessee Walking Horse, Another Masterpiece. (Courtesy of the Tennessee Walking Horse Breeders and Exhibitors Association)

FIGURE 35-17. The Thoroughbred is noted for its ability to race long distances. Shown here is Ruffian. (Courtesy of the New York Racing Association, Inc.)

Thoroughbred

The Thoroughbred (Figure 35-17) was developed by the royal families of England before being imported to America almost on the heels of the first settlers. Its use in England gave rise to the term "the sport of kings" in reference to horse racing because English noblemen and women bred and raced this magnificent horse. Besides great heart and intelligence, its outstanding characteristics are speed and stamina as evidenced for hundreds of years in flat and jumping races such as the Kentucky Derby and English Grand National Steeplechase.

FIGURE 35-18. Welsh Pony, also known as Welsh Mountain. When crossed with the Welsh Cob, they produce the famous trekking pony, or Welsh Pony of Cob type. (Courtesy of Welsh Pony Society of America)

Welsh Pony

The Welsh breed (Figure 35-18) had its origin in Wales. Rough, mountainous terrain and sparse vegetation led to a natural selection system from which only the most fit survived. This rugged, agile breed is somewhat larger than the Shetland and is unexcelled as a mount for older, more experienced children. It is also popular as a show pony in the East, often crossed with the Thoroughbred. Often described as a miniature coach horse, it is also used, within size limitations, for roadsters, racing, trail riding, stock work, and hunting.

DRAFT HORSES

American Cream Horse

The American Cream Horse (Figure 35-19) had its origin in Iowa and descended from one mare of unknown ancestry that was cream colored and produced all cream-colored offspring. This is an unusual breed because it stems not from a stallion but from a mare that had a great deal of prepotency (ability to stamp a characteristic on offspring). Being the only U.S.-developed draft horse is the outstanding characteristic of this medium-sized breed.

FIGURE 35-19. The only U.S.-developed draft breed is the American Cream Horse. (Courtesy of the American Albino Association, Inc.)

Belgian

The Belgian (Figure 35-20) descended directly from a great wild horse of the Flanders area of Belgium and was first imported for draft use in America in 1866. The Belgian is the most massive of the draft horses, weighing 1900 to 2200 lb or more.

Clydesdale

The Clydesdale (Figure 35-21), which has become familiar to us as the beer wagon horse, came from the valley of the River Clyde in Scotland. This breed

FIGURE 35-20. The Belgian breed is the most popular draft horse in the United States and Canada. (Courtesy of Big Ed's Photos, Davenport, Iowa)

FIGURE 35-21. Among draft horses, the Clydesdale is noted for style, beauty, and action. (Courtesy of Big Ed's Photos, Davenport, Iowa)

is now quite well known in the United States because of its massive size, carrying style, distinctive coloring, and great action. The Clydesdale is considered a real show horse in the area of draft animals.

Percheron

The Percheron (Figure 35-22) horse originated in the northern French district of La Perche. Flemish and Arabian blood appear to have had some influence on the Percheron, and its quality was improved later through selection. Almost exclusively black or gray in color and a popular draft breed, the Percheron

FIGURE 35-22.   The Percheron is noted for its clean head and neck, a long croup, big round hips, and extra heavy muscling in the thighs. (Courtesy of Big Ed's Photos, Davenport, Iowa)

ranks generally behind the Clydesdale and Belgian in use today. Like all the draft breeds, its numbers have declined since farm mechanization took place.

Shire

The Shire (Figure 34-23) came from Lincolnshire, England. Robert Bakewell, the famous English breeder, often called the father of animal science, became interested in the breed and contributed greatly to its development. The unusual characteristic of the Shire is its height, being the tallest of the draft breeds.

FIGURE 35-23. The Shire is characterized by its great size and heavily feathered legs. (Photo by the author)

Suffolk

The Suffolk (Figure 35-24) came from the county of Suffolk, England. The Suffolk is all chestnut in color and noted for its courage and strength. Perhaps this horse exemplifies the desire an animal can have to please its master. A characteristic contest for this breed developed in England. The horses were hitched to a tree, and the winner of the contest was determined by the number of efforts it would make in trying to move the immovable object on voice command. Horses were often on the verge of exhaustion before they would give up in this attempt to respond.

FIGURE 35-24. The Suffolk or Suffolk Punch is a rotund, muscular, "punched-up" breed of English origin. (Coutresy of Kentucky Horse Park, Lexington, Kentucky).

FIGURE 35-25. A superior standard donkey jack. (Courtesy of Don Shugart Photography, Grapevine, Texas)

DONKEYS: JACKS AND JENNETS

A jack is the male of the *Equus asinus* species while a jennet is the female of the species. Compared with the horse, the donkey (Figure 34-25) is much smaller in stature, less subject to founder or injury, much more hardy, less subject to hysteria under stress, and has a longer gestation period (totaling 12 months).

MULES

It has been said that the mule is without pride of ancestry or hope of posterity—this is true. Like most hybrids, it is seldom fertile. However, there have been a few hybrid cases reported in which mare mules had offspring. However, this is the exception and generally it can be said that a mule is sterile, that is, it will not reproduce. The use of the mule was first popularized by George Washington and Henry Clay. George Washington first imported a jack in 1787 as a gift of the king of Spain. This jack had a very docile temperament, and using another jack that had a mean disposition, Washington was able to combine the qualities of both of these superior jacks through his jennets, developing a strain of donkey that was eventually crossed with a stallion to develop the first good-quality mules in the United States. The state of Missouri is famous for its production of high-quality mules and is one of the few states today in which mules are actively shown. Compared with the horse, the mule (Figure 35-26) can withstand much higher temperatures, can eat irregular meals or be self-fed without digestive dangers, is much sounder of foot, and will work in certain areas where most horses would panic. For instance, an

FIGURE 35-26. An excellent example of the standard mule. (Courtesy of Don Shugart Photography, Grapevine, Texas)

object touching the head of a horse can cause it to rear up or to react in such a manner as to injure itself and perhaps its handler, whereas an object touching the head of a mule usually causes it to react in the opposite way—lowering its head very casually away from the sensation of touch.

TYPES AND CLASSES OF HORSES

Horses are classified as light horses (14 hands 2 in. to 17 hands high, usually 900 to 1400 lb), draft horses (14 hands 2 in. to 17 hands 2 in. high, usually weighing more than 1400 lb), and ponies (14 hands 2 in. and below and usually 500 to 900 lb). The hand measurement had its origin as the width of a human hand. It is now a standard size of 4 inches in width. Because draft horses and ponies are not as popular as the light horse industry in the United States, this discussion will be confined to the breakdown of light horses. Light horses can be broken down into seven categories.

The most numerous type of light horse is the *three-gaited horse*, so called because it has been trained to develop a flashy walk, trot, and canter. It is shown with a roached mane and a trimmed tail and can be easily spotted at horse shows. Three-gaited horses are usually American Saddle Horses, Morgans, or Arabians by registry. However, any horse with sufficiently high action may be shown or considered in this category.

The *five-gaited horse* has, in addition to the walk, trot and canter, been trained to exhibit a slow gait and a rack. The latter two gaits are very difficult to teach; thus the five-gaited horse is generally a more expensive animal. It is always shown with a full flowing mane and tail. The five-gaited horse is generally an American Saddle Horse by registry.

Walking horses are mostly of one breed. The Tennessee Walking Horse, developed by plantation owners, holds a unique place in horse shows and fairs and is one of the most popular competitors because of its beauty and gliding appearance.

The first *stock horse* descended from the mustang and had no particular breeding. It is the largest class of light horses, amounting to over 400,000 in the United States and continues to lend its talents to the cattle industry and rodeos. Many stock horses are now registered Quarter Horses or at least are half or quarter bred. Many also carry some Morgan blood. *Polo ponies* are predominantly of Thoroughbred breeding, and few are actually ponies. Once very popular, they are now restricted to special classes of people who are able to afford such an expensive sport. Transportation, training, feed, and related expenses have relegated the popularity to the wealthy. Because polo is played almost constantly at full speed, a player may need three mounts to compete in a match. Some Quarter Horse blood can be found in polo ponies, primarily in western United States.

Hunters and jumpers (Figure 35-27) are usually full or part Thoroughbred, if horses; pony hunters and jumpers are primarily Shetlands, Welsh ponies, small Thoroughbreds, or any combination thereof. Almost every state in the union has an active fox hunting organization. The Thoroughbred or half Thoroughbred is ideally suited for the galloping and jumping fox hunting involves. Horse shows offer competitions that simulate the hunt field as well as other jumping classes where only how high and fast a horse can jump is counted. Another type of jumping competition where the full or part Thor-

FIGURE 35-27.   Hunters and Jumpers are still used for fox hunts in England and the United States.

oughbred is used is the combined training or three-day event where the horse must gallop and jump natural, cross-country obstacles and jump a course of stadium jumps and perform a dressage program on the flat.

Race horses are big business in the United States as well as other parts of the world. In the United States, Thoroughbred race horses compete at distances from six furlongs (three-fourths of a mile) to 2 miles at the most on the flat and up to 4 miles over timber or brush jumps. Standardbreds and Morgans are raced in harness over distances of 1 mile. Quarter Horses are only raced over a quarter of a mile.

MINIATURE HORSES

One of the most unusual breeds of horses is the true Miniature (Figure 35-28), not to be confused with Shetland ponies or the dwarf. A revival of interests by hobbyists in the Miniature Horse has lead to a recent rediscovery of these small animals, which are found in increasing numbers throughout the United States, Republic of South Africa, England, Canada, and Australia.

Stables of Miniature Horses are kept by such notable personalities as Queen Elizabeth II, King Juan Carlos, the Kennedy Family, and entertainer Dean Martin. There are two general types of miniature horses today: the Draft horse and the finer-boned Arabian type.

A registered Miniature Horse with permanent papers cannot exceed 34 in. in height at the withers when fully grown (36 months). The characteristics and confrontation of a full-sized horse, only in miniature, are stressed.

One peculiar aspect related to the world of Miniatures is that the smaller they are, the more valuable they become. Only about 230 horses worldwide are less than 32 in. at the withers, and those that are below this height vary from "expensive to priceless." They vary in size from 26 in. to 34 in.

One of the outstanding characteristics of this tiny pet is that it may be kept, studied, and worked with, just as a mature horse but for only a fraction

FIGURE 35-28. A grand champion miniature stallion. (Courtesy of Don Shugart Photography, Grapevine, Texas)

of the expense. A dozen or so horses of the Miniature variety can be kept for the cost of one standard horse. A bale of hay can last, for one horse, up to three months. They eat everything the larger horses do only in smaller quantities.

This horse is strictly a pet, a hobby for those who are intrigued by its unusual size. Although a few breeders advocate very small children riding them, most owners think that the horse is simply too small to bear much weight on its back. They have been used to pull specially made carts for children, or for miniature harness racing, a more commonly accepted practice.

This horse is ideal for those who have little acreage available for pasture; even a backyard will suffice. Two or three may be stocked per acre and fences need only be about 2 ft high to contain them. They are often hauled to Miniature Horse Shows in the back of station wagons, vans, even riding in the back seat of Cadillacs between two adults.

One of the more notable uses of the Miniature Horse is to draw attention from the news media. "TD," a registered Miniature Horse, has been used as the mascot for a professional football team, the Denver Broncos.

Foaling difficulties are to be expected with the Miniature. As foaling time approaches, mares are often kept in small stables, even baby beds, and watched closely in order to save the foal. They are watched around the clock at foaling time because assistance is inevitably needed.

Why would anyone want such a tiny horse? One breeder explains his hobby of raising the Miniature by stating: "It's like a Porsche, nobody needs one but everyone wants one."

STUDY QUESTIONS

1. The horse probably originated in America. (True, false) _____.
2. The horse is a ruminating mammal. (True, false) _____.
3. The species *Equus asinus* refers to the _____.
4. The proper terminology for a breeding male is a _____.
5. An intact male no older than 3 years is a _____.
6. A young horse under the age of 1 year of either sex is called a _____.
7. The Nez Perce Indians developed the _____.
8. Arabians are known for _____ and _____.
9. The _____ is the leading heavy harness horse.
10. The _____ horse is the product of one outstanding stallion from New England.
11. Spanish royalty originated the _____.
12. A _____, _____ cross will often produce the Palomino color.
13. A "glass eye" is not discriminated against in the _____ horse.

14. There are two types of Quarter Horses, the _____ and _____.

15. The racing Quarter Horse type has some _____ blood.

16. The Shetland Pony was first used for _____.

17. Standardbreds are raced as _____ or _____.

18. The Thoroughbred originated in _____.

19. The largest type of horse is the _____.

20. An Englishman named _____ is credited with developing _____ breed.

21. Crossing a jack on a mare will produce a _____.

22. Light horses usually weigh a minimum of _____ lb.

23. Draft horses exceed _____ hands _____ inches.

24. Five-gaited horses are always shown with a _____ mane and _____ tail.

DISCUSSION QUESTIONS

1. What were horses first used for, what additional uses arose through the early twentieth century, and how are they used today? Really give this some thought (there may be hundreds of uses).

2. What distinguishes a draft horse from a light horse? A light horse from a pony?

3. What is the relationship between the Hackney, Quarter Horse, and Standardbred?

4. Name and describe the five gaits of horses.

5. Compare advantages of using mules versus horses.

6. What is the major difference between Polo Ponies and Shetland Ponies?

7. How does the horse population in your state (or country) compare with U.S. and world figures?

DIRECTORY OF BREED ASSOCIATIONS AND OTHER INFORMATION SOURCES

TENNESSEE WALKING HORSE
  BREEDERS' AND
  EXHIBITORS ASSOCIATION
250 N. Ellington Pky.
P.O. Box 286
Lewisburg, TN 37091

AMERICAN BASHKIR CURLY
  REGISTRY
Mrs. Sunny Martin, Secretary
Box 453
Ely, Nevada 89301

AMERICAN BAY HORSE
ASSOCIATION
P.O. Box 884 F
Wheeling, Illinois 60090

AMERICAN PART-BLOODED
HORSE REGISTRY
Ms. Barbara Bell
4120 S.E. River Drive
Portland, Oregon 97222

KENTUCKY HORSE PARK
R.R. 6, Iron Works Road
Lexington, Kentucky 40511

AMERICAN SADDLE HORSE
BREEDERS' ASSOCIATION,
INC.
C. J. Cronan, Jr.
929 South Fourth Street
Louisville, Kentucky 40203

AMERICAN QUARTER PONY
ASSOCIATION
Linda Grim, Secretary
New Sharon, Iowa 50207

NATIONAL QUARTER PONY
ASSOCIATION, INC.
Edward Ufferman
Rt. 1
Marengo, Ohio 43334

AMERICAN WALKING PONY
ASSOCIATION, INC.
Mrs. Joan H. Brown
Rt. 5, Box 88, Upper River Road
Macon, Georgia 31201

THE SUGARMAN RANCH,
ANDALUSIANS
William C. Hass, VP, Business
Affairs
300 West Potrero Road
Hidden Valley
Thousand Oaks, California 91360

INTERNATIONAL
ARABIAN HORSE
ASSOCIATION
224 E. Olive Avenue
P.O. Box 4502
Burbank, California 91503

INTERNATIONAL BUCKSKIN
HORSE ASSOCIATION, INC.
P.O. Box 357
St. John, Indiana 46373

AMERICAN BUCKSKIN
REGISTRY ASSOCIATION,
INC.
Randi Spears, Executive Secretary
P.O. Box 1125
Anderson, California 96007

HORSES
The Ridge Press
New York

DRAFT HORSE JOURNAL
Maurice Telleen, Editor
Rt. 3
Waverly, Iowa 50677

COLORADO RANGER HORSE
ASSOCIATION
John E. Morris, President
7023 Eden Mill Road
Woodbine, Maryland 21797

AMERICAN CONNEMARA
PONY SOCIETY
Mrs. John E. O'Brien, Secretary
Hoshiekon Farm R.D. 1
Goshen, Connecticut 06756

GALICENO HORSE BREEDERS
ASSOCIATION
Roxanne McDole, Secretary
111 East Elm Street
Tyler, Texas 75702

THE AMERICAN HORSE
COUNCIL
1700 K Street, N.W.
Washington, DC 20006

AMERICAN GOTLAND HORSE
ASSOCIATION
R.R. 2, Box 181
Elkland, Missouri 65644

THE HALF SADDLEBRED
REGISTRY OF AMERICA
Roberta N. Busch, Secretary
660 Poplar Street
Coshocton, Ohio 43812

AMERICAN INDIAN HORSE
REGISTRY, INC.
Rt. 1, Box 64
Lockhart, Texas 78644

THE ROYAL INTERNATIONAL
LIPIZZANER CLUB
Mrs. Margaret D. Fleming,
Secretary
Rt. 7
Columbia, Tennessee 38401

INTERNATIONAL MINIATURE
HORSE REGISTRY
P.O. Box 907
Palos Verdes Estates, California
90274

MISSOURI FOX TROTTING
HORSE BREED ASSOCIATION
Ava, Missouri 65608

FREDA BEISNER, MISSOURI
FOX TROTTING HORSES
United Missouri Bank of Hickman
Mills
11702 Hickman Mills Drive
Kansas City, Missouri 64134

THE AMERICAN MORAB
HORSE ASSOCIATION
P.O. Box 687
Clovis, California 93613

BOSTON MORGANS
Mrs. J. William Crawford
7400 S.W. Boeckman Road
Wilsonville, Oregon 97070

AMERICAN MORGAN HORSE
ASSOCIATION, INC.
Oneida Co. Airport Industrial Park
Box 1
Westmoreland, New York 13490

AMERICAN MUSTANG
ASSOCIATION, INC.
P.O. Box 338
Yucaipa, California 92399

THE NORWEIGIAN FJORD
HORSE ASSOCIATION OF
NORTH AMERICA
R.R. 1, Box 370
Round Lake, Illinois 60073

PONY OF THE AMERICAS
CLUB
P.O. Box 1447
Mason City, Iowa 50401

PASO FINO OWNERS AND
BREEDERS ASSOCIATION,
INC.
Rosalie MacWilliams, Registrar
P.O. Box 764
Columbus, North Carolina 28722

TIOVIVO RANCH, PASO FINOS
Elma H. White, Ph.D.
Rt. 1, Box 185
Weimar, Texas 78962

AMERICAN PASO FINO
HORSE ASSOCIATION, INC.
P.O. Box 2363
Pittsburgh, Pennsylvania 15230

PERCHERON HORSE
ASSOCIATION OF AMERICA
R.R. 1
Belmont, Ohio 43718

THE AMERICAN PERUVIAN
PASO HORSE REGISTRY
Rt. 5, Box 361
Harrison, Arkansas 72601

AMERICAN ASSOCIATION OF
OWNERS AND BREEDERS OF
PERUVIAN PASO HORSES
P.O. Box 2035
California City, California 93505

INTERNATIONAL PERUVIAN
PASO HORSE ASSOCIATION
P.O. Box 1194
Greenville, Texas 75401

PINTO HORSE ASSOCIATION
OF AMERICA, INC.
7525 Mission Gorge Road, Suite C
San Diego, California 92120

AMERICAN QUARTER HORSE
ASSOCIATION
Box 200
Amarillo, Texas 79605

AMERICAN SHETLAND PONY
CLUB
P.O. Box 468
Fowler, Indiana 47944

THE UNITED STATES
TROTTING ASSOCIATION
750 Michigan Avenue
Columbus, Ohio 43215

AMERICAN SUFFOLK HORSE
ASSOCIATION
Mary Margaret Read, Secretary
15 B Roden
Wichita Falls, Texas 76311

AMERICAN TARPAN STUD
BOOK ASSOCIATION
Ellen J. Thrall
1760 Rehoboth Church Road
Griffin, Georgia 30223

THE JOCKEY CLUB
300 Park Avenue
New York, New York 10022

MARGARET S. GAFFORD,
TRAKEHNERS
Sankt Georg Farm
Rt. 4, Box 177
Petersburg, Virginia 23803

AMERICAN TRAKEHNER
ASSOCIATION, INC.
Leo H. Whinery
P.O. Box 268
Norman, Oklahoma 73070

WELSH PONY SOCIETY OF
AMERICA
P.O. Box 2977
Winchester, Virginia 22601

AMERICAN ALBINO
ASSOCIATION, INC.
Box 79
Crabtree, Oregon 97335

SPANISH-BARB BREEDERS
ASSOCIATION
Box 7479
Colorado Springs, Colorado 80907

RACKING HORSE BREEDERS'
ASSOCIATION OF AMERICA
Helena, Alabama 35080

APPALOOSA HORSE CLUB,
INC.
Box 8403
Moscow, Idaho 83843

THE PALOMINO HORSE
ASSOCIATION, INC.
P.O. Box 324
Jefferson City, Missouri 65101

AMERICAN PAINT HORSE
ASSOCIATION
P.O. Box 18519
Fort Worth, Texas 76118

NATIONAL CUTTING HORSE
ASSOCIATION
Box 12155
Fort Worth, Texas 76116

Chapter Thirty-Six

HORSE REPRODUCTION AND MANAGEMENT

The horse is by nature a high spirited, freedom loving, nomadic animal. In the wild, the horse's reproduction approaches 90% efficiency or more. Under the domestic conditions imposed by humans, the rate falls far short of this mark. Lack of proper exercise, exposure to disease, and many other factors contribute to a poor showing in the race for efficient conception and foaling rates. Therefore, knowledge of reproductive traits is especially important if the horse breeder is to be successful.

BREEDING HINTS

A filly normally will reach puberty at 12 to 15 months of age but should not be bred before the age of 2 years and preferably not before 3 years. A filly bred at the earlier age traditionally has a poor conception rate the following year. When bred at 3 years and properly cared for, a mare can be expected to produce 10 to 12 foals during its lifetime. A mare 20 years old or older can often produce foals.

Estrus

A mare's estrous cycle is normally 21 days, although it may range from 10 to 37 days. The estrous period averages 4 to 6 days but it too may range from one day to continuous heat for some maiden mares. Estrous signs should be observed and breeding initiated only when a normal cycle is obvious to achieve best results. Signs of estrus are nervousness, a desire for company, frequent urination, nickering, and a swelling and movement of the vulva.

Teasing

Various schedules and techniques are used in the unique handling of mares to ensure maximum conception when breeding. To more effectively detect true estrus, the mare suspected of "coming in" is allowed close proximity to a teaser stallion. The teaser may be on one side of a teasing board, restrained by halter in the field, or just led through the barn. Some establishments use Shetland stallions as teasers because there is less chance, because of size differences, of an accidental mating.

Actions of the mare as observed by an experienced handler indicate the most logical time to breed. The mare will empty its bladder frequently and exhibit restlessness. Most handlers think that talking to a mare during this nervous time helps to calm it. This can be especially important in valuable brood mares to prevent injuries, so the phrase "Whoa Nellie" is quite common around the stud barn. By teasing mares to detect estrus, a good handler may get 75% of the mares in foal.

Breeding

Because conception is low in domestic horses to begin with, the strict observance of sanitary measures and other precautions is important in successful breeding. The stallion and mare's genitals are washed with warm water and soap. A tail bandage made of flannel or gauze is applied to the mare and she may be restrained by twitch and hobbles. These precautionary measures are needed only about 10% of the time but are recommended to prevent possible harm to the male from biting or kicking.

Time to Breed

Ovulation occurs in the latter portion of the estrous period. The egg is viable for about 6 hours and sperm from the stallion will live about 30 hours in the reproductive tract. It is recommended therefore, that mares be served daily or every other day beginning with the third day of estrus.

Gestation

The average gestation is 336 days, with normal variations reported from 315 to 350 days. Mares tend to follow a pattern, some consistently foaling early, others consistently late. Once the pattern is determined, an experienced horse-person can make more accurate predictions.

FOALING

A critical time for both the mare and foal occurs at parturition. While some births require no assistance at all, other cases develop where life hangs by a thread and human assistance prevents a fatal situation. For unknown reasons, nearly all mares foal during dark hours, but as inconvenient as it may be, the experienced horse breeder is on hand to assist the mare.

Signs of Approaching Parturition

One of the first signs that foaling is near is the the mare's beginning to "make bag," the gradual enlargement of the mammary glands. Typically, *waxing* occurs (a waxy substance appears at the end of each teat) prior to foaling. Usually within 12 to 24 hours of birth, the wax softens, falls away, and milk may drip, even stream from the teats. It should be mentioned that these signs are not always on prompt schedule. Occasionally, a mare will wax two or three times and may have milk streaming from the teats 10 days before parturition. At about the same time, the mare will show a soft swelling and noticeable relaxation of the muscles of the vulva. The muscles and ligaments associated with the pelvis relax, making the mare appear loose about the hips.

The mare will usually leave other horses, if in the pasture, seeking solitude. It may appear in a bad temper, pinning its ears back, kicking when other horses approach. The mare carries its tail away from the body slightly, urinates frequently, may bite at its sides, and alternates lying and standing. Sweating profusely is normal at this time. The water bag (placenta) usually breaks at this point, discharging 2 to 5 gallons of fluid, helping to lubricate the birth canal. If the foal is in the right position (Figure 36-1), involuntary muscle contractions of the uterus and abdominal muscles (labor) may start and birth can occur with surprising ease.

Care at Parturition

Only a few items need be on hand at birth—a clean, well-bedded box stall; a source of light; clean, hot water and soap (Ivory is suggested); tail bandages; approved disinfectant; a suitable navel dressing (iodine, merthiolate); an enema bag, and a laxative such as Milk of Magnesia.

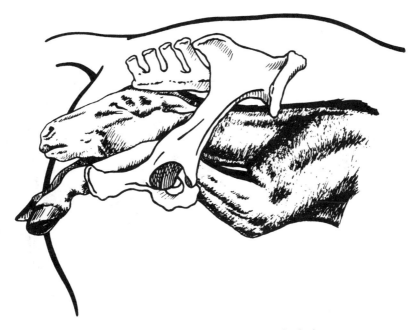

FIGURE 36-1. Normal presentation of a foal.

When there is no doubt that the mare is about to foal, the attendant applies a tail bandage (flannel or gauze) and waits hopefully for nature to do the rest. No assistance should be given until it is obvious that help is needed. It is often necessary to tie a soft, thick rope around the foal's forefeet and pull to assist the mare. If both feet and the muzzle of the foal can be seen, then all is well. Birth may take 10 minutes to an hour.

If after a reasonable amount of time is allowed and the muzzle or one or more legs cannot be seen, the attendant may need to wash his or her hands and the vulva with soap, water, and antiseptic solution and make an examination for complications. Some common difficulties are illustrated by Figure 36-2 (head turned to one side), Figure 36-3 (the dorso-transverse position), and Figure 36-4 (breech position). The foal must be rotated to the normal position (Figure 36-1) or suffocation may result. Nature supplies the foal with an umbilical cord of sufficient length to completely clear the birth canal only if birth occurs in the normal position. If the cord is crushed by a backward birth, oxygen is cut off and breathing may be stimulated in an anerobic (without air) environment, resulting in suffocation. Suffocation may also occur in other positions if delivery of the head is delayed.

The new foal born with or without assistance should be checked for breathing restrictions. Any membranes or fluids covering the mouth or nostrils should be cleared. Allow 2 or 3 hours for the foal to gain strength and see that it nurses. It is absolutely necessary that the foal receive colostrum milk to supply its system with the necessary antibodies, vitamins, and energy required to start and maintain life. The antibodies that provide protection disap-

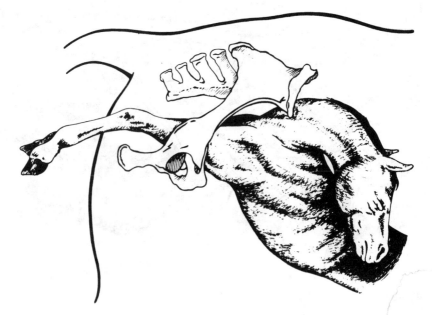

FIGURE 36-2. A foal with its head turned to one side.

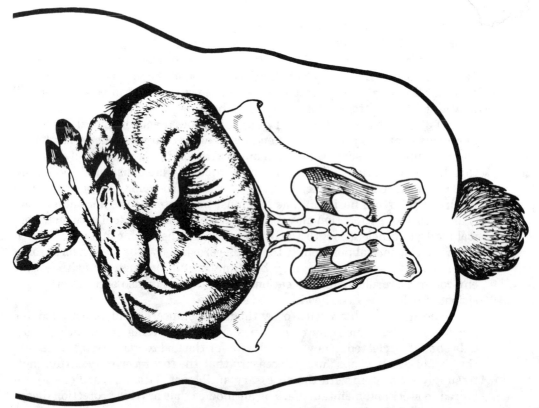

FIGURE 36.3. The dorso-transverse position of a foal.

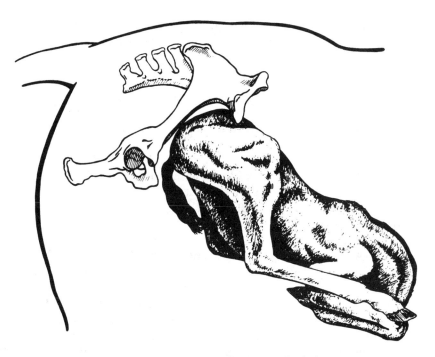

FIGURE 36-4. The breech position of a foal.

pear from the milk within 24 to 36 hours. The foal, unlike the calf that receives some immunity through blood circulation, must rely almost entirely on colostrum. It is also a common practice to administer a tetanus (lockjaw) antitoxin and a penicillin–streptomycin package loaded in a disposable syringe.

An enema with warm, soapy water should be given the foal and a dose of Milk of Magnesia. This is standard practice to ensure the passing of *meconium*, a toxic substance developed in the large intestine just prior to birth and sometimes absorbed if constipation occurs. After this passage, the problem does not develop again. Colic may also develop if early elimination is not induced.

The navel cord should break by itself; it should not be tied because of the possibility of causing a fatal case of navel ill. To prevent entry of harmful organisms to the bloodstream, as soon as the break occurs the navel stump and surrounding area are treated with iodine or merthiolate daily until they are dried.

Post-Foaling Care

The placenta, ideally shed within 3 hours, should be spread out and checked for missing pieces because retention of even a piece the size of a hand can cause founder or infection of the mare. If all or part of the placenta is not shed within 6 to 8 hours, removal by a veterinarian is recommended.

Moderate exercise is important in stimulating the mare's uterus to tone up and in developing the foal's protective mechanism. A paddock should be provided for this purpose, and eventually the pair can be turned out to pasture during the day or night.

RE-BREEDING

Horses exhibit *foal heat*, more often called *nine-day heat*, the first estrus at which re-breeding is possible. In actual practice, this foal heat may occur at 5 to 10 days after foaling or even at a later date. On the average, 9 days has been established by breeders as the thumb rule to follow and professional establishments plan to breed the mare at this time barring complications of the genital tract.

A mare that shed the placenta within 3 hours of foaling and that has no visible signs of genital lacerations or bruises due to foaling and no signs of infection qualifies because the primary objective of breeding is to get her in foal. If this estrus is not used to advantage, it may be 50 to 60 days or longer before the next estrus. Also, there are occasional mares that will not appear to be in estrus even to the teaser while nursing a foal, except at foal heat. Conception rate at this time averages only about 25%, far from desirable, but any advantage is appreciated by breeders, no matter how slight. Mares not conceiving at foal heat continue in the regular breeding program cycle previously described.

ABORTION

Mares are subject to a higher incidence of abortion compared to other species, further compounding the difficulties faced by breeders. It is important to have a veterinarian determine the cause of any aborted fetus (foal) by examination of it because some forms are caused by bacteria or virus and thus subject to spread throughout the herd. Preventative vaccines or other sanitary measures should follow accurate diagnosis.

Two causes of abortion are not contagious and not uncommon. A twisted navel cord caused by fetal rotation that shuts off blood circulation is fatal. It is interesting to note that the other cause is of little concern in other species—the incidence of twinning. However, for reasons not clearly understood, 90% of equine twins are aborted.

PREGNANCY EXAMINATION

To get the highest efficiency possible, the experienced breeder checks mares suspected of conceiving by rectal palpation at about 60 days of gestation. Although other biological tests using rabbits, rats, or frogs are sometimes employed, they are more time-consuming and less accurate at early stages of pregnancy. This step is necessary because mares safely in foal sometimes react to the teaser and if bred may abort. Of course those females diagnosed open or not in foal continue in the routine until conception is verified.

ARTIFICIAL INSEMINATION

Artificial insemination has been a common practice in breeding horses since 1938. When properly fed and managed, a young stallion is ready for service at 24 months of age. Exercise is most important in the stallion for maintenance of semen quality, and the equivalent of a half-day's work under the saddle or harness is recommended. The artificial vagina is the normal method of semen collection from stallions. Semen is diluted, frozen, and stored as for cattle. Thawing and actual insemination are similar to the techniques used in cattle insemination.

STUDY QUESTIONS

1. Breeding efficiency in the wild is much _____ than domestic breeding.
2. The age of puberty in the filly is _____ months of age.
3. It is generally recommended not to breed fillies before _____ years of age.
4. The estrous cycle averages _____ days and the estrous period averages about _____ days.
5. Good signs of approaching estrus are frequent urination, nervousness, and a large swelling of the mammary glands. (True, false) _____.
6. The most accepted method of detecting estrus is through the use of a _____ stallion.
7. Ovulation occurs near the _____ day of estrus.
8. Mares should be served _____ beginning with the _____ day of estrus.
9. The average length of pregnancy is _____ days.
10. The act of giving birth is called _____.
11. A sign of parturition within 24 hours is _____ of the teats.
12. After parturition has begun, the _____ will break to provide lubrication.
13. A backward birth is dangerous because _____ can occur because of a crushed _____.

14. If a foal is in the normal position, the _____ and _____ should be visible in the early stages of birth.
15. _____ or "first milk" is necessary for the young foal to receive _____ necessary to provide natural immunities to disease.
16. Iodine is used to treat the navel cord to prevent _____.
17. A piece of retained placenta in the mare can cause _____, _____.
18. To ensure the passage of meconium from the foal, a _____ is administered.
19. Foal heat occurs _____ days after parturition.
20. The incidence of twinning in mares will almost always result in _____.
21. The most common way to test pregnancy in mares is _____ _____ at about _____ days of gestation.

DISCUSSION QUESTIONS

1. Define the following:
   a. foal heat   d. teasing
   b. meconium   e. estrus
   c. parturition   f. estrous period
2. What preventive measures should be taken with the newborn foal to ensure good health and life?
3. Give age recommendations for breeding light horse fillies, signs of approaching estrus, when to breed and how often, parturition signs, and recommended care at foaling.
4. What are the common causes of noncontagious abortion? Of contagious abortion?
5. Why should the placenta be observed after foaling? How could you tell if a problem could occur and what would you do to prevent it?

Chapter Thirty-Seven

FEEDING HORSES

Nutrient requirements of horses are not unlike those of cattle (Chapter 4). However, equines are used almost exclusively for sport, work, and recreation rather than for meat or milk. Owners should also be cautioned against feeding prepared cattle feeds to horses because of the possibility of nonprotein nitrogen (NPN) content. Cattle and other ruminants can utilize some NPN, which cheapens the ration, but horses have a very low tolerance for it and may develop toxic reactions. Special prepared horse rations are commercially available to meet every safety and dietary need. Considerable skill and judgment are required to feed according to nutritive needs under fluctuating requirements.

The horse, although it eats hay and grass, is not a ruminant. Its digestive system is unique among domestic farm animals (Figure 37-1).

The horse does not ruminate and is physiologically unable to regurgitate. It is able to make use of forages because of the large cecum that houses microorganisms that break down the fibrous feed, releasing absorptive nutrients. Because of the location of the cecum near the end of the system, it is understandable that digestive efficiency is not nearly as good as in ruminants. This is evident in the droppings of horses, which are comparatively drier and bulkier because of the passage of undigested fiber.

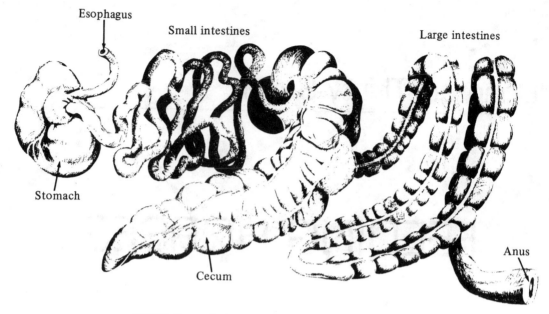

FIGURE 37-1.   The horse's digestive system. Note the size of the cercum.

Because the horse is less efficient as a forage converter, it is necessary to provide nutrients through concentrates (grains) that provide energy that is absorbed through the stomach. If grain is not fed, horses tend to eat more hay or grass to compensate for a lack of energy, increasing the size of the cecum and consequently, the barrel. Some horses have a greater tendency than others to expand in this manner and are often referred to as hay burners.

Founder and colic (see Chapter 39) are common disorders related to diet because horses require high energy but need some bulk to prevent compaction in the relatively small stomach, which leads to colic and/or founder. For this reason, horse breeders have long recognized oats as a safe, superior grain source for their animals because it is high in nutrition and moderately bulky. The horse owner's admiration for the grain is excelled only by the horse's desire to consume it. Any breeder or groom will testify that "mares eat oats" and few digestive problems occur when used with good judgment.

NUTRIENT NEEDS OF HORSES

Specific feed requirements vary depending on the use of the horse. Idle horses require less energy than working horses, lactating mares require more protein than open brood mares, and nearly all requirements are higher for young as compared to mature horses.

Protein

Although not a ruminant, the horse does synthesize considerable energy in the cecum, and protein utilization is doubtful, making high quality or a variety of protein sources desirable. The most popular source of protein for horses is undoubtedly linseed oil meal because of its resulting bloom (slick, shiny coat). Cottonseed meal should be used sparingly because if fed at rates exceeding the requirement, toxicity problems could occur because of gossypol content. Horses are resistant to gossypol only at lower levels. The total protein requirements vary for foals, gestating brood mares, and mature horses (Tables 37-1 to 37-4).

Energy

The major consideration in growing, lactating, or working horses is energy. Pasture or hay meets this need to a limited extent. Grain is the cornerstone to building energy levels that will meet body demands. Although oats is the top grain, corn and cob meal (ground ear corn), barley, and wheat have a place in building a ration. Tables 37-1 to 37-4 give requirements in terms of digestible energy expressed in megacalories.

Minerals

Function and deficiency symptoms are about the same as previously discussed for cattle. Calcium and phosphorus requirements are especially critical in horses because strong bones are so necessary in work. A lack of either may cause rickets in the young or osteomalacia in mature mounts. The calcium/phosphorus ratio is recommended to be maintained at 1.1:1 to 1.4:1 to prevent the occurrence of this condition. Horses require rations with 0.60 to 0.70% calcium and phosphorus. Bonemeal is a recommended source and may be mixed 2:1 with salt and fed free choice. A more complex mineral mixture is illustrated in Table 37-5.

The feeding of iodized, trace mineral salt should stave off problems caused by other mineral-induced disorders. Salt should also make up 0.5 to 1.0% of the feed. Because the horse perspires to cool body temperature, salt is vital.

Vitamins

Luxuriant pasture from fertile soils offers the best insurance against vitamin shortage. In the absence of pasture, green, leafy forages from well-fertilized lands, offer a palatable substitute. Most fat-soluble (A, D, E, and K) vitamins as well as the B vitamins may be derived from natural sources. However, Table 37-6 shows a sure source of vitamins.

TABLE 37-1. NUTRIENT REQUIREMENTS OF HORSES (Daily Nutrients per Horse)

Body Weight (lb)	Daily Feed (lb)	Digestible Energy (Mcal)	Digestible Protein (lb)	Calcium (g)	Phosphorus (g)
For Maintenance					
800	12.9	12.9	0.55	18.1	12.3
900	14.1	14.1	0.60	19.8	13.4
1000	15.3	15.3	0.65	21.5	14.6
1100	16.4	16.4	0.71	23.1	15.6
1200	17.5	17.5	0.75	24.6	16.7
For Light Work (1 to 3 hr of riding or driving per day)					
800	16.9	18.6	0.71	23.8	16.1
900	18.5	20.3	0.78	26.0	17.6
1000	20.1	22.1	0.84	28.3	19.2
1100	21.5	23.7	0.90	30.2	20.5
1200	23.0	25.3	0.97	32.3	21.9
For Medium Work (3 to 5 hr of riding or driving per day)					
800	17.9	21.5	0.73	25.2	17.1
900	19.6	23.5	0.80	27.6	18.7
1000	21.2	25.4	0.87	29.8	20.2
1100	22.8	27.3	0.93	32.1	21.7
1200	24.3	29.1	1.00	34.2	23.2
For Hard Work (above 5 hr of riding or driving per day)					
800	18.7	24.3	0.75	26.3	17.8
900	20.5	26.6	0.82	28.8	19.5
1000	22.2	28.8	0.89	31.2	21.2
1100	23.8	30.9	0.95	33.5	22.7
1200	25.4	33.0	1.02	35.7	24.2

Source: Courtesy of John P. Baker, University of Kentucky.

FEEDING HINTS

Table 37-7 gives example rations for the professional horse owner. The average horseperson doing average recreational riding may do just as well on a simple ration of 95% oats, 5% linseed meal for a concentrate source or half oats, half sweet feed (premixed corn, soybean, linseed, wheat, and vitamins and mineral supplements) and a bright, clean, leafy hay for roughage.

Owners should be aware of sand colic, a very common disorder caused by feeding hay on the ground. This is a digestive disturbance that can easily be prevented by providing hay in bunks or in the characteristic hay nets seen at eye level in many horse stalls.

Each horse should be fed as an individual because some are gluttons, others shy eaters. Feed boxes should always be kept clean of moldy feed. A horse should clean up its daily concentrate allowance in about 30 minutes, and may be fed once, twice, or three times daily but always at the same time, even

TABLE 37-2. NUTRIENT REQUIREMENTS OF HORSES (Daily Nutrients per Horse)

Body Weight (lb)	Average Daily Gain (lb)	Daily Feed (lb)	Digestible Energy (Mcal)	Digestible Protein (lb)	Calcium (g)	Phosphorus (g)
For Growth (1200 lb mature wt.)						
200	2.2	7.3	10.2	1.13	26.8	17.2
400	2.1	11.2	14.5	1.20	38.1	23.9
600	1.6	13.7	16.4	1.16	40.4	25.5
800	0.8	14.7	16.2	0.90	33.3	22.0
1000	0.4	17.0	17.0	0.82	30.8	20.8
1200	—	17.5	17.5	0.75	24.6	16.7
For Mares, Last 90 Days of Pregnancy						
800	—	12.5	13.7	0.76	21.6	16.4
900	—	13.6	15.0	0.83	23.4	17.9
1000	—	14.7	16.2	0.89	25.3	19.3
1100	—	15.8	17.4	0.96	27.2	20.8
1200	—	16.9	18.6	1.03	29.1	22.2
For Mares, Peak of Lactation						
800	—	19.7	23.7	1.50	42.0	33.1
900	—	21.3	25.6	1.62	45.4	35.7
1000	—	22.9	27.5	1.74	48.8	38.4
1100	—	24.5	29.4	1.85	52.2	41.1
1200	—	26.0	31.2	1.97	55.4	43.6

Source: Courtesy of John P. Baker, University of Kentucky.

on weekends. Not eating is the first sign of illness. Fast eaters may be slowed down (to prevent digestive problems) by placing a few large round stones in the feed box so that the horse has to eat around them.

Rations should never be changed abruptly; allow a week to gradually switch to a new mixture to prevent digestive problems. Feeding is as much an art as it is a science.

Exercise is important in overall condition. Horses that are penned in a small place often resort to *cribbing* (eating and sucking on fences), which is not a sign of nutritional deficiency but boredom. A spirited animal must be allowed adequate exercise.

Care should be taken not to let horses take on a full fill of water after vigorous exercise because of the possibility of founder or colic. Always walk a horse to let it cool off before drinking.

As stated previously, each horse must be fed according to its needs as determined by observation of its condition, preferably by the same groom or stable manager over an extended period of time. Some helpful rules of thumb to serve as a base follow:

1. Horses doing light work (under 3 hours)—0.5% of body weight in concentrate (see the sample rations in Table 37-7), 1 to 1.25% hay.

TABLE 37-3. NUTRIENT REQUIREMENTS OF HORSES (Nutrient Concentration in Ration)

Body Weight (lb)	Daily Feed (lb)	Digestible Energy per Pound of Feed (Mcal)	Crude Protein in Feed (%)	Digestible Protein in Feed (%)	Calcium in Feed (%)	Phosphorus in Feed (%)	Vitamin A per Pound of Feed (IU)
For Maintenance							
800	12.9	1.0	9.0	4.3	0.31	0.21	750
900	14.1	1.0	9.0	4.3	0.31	0.21	750
1000	15.3	1.0	9.0	4.3	0.31	0.21	750
1100	16.4	1.0	9.0	4.3	0.31	0.21	750
1200	17.5	1.0	9.0	4.3	0.31	0.21	750
For Light Work (1 to 3 hr of riding or driving per day)							
800	16.9	1.1	8.1	4.2	0.31	0.21	750
900	18.5	1.1	8.1	4.2	0.31	0.21	750
1000	20.1	1.1	8.1	4.2	0.31	0.21	750
1100	21.5	1.1	8.1	4.2	0.31	0.21	750
1200	23.0	1.1	8.1	4.2	0.31	0.21	750
For Medium Work (3 to 5 hr of riding or driving per day)							
800	17.9	1.2	7.7	4.1	0.31	0.21	750
900	19.6	1.2	7.7	4.1	0.31	0.21	750
1000	21.2	1.2	7.7	4.1	0.31	0.21	750
1100	22.8	1.2	7.7	4.1	0.31	0.21	750
1200	24.3	1.2	7.7	4.1	0.31	0.21	750
For Hard Work (above 5 hr of riding or driving per day)							
800	18.7	1.3	7.1	4.0	0.31	0.21	750
900	20.5	1.3	7.1	4.0	0.31	0.21	750
1000	22.2	1.3	7.1	4.0	0.31	0.21	750
1100	23.8	1.3	7.1	4.0	0.31	0.21	750
1200	25.4	1.3	7.1	4.0	0.31	0.21	750

Source: Courtesy of John P. Baker, University of Kentucky.

2. Horses doing medium work (3 to 5 hours)—1% of body weight in concentrate, same amount of hay.

3. Horses doing hard work (over 5 hours)—1¼% of body weight in concentrates, 1% in hay. Breeding stallions should be fed as a horse doing hard work and should be exercised thoroughly prior to and during the breeding season.

4. Pregnant mares—0.75 to 1.5% of body weight in concentrate, 0.75 to 1.5% of body weight in hay.

5. The total combined consumption of concentrate and hay should be kept in the range of 2 to 2.5% of body weight.

6. Foals—The mare's milk will sustain adequate growth until about 6 weeks of age, when it is advisable to begin creep feeding. The foal should be consuming about 0.75% of its body weight in a good legume or grass hay and an equal amount of concentrate creep feed. The concentrate mix

TABLE 37-4. NUTRIENT REQUIREMENTS OF HORSES (Nutrient Concentration in Ration)

Body Weight (lb)	Average Daily Gain (lb)	Daily Feed (lb)	Digestible Energy per Pound of Feed (Mcal)	Crude Protein in Feed (%)	Digestible Protein in Feed (%)	Calcium in Feed (%)	Phosphorus in Feed (%)	Vitamin A per Pound of Feed (I.U.)
For Growth (1200 lb mature weight)								
200	2.2	7.3	1.4	19.4	15.5	0.81	0.52	600
400	2.1	11.2	1.3	16.5	10.7	0.75	0.47	700
600	1.6	13.7	1.2	14.2	8.5	0.65	0.41	800
800	0.8	14.7	1.1	12.2	6.1	0.50	0.33	1000
1000	0.4	17.0	1.0	10.2	4.8	0.40	0.27	1000
1200	—	17.5	1.0	9.0	4.3	0.31	0.21	750
For Mares, Last 90 Days of Pregnancy								
800	—	12.5	1.1	11.7	6.1	0.38	0.29	1500
900	—	13.6	1.1	11.7	6.1	0.38	0.29	1500
1000	—	14.7	1.1	11.7	6.1	0.38	0.29	1500
1100	—	15.8	1.1	11.7	6.1	0.38	0.29	1500
1200	—	16.9	1.1	11.7	6.1	0.38	0.29	1500
For Mares, Peak of Lactation								
800	—	19.7	1.2	12.0	7.6	0.47	0.37	1200
900	—	21.3	1.2	12.0	7.6	0.47	0.37	1200
1000	—	22.9	1.2	12.0	7.6	0.47	0.37	1200
1100	—	24.5	1.2	12.0	7.6	0.47	0.37	1200
1200	—	26.0	1.2	12.0	7.6	0.47	0.37	1200

Source: Courtesy of John P. Baker, University of Kentucky.

should contain a minimum of 12% crude protein and at least 5% fiber. A standard ratio is 9 parts grain to 1 part oil meal. No more than one-half the grain mix should be corn, milo, or wheat combined with oats or barley because of the possibility of compaction due to the heavier concentrates. This same mix can be continued after weaning to about 1 year of age. Consumption by this time will have increased to 1% of body weight in concentrates and 1.5% in hay. A suitable mineral mixture should be supplied free choice as for other horses.

TABLE 37-5. TRACE MINERALIZED SALT

Mineral	Trace Mineral Content	
	T.M. Salt (%)	Amount per Pound of Grain Mixture
Iodine	0.007	318 mcg
Iron	0.80	36 mg
Copper	0.16	7 mg
Zinc	1.00	45 mg
Manganese	0.40	18 mg

Source: Courtesy of John P. Baker, University of Kentucky.

TABLE 37-6. VITAMIN PREMIX FOR HORSES

Vitamin	Amount per Pound of Premix	Amount per Pound of Feed When Premix Added at:	
		5 lb/ton	1 lb/ton
Vitamin A	1,000,000 IU	2,500 IU	500 IU
Vitamin D	100,000 IU	250 IU	50 IU
Vitamin E	5,000 IU	12.5 IU	2.5 IU
Thiamine	1.2 g	3.0 mg	0.6 mg
Riboflavin	800 mg	2.0 mg	0.4 mg
Pantothenic acid	800 mg	2.0 mg	0.4 mg
Vitamin B_{12}	5 mg	12.5 μg	2.5 μg

Source: Courtesy of John P. Baker, University of Kentucky.

TABLE 37-7. SOME EXAMPLE GRAIN MIXTURES[a] FOR HORSES

	Nutrient Concentration (%) of Mixture:[b]					
	1	2	3	4	5	6
Rolled oats	15.0	25.0	31.0	45.0	35.0	41.0
Cracked corn	15.5	20.5	29.5	25.5	25.0	35.0
Oat groats	15.0	—	—	—	—	—
Soybean meal	15.0	10.0	10.0	5.0	10.0	5.0
Linseed meal	10.0	10.0	5.0	5.0	10.0	—
Dried skimmed milk	5.0	5.0	5.0	—	—	—
Dehydrated alfalfa	5.0	10.0	10.0	10.0	5.0	—
Wheat bran	10.0	10.0	—	—	5.0	10.0
Molasses	7.0	7.0	7.0	7.0	7.0	7.0
Dicalcium phosphate	—	0.5	1.0	1.0	1.0	—
Ground limestone	1.5	1.0	0.5	0.5	1.0	1.0
Trace mineral salt[c]	1.0	1.0	1.0	1.0	1.0	1.0
Vitamin premix[d]	+	+	+	+	+	+
	100.0	100.0	100.0	100.0	100.0	100.0
Digestible energy meal/lb	1.4	1.4	1.4	1.4	1.4	1.4
Crude protein (%)	20.2	17.8	15.9	13.2	16.1	11.9
Digestible protein (%)	15.8	13.7	11.9	9.4	12.0	8.0
Calcium (%)	0.82	0.81	0.72	0.65	0.79	0.46
Phosphorus (%)	0.53	0.57	0.54	0.49	0.58	0.37

Source: Courtesy of John P. Baker, University of Kentucky.

[a]To be fed with grass or grass-legume hay.

[b]Type of ration: (1) milk replacer; (2) creep feed; (3) postweaning; (4) grower; (5) broodmare; (6) maintenance and working.

[c]See Table 37-5.

[d]See Table 37-6.

MANAGEMENT SUGGESTIONS

1. Withhold half of the grain ration and increase hay on days that working horses are idle.
2. Use only dust-free and mold-free feeds.
3. Water before feeding. If horse is heated, avoid excessive watering.
4. Feed hay before grain.
5. Do not feed a tired horse its total allotted diet of grain at one time. Feed half the grain, then feed the rest 1 hour later.
6. Do not work a horse hard after feeding a full grain allotment.
7. Feed and water regularly and not less than twice daily.
8. Observe the condition of the horse and feed accordingly.

STUDY QUESTIONS

1. The horse is a forage eating ruminant. (True, false) _____ .
2. The part of the horse's digestive system that allows for microbial breakdown of fiber is the cecum. (True, false) _____ .
3. The horse's digestive system is more efficient in fiber conversion than the cow's. (True, false) _____ .
4. Horses need grain to provide _____ .
5. _____ are the favorite grain of horses and the safest because they are _____ .
6. The most common digestive problem in horses is _____ .
7. _____ _____ is a recommended source of calcium and phosphorus and is usually mixed with _____ .
8. Salt should be a maximum of _____ % of the ration.
9. A horse should clean up its feed in about _____ minutes.
10. A fast eater should be slowed down because it may get _____ .
11. Cribbing is a nutritional disorder causing the horse to eat wood. (True, false) _____ .
12. Too much feed or water after vigorous exercise can cause either _____ or _____ .
13. A horse being used 2 or 3 hours per day is classed as doing _____ work.
14. A 900-lb horse doing light work will eat about _____ lbs of concentrate and free-choice hay.
15. According to Table 37-3, a 1000-lb horse working 4 hours daily should be fed _____ lb of feed.
16. Pregnant mares have a higher protein requirement than weanlings. (True, false) _____ .

17. A 1200-lb horse doing moderate work requires a ration containing at least _____ % of digestible protein.

18. A trace mineralized salt should contain _____ , _____ , _____ , _____ , and _____ .

DISCUSSION QUESTIONS

1. Compare the anatomy of the horse's digestive system to that of cattle and discuss digestive limitations and the reasons for them.

2. Why should prepared cattle feeds not be fed to horses? Are there possible exceptions to this rule, and if so, what are they?

3. What are protein requirements of the following: (Give figures in pounds of digestible protein, percent of digestible protein, and percent of crude protein):
   a. 1100-lb gelding, hard work
   b. 900-lb mare at 200 days of gestation
   c. 1200-lb mare, just foaled 2 weeks ago.
   d. 600-lb pony, light work
   e. 200-lb foal

4. Suppose that you own the five horses in Question 3. How much total feed would be needed to meet their nutritional needs for 1 year?

5. Define the following:
   a. cribbing   d. founder
   b. sand colic   e. cecum
   c. gossypol   f. protein quality

6. Calculate the daily amount of feed (concentrate) and hay needed for the following:
   a. 950-lb pregnant mare
   b. 1120-lb gelding, light work
   c. 1610-lb draft horse, medium work

Chapter Thirty-Eight

HORSE SELECTION

Judging a horse should be like looking at a beautiful painting; everything there should be pleasing to the eye. We can even go a little further in the selection of horses and say that everything we see has a direct relationship to the performance of that horse. It is with this thought in mind that the following examples are used to illustrate the direct relationship between conformation, type, and desired function.

SELECTION OF HORSES AND HORSE TYPES

There are basically three ways to select a horse—by pedigree, by performance, or by visual observation. The best method would be to utilize all three types. In this chapter we limit our discussion to visual observations that are normally referred to at horse shows as *halter, model,* or *breeding classes.* The following discussion of selection will cover the two most popular breeds in the United States—the Quarter Horse and the Thoroughbred.

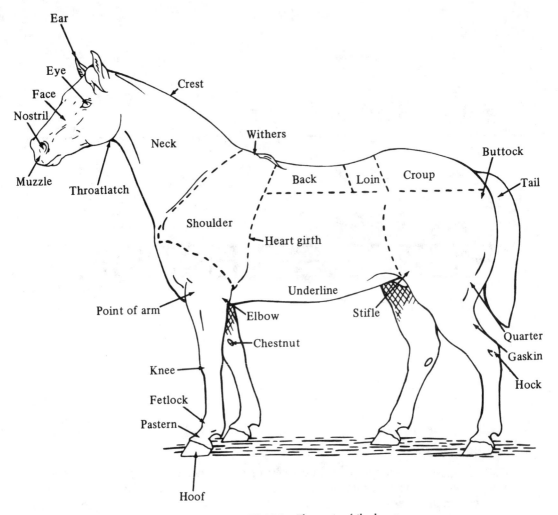

FIGURE 38-1. The parts of the horse.

HORSE TERMINOLOGY

You have a name and so does every part of your body, but nature has made the equine different; thus terms that are unique to the horse are quite necessary in expressing observations (Figure 38-1).

JUDGING THE QUARTER HORSE

The horse or a class of horses of any breed should always be judged in a logical sequence. A common way is to look at a horse first from the side, then from the front, and lastly from the rear. Then the accomplished judge will want to judge the moving horse. After taking a quick, overall view of the horse from all three positions, one can start looking for the finer points.

The Head and Neck

A good-quality head of a Quarter Horse is shown in Figure 38-2. The ears should be short, erect, and "fox like." Ears are a good indication of temperament and intellect in a horse. Short, erect ears that often point forward indicate a horse that is in good condition, alert, and paying attention to what is going on, which is necessary in a cutting horse, roping horse, or stock horse. The eyes should be prominent and well set apart. A fairly broad area between the eyes is desirable in Quarter Horse conformation and gives the horse visual latitude (which means it can see forward and backward without moving its head). This allows a horse to react quicker because it is able to see something

FIGURE 38-2.   The Quarter Horse head.

that is coming from behind, the side, or front. The nostrils are also important. It is desirable to have a very large nostril because intake of air is necessary for a hardworking horse. Most Quarter Horses, whether they are racing or working, must have endurance. If they have sufficient lung capacity but are restricted at the nose, the lungs cannot be filled, thus reducing their endurance.

Figure 38-3 shows the various parts of the head from the side view. The jaw is a good indication of strength of bone and constitution (hardiness). The desirable type is called a "dinner plate" jaw, being large and round. The mouth should be relatively shallow so that a bit placed in the mouth is able to control the horse with a minimum of pressure. A horse with a very deep mouth may be one that is difficult to control. The neck should be relatively thin because this determines whether or not the horse has the ability to turn quickly when following a calf or doing other types of work characteristic of the needs for a typical Quarter Horse. The throat latch must be clean with no excess muscling or finish because this is the pivot point for the head and as the head goes, so goes the horse.

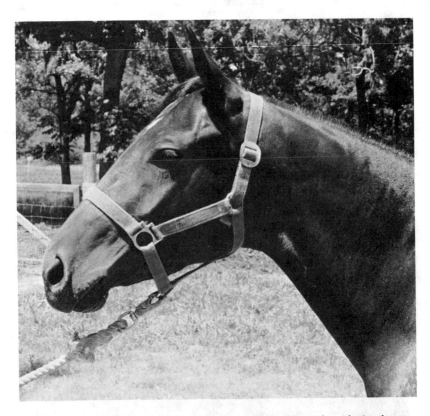

FIGURE 38-3. A "dinner plate" jaw, shallow mouth, thin neck, and trim throat latch are desired.

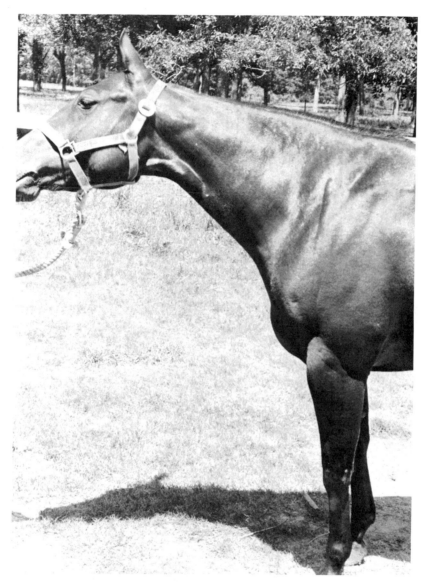

FIGURE 38-4. The slope of the shoulder and pastern should approximate 45 degrees. A long forearm muscle is desired.

The Shoulder, Forehand, and Withers

The slope of the shoulder and the slope of the pastern are very important (Figure 38-4). This is a point missed by the average student Quarter Horse judge. The slope of the shoulder should be approximately 45 degrees, and this should match the slope of the pastern. The pastern acts as a shock absorber to the front end and gives a smoother ride. If this angle is much more than 45 degrees, the horse has a tendency to tire easily or go lame. If the slope is too straight it may develop into a bone jarring type of ride, which no one desires for very long. Also illustrated is the forearm muscle. This, of course, is a necessary part of moving the foreleg, and the more highly developed the forearm muscle, generally the better movement the horse will have in the front end. A longer attachment is generally preferred to the heavier, more closely attached forearm muscle.

The Quarter Horse's front legs form an inverted "V" from the front view (Figure 38-5). This is the most important aspect of the front. We do not want this area of muscling to be flat but rather a well-balanced "rooftop" type of muscling. This is very important because it enables the horse to cross its front

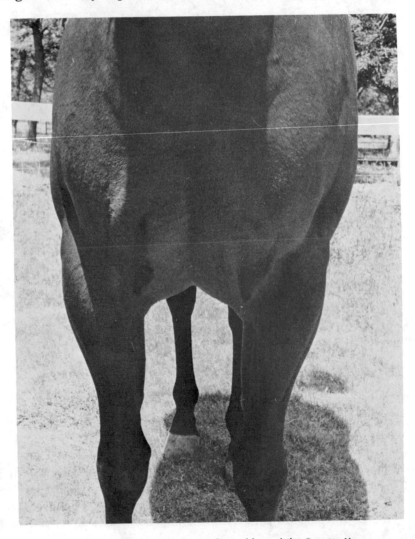

FIGURE 38-5. The "V" muscling and set of legs of the Quarter Horse.

Chap. 38 HORSE SELECTION

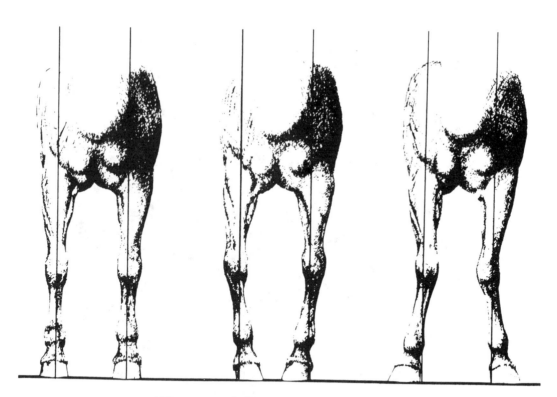

FIGURE 38-6. The legs of a horse should be mentally aligned.

legs when running which gives it the necessary agility. If an athlete, for example a tackle, is built square and not able to change direction quickly enough, the coach will select another player who has the ability to move quickly to carry the ball. This agility correlates very closely with the type of muscling and the conformation of the athlete—in this case, the Quarter Horse.

Figure 38-6 indicates what we should be looking for in a set of legs. Imagine dropping a plumb line down through the middle of the horse's leg, knee, and hoof and study the examples in Figure 38-6 to determine the common deviations and the correct placement.

The side view (Figure 38-7) shows the withers, which should be prominent. The beginner may think this prominence unsightly. However, its greatest function is to give the saddle a place to hang and prevent the saddle from turning. So we do not like a horse that has full, flat withers. The top line of a Quarter Horse should be short or "close coupled" and the bottom should be long. This gives the horse the unique ability to be "wound up tight" on top and able to stretch out underneath. This is very important in Quarter Horses. The croup should have a slight slope from the hip to the rear. This enables the horse to get its rear feet up and under it which is the source of great, quick starts and power over a short distance. Also from the side view note the muscling commonly referred to as the quarter. It should be well developed with a bulge that lets down to the area of the stifle muscle.

FIGURE 38-7. Prominent withers, short back, and long underline.

The leg settings from the side view as illustrated in Figure 38-8 are also observed using the same imaginary plumb line.

Hindquarters

Perhaps the most important view of all is the rear view as illustrated in Figure 38-9. One should visualize the conformation lines of an apple turned with the stem down. Muscle that is quite similar to this imaginary example is desired. The horse should be very wide through the stifle muscle and this area should be the widest part seen from the rear. The gaskin muscle is the most prominent feature on the way down. It should be full and bulging outward. Also directly inside this muscle is another gaskin referred to as the inside gaskin muscle. Most breeders agree this is one of the most important muscles on a horse and one of the most difficult to develop through selection and breeding.

Figure 38-10 shows the rear view leg settings, both good and bad, using the techniques previously developed in Figures 38-6 and 38-8. Every breed of horse should conform to the ideals illustrated in these figures.

The Moving Quarter Horse

The standard procedure in halter classes is to view the horse walking away from the judge and jogging back. The judge is looking for its "way of going or moving." This means that its feet and legs move as smooth and straight

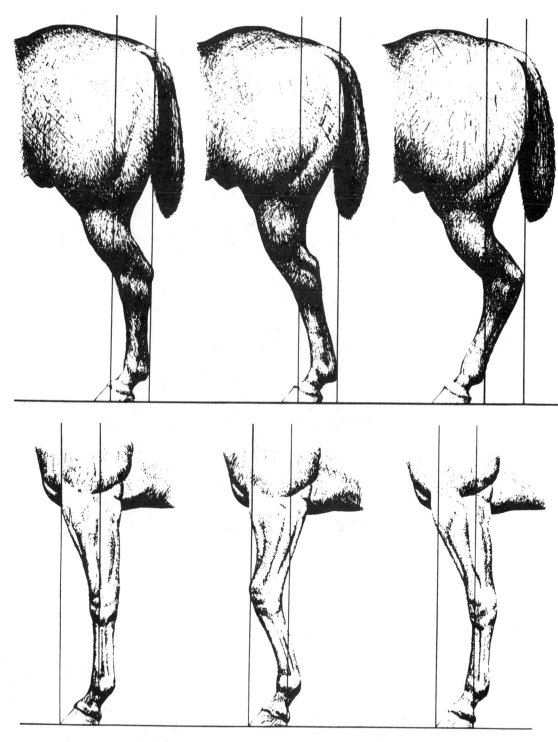

FIGURE 38-8. Leg settings should not vary drastically from the normal.

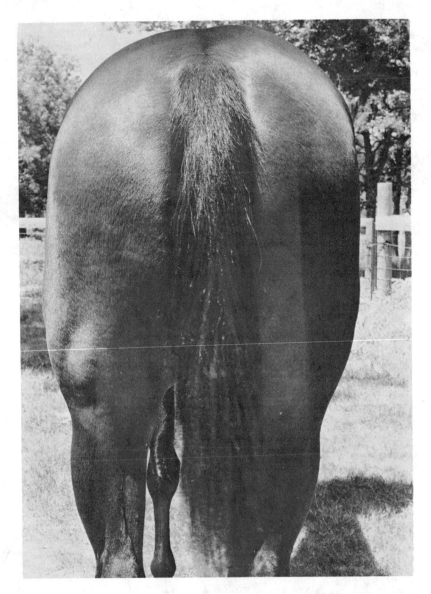

FIGURE 38-9. Lines of the rear quarter resemble an upside down apple.

as expected from the standing "plumb line" observations. Imagine pushing someone in a swing. If well balanced, the movement will be like a pendulum on a clock. Likewise, the legs of a horse (when viewed from front and rear) should move in a similar manner. Defects not detected at slower speeds may show up at faster movement, thus the reason for observations when walking and jogging. Some common defects include "winging out" or "paddling" (slinging the feet outward), "scalping" (front legs rubbing together), "toeing in" (slinging the feet inward), and "forging" (hind feet stepping on front feet).

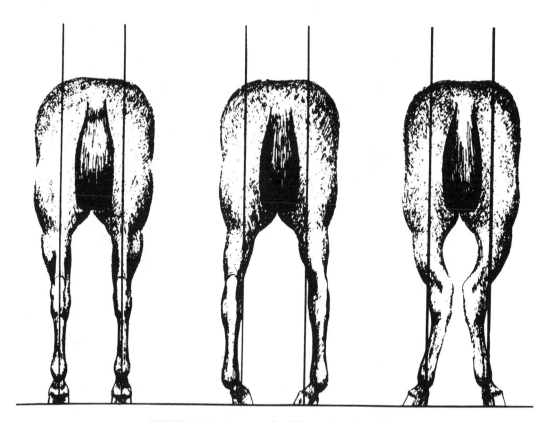

FIGURE 38-10. An example of leg settings from the rear view.

POINTS OF CONFORMATION OF THE THOROUGHBRED

There are not as many Thoroughbred breeding classes as there are Quarter Horse halter classes, but the procedures followed by the judge of Thoroughbreds looking for quality conformation can be utilized by the owner, breeder, or potential purchaser of Thoroughbreds. The general overall look of the Thoroughbred is one of great refinement and quality. Its coat is fine and close, its skin thin, and its veins prominent. In general, the Thoroughbred is more excitable and high strung than other horses, but also can be more easily called upon to push its nose across the finish line first or carry itself and a rider cleanly over a seven-foot wall. The Thoroughbred is lithe and clean and shows not a drop of any cold blood in its veins. Each individual segment of its conformation as discussed below goes to make up this elegance and class.

The Head and Neck

The head is small, well proportioned, and intelligent looking (Figure 38-11). In profile, the nose is either straight or slightly dished. The muzzle is fine and

FIGURE 38-11. The head of a typical Thoroughbred. Note the fine, intelligent look.

cleanly cut, the nostrils capable of opening wide to pull in great breaths of air while the horse is running or jumping—the two activities at which it is unsurpassed by other breeds. The ears of the Thoroughbred are ideally small and neat; the eyes are large, prominent, and deep. The jaw line is not so heavily muscled as the Quarter Horse and the Arab, and the throat is fine and clean, attaching to a long, slender but well-muscled neck. This neck is not only arched, but also very flexible, enabling the horse to use it to balance its actions. The mane is usually fine and thin haired.

The Withers, Shoulders, and Forehand

The withers are pronounced and rightly so because they are the top of the shoulder that must be so long and sloping (Figure 38-12). The supposed ability of a Thoroughbred to jump well and stride out while galloping is often judged by the slope of the shoulder. Without a slope of about 45 degrees, the horse will not find it as easy to tuck its forelegs up and under its chin in clearing an obstacle or in reaching out while galloping. As with most breeds, the slope of the horse's shoulder relates directly to how comfortable the horse is to ride. The chest is broad and deep, accommodating the big heart and expansive

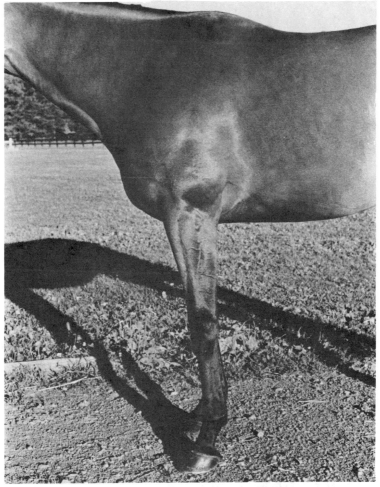

FIGURE 38-12. A good example of the Thoroughbred's forehand with solid clean legs and a large, sloping shoulder.

lungs (Figure 38-13). The forelegs should be well set on the chest, not too close together or out too far.

The Thoroughbred's forelegs are clean, hard, and of good bone. Long and slender, the legs make up the bulk of the Thoroughbred's height. The forearm should be well muscled and powerful but not stubby looking; the knee is flat; the cannon bone short but not as short as that of the Quarter Horse. A too-long cannon bone hints of early fatigue as do pasterns that are too sloping or too straight. In addition, a pastern of other than medium length and slope can more easily become lame. The Thoroughbred's forelegs can easily be damaged if not of good quality or if the horse is pushed along in training too hard and too quickly. Broken at 1 year of age and often raced the next or at least by

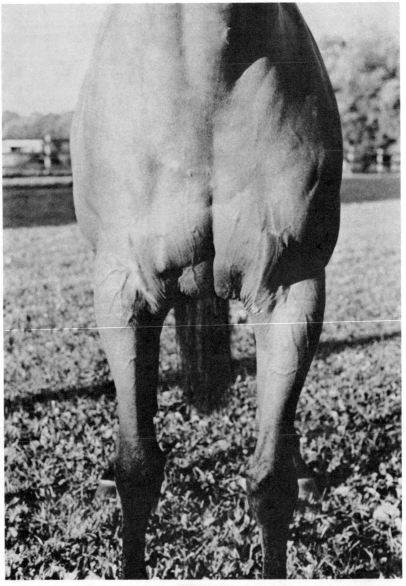

Photograph by Paul Rosen

FIGURE 38-13.   The Thoroughbred's chest is broad and deep enough to accommodate its strong heart and large lungs.

the third year, the Thoroughbred can break down unless it at least has the
benefit of good, strong conformation in the forehand.

The hooves of the horse should be fine and not disproportionately large
for the size of the horse. In fact, the Thoroughbred should appear light on its
feet even when standing still. Strong feet and legs are especially vital to a

good Thoroughbred because a 3-year-old can already be as tall as 16 hands or more and its legs may not be closed (strong, hard, and fully matured) enough at that age to maneuver a weight of 900 to 1200 lb, not to mention a rider, without injury. The previously discussed idea of visualizing a plumb line dropped down the foreleg and the illustrations of Figure 38-6 apply to the Thoroughbred (and any breed), too.

The Side View

Figure 38-14 is the side view of a typical Thoroughbred and shows the head, neck, shoulder, and foreleg as already described. Notice the flat and non-bunchy muscles of the shoulder, neck, and hindquarters. The back is on the long side, but not as long as the American Saddle Horse or Standardbred. The underline is slightly up-sloping on the fit horse—giving rise to the term "wasp-waisted" that is applied to racing-fit horses carrying no fat. Not only is the Thoroughbred's chest wide as already mentioned, but the body is also—the ribs do not spring outward as in the Arab or Quarter Horse, rather more out and down, but there should be good horizontal depth through the body at the topline and the underline.

The croup is higher than the dock, giving rise to what some steeplechase buffs call the jumping bump. It is often taken as an indication of a good jumper, and a horse that has been well schooled and jumped for several years will develop the croup even higher.

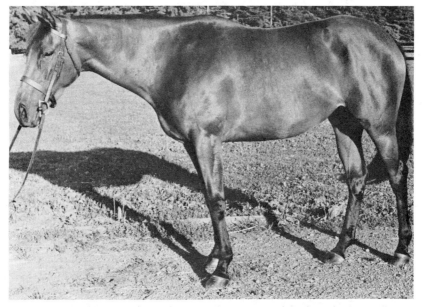

Photograph by Paul Rosen

FIGURE 38-14. A typical 3-year-old Thoroughbred filly.

Hindquarters

The hindquarters of the Thoroughbred are its most utilitarian and interesting aspect. Notice the great distance between the hip and the hock (Figure 38-15). This distance coupled with the overall length of the leg and the angles at the hip, stifle, and hock comprise a great source of power. This is the point of propulsion for the horse and indicates the length of stride and power the horse will muster whether on the track, in the hunt field, or over a course of jumps. Along with this hip to hock length should come good muscling—a hindquarter less well developed than a forehand is undesirable. There should be balance between fore and hind, but it should be obvious that the power lies behind.

When it is viewed from the rear, the Thoroughbred is pear-shaped rather than square or apple-shaped like the Quarter Horse. Notice the fineness of the tail that ideally is kept at a length halfway between the hocks and fetlocks.

Photograph by Paul Rosen

FIGURE 38-15. The long distance from hip to hock gives the Thoroughbred its great propulsion.

The tail is braided to the end of the bone for horse shows to better show off the fine conformation of the hindquarters and to present a tidy, put-together appearance (for the same reason, the mane is braided, showing off the quality of the neck).

Again, as with all other breeds, the hind legs should be straight and sturdy when viewed from the back. One of the worst faults is being cow hocked where the legs converge to meet at the hocks and splay out at the feet. Also, the hind feet should not be too close together; the insides of both the hind and fore feet can injure the leg opposite if set too closely. Refer again to Figures 38-8 and 38-10 to study the proper alignment of the hind legs.

Chap. 38   HORSE SELECTION

Way of Moving

A judge in a breeding class, under saddle class, or hunter over fences class will weigh a horse's way of going and soundness heavily. A horse can be sound but still be a poor mover. Traveling problems that can arise in any breed including the Thoroughbred have been discussed previously and apply here also.

The judge of a non-broodmare breeding class (unsoundness unless from severe conformation fault does not count in a broodmare class) will want to see the entries at the trot primarily, but will probably observe them informally at the walk also. The hunter under saddle judge will judge the entries on way of going, soundness, and discipline at all gaits; and the judge of hunters over fences will watch their way of going at the canter or hand gallop and jumping quality as well as their soundness. *Theoretically,* a jumper or a race horse could

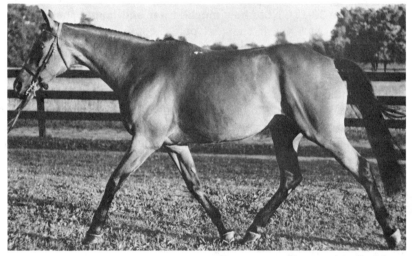

Photograph by Paul Rosen

FIGURE 38-16. The Thoroughbred at the trot.

be a bad mover or actually lame and it would not necessarily affect whether it won or lost.

The Thoroughbred at the walk should move alertly, evenly, and with a long striding action. At the trot (Figure 38-16), it should clip daisies—moving its legs low, even, and with no excess up and down motion. At the canter, the legs also move low, either long or short depending on the cantering speed desired, and with no chop to the stride, thereby being easy to sit to even at an extended canter. At the gallop, the stride lengthens tremendously and the body stretches out with the hind quarters coming up and under the horse and the forelegs reaching out straight, meeting the ground and pulling it behind.

Nearly everyone has seen the horse trader on the late, late western cast a suspicious eye on a prospective horse being traded or bought. After examining the teeth by opening the mouth, the trader calmly announces that the horse is much older than advertised. What did he see and what made him so sure? Horses' teeth come in with such predictable accuracy that it is now quite common for race tracks to employ a specialist to determine age, thus preventing unfair practices of running more mature horses against younger competitors.

Although age determination within a few months can be quite simple for the expert, a more simplified approach is taken in this text. A young horse will have baby teeth that are replaced by permanent teeth. There are six upper front teeth and six lower front teeth. The permanent teeth start erupting in pairs, starting with the two middle incisors (the front teeth) around 2 to $2\frac{1}{2}$ years of age. Both top and bottom middle incisors are complete by 3 years of age (Figure 38-17). Note how much larger and longer these teeth are than the baby teeth. At 4 years, the next pair is complete (Figure 38-18), leaving only one pair of temporary incisors.

Figure 38-19 illustrates the complete set of permanent teeth that exists at 5 years. An interesting point is the development of canine teeth at about this age (although it may be as early as $3\frac{1}{2}$). These canine, or wolf teeth, are always seen in a stallion or gelding but only rarely in a mare. Figure 38-20 is a rare picture of an 8-year-old mare's mouth with a protruding canine tooth.

In summary, the average horse handler can tell the age of horses that are

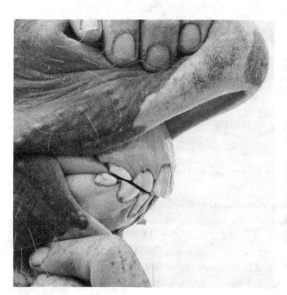
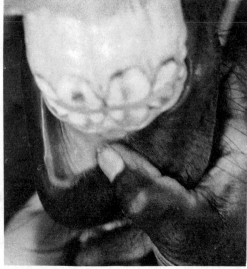

FIGURE 38-17. The 3-year-old's mouth.

Chap. 38   HORSE SELECTION

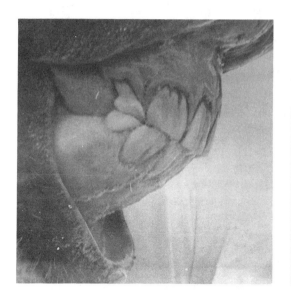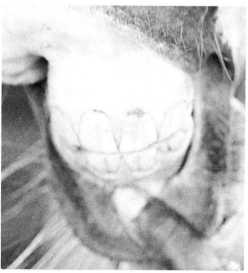

FIGURE 38-18.   The 4-year-old's mouth.

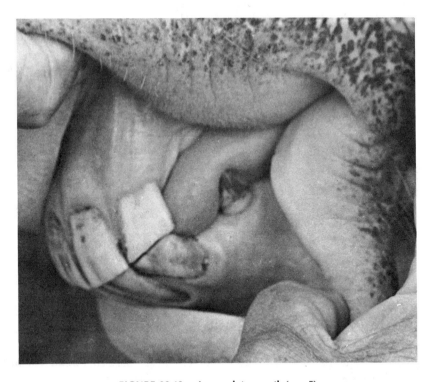

FIGURE 38-19.   A complete mouth (age 5).

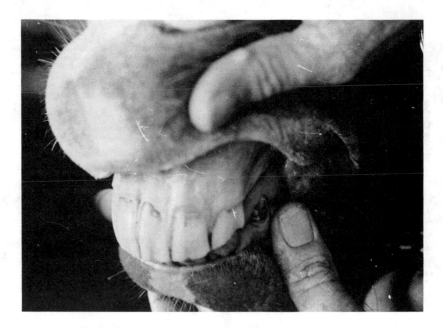

FIGURE 38-20.   An 8-year-old mare's mouth with rare canine tooth.

3, 4, and 5 with a little practice. Beyond 5, more detailed observations are made with reference to the wearing of cusps and the angle of incisors. An excellent reference with great detail is published by the American Association of Equine Practitioners, Golden, Colorado 80401 for the serious student of this neglected art.

STUDY QUESTIONS

1. Three ways to select horses are _____, _____, and _____ _____.
2. The selection of horses based only on visual observation is commonly conducted in _____, _____, _____ classes at horse shows.
3. Judging horses is unique in that everything we look for in a horse may have a definite _____ on its performance.
4. There are two types of Quarter Horse: _____, _____.
5. Attention is evidenced by the _____.
6. The general appearance of the Thoroughbred is one of _____, _____.
7. Wide set eyes are desirable so that a horse has good vision to the front, side, and _____.

8. Small _____ are undesirable because they restrict lung capacity.

9. The _____ _____ of the Thoroughbred is not so heavily muscled as that of the Quarter Horse.

10. A large dinner plate jaw in the Quarter Horse indicates good _____.

11. A Quarter Horse that responds well to the bit should have a _____ mouth.

12. A horse with a heavy muscled, thick neck is desirable in a working Quarter Horse. (True, false) _____.

13. A trim throat latch is very desirable in both the Quarter Horse and the Thoroughbred. (True, false) _____.

14. The slope of the shoulder should be approximately _____ degrees.

15. A rough ride may be anticipated from a shoulder slope that is too _____.

16. A severe slope in the pastern could cause the horse to become _____.

17. The Thoroughbred's _____ if not at least of basically good conformation can more easily break down under heavy training or work.

18. The rear end of a Quarter Horse should look like an inverted _____.

19. The rear end of the Thoroughbred is _____ shaped.

20. Desirable "V" muscling is characteristic of Quarter Horses with good _____.

21. The widest point through the rear quarter of a Quarter Horse should be through the _____ muscle.

22. The Thoroughbred's great source of propulsion is found in the length from _____ _____ _____.

23. The most difficult muscle to "breed on" a horse is the _____ _____.

24. A prominent _____ allows good saddle "set" and indicates a large shoulder.

25. A moderate slope to the _____ allows the Quarter Horse to develop power from the hind legs by getting them further under the body.

26. The swiftness of the Quarter Horse can be partially attributed to a short _____ and a long _____.

27. Horses are judged for conformation _____, _____, _____ classes.

28. The judge watches the moving horse for its way of _____.

29. At a trot, the Thoroughbred should move _____.

30. "Winging out" is a defect noted in some horses when walking or jogging. It means slinging the _____ _____.
31. The permanent front teeth are called _____.
32. A 3-year-old horse will have permanent _____ incisors.
33. Six permanent incisors mean a _____-year-old mouth.
34. The complete set of permanent teeth exists at _____.
35. At 5 or earlier, _____ teeth will develop in _____ and _____.
36. Canine teeth _____ (rarely, never) appear in mares.

DISCUSSION QUESTIONS

1. Name 50 common terms relating to parts or anatomy of the horse.
2. What is the relationship between slope of the shoulder and slope of the pastern, what should it be, and why?
3. What are the chief differences between judging standards of Quarter Horses versus Thoroughbreds?
4. Why should a Quarter Horse be close coupled on the back and long in the underline? How does this differ from Thoroughbreds?
5. Describe the major foot and leg deviations that can be observed by watching a horse traveling and at halter.
6. Describe in detail how to tell the age of horses up to 5 years by teething.

Chapter Thirty-Nine

DISEASES OF HORSES

Health programs include disease prevention, worming, and first aid practice. It is important to make the correct diagnosis and have a working knowledge of medication. The best horse owners and breeders rely heavily on veterinarians. A brief description of the more common diseases, parasites, injuries, and health problems should aid in the successful management of horses. The first rule of management is cleanliness of both the stable and the horse itself. Feed boxes, bedding, and stable areas must be managed properly to prevent problems. The temperature of the stable should be close to the outside temperature to reduce the possibility of respiratory diseases.

One of the first signs of any problem is a poor appetite or failure to eat at all because a healthy horse is almost always hungry. A simple check of temperature by rectal thermometer should verify any suspicions of illness. Any deviation above or below the average range 99 to 100.8 °F could indicate a serious problem, and proper diagnostic measures should be conducted by a qualified person. A description of some common diseases, parasites, and injuries follows.

Azoturia (Monday Morning Disease)

Horses that are normally worked hard every day, fed well, and rested on Sunday with the same level of feeding may develop lameness on Monday morning when work is resumed. Muscles of the quarter are extremely tight, legs are stiff, and extreme sweating occurs. To prevent the condition from occurring, idle horses should be fed at moderate levels. This condition is often seen in light horses.

If the disease strikes, immediate, absolute rest in a standing position is considered the best treatment. The animal should not be moved from the spot where the condition occurs. Temporary shelter and slings may be necessary in a severe case. Medication varies with the intensity of the situation and it is essential that a veterinarian prescribe sedatives, laxatives, alkalizing drugs, and so on.

The cause is thought to be overproduction of lactic acid as a byproduct in the metabolism of glycogen (animal starch). Excessive lactic acid production due to inactivity but continued glycogen metabolism is thought to cause deranged muscular activity. The condition is similar to *tying-up syndrome*, but treatment is considerably different.

Bowed Tendon (Tendinitis)

This condition refers to enlarged tendons behind the cannon bone in both the front and the hind legs. It is most common in the forelegs. Location of the injury is described as "high" (just under the knee), "low" (just above the fetlock), or "middle" (in between). Tendinitis may occur in the high or low areas alone, but usually not in the middle area alone. When it does occur in the hind leg (rarely), it occurs in the low area only. As viewed from the side of the cannon, the swollen tendon is bulged or bowed out giving rise to the name "bowed tendon."

A severe strain is the cause, brought about by long, weak pasterns; toes that are too long; exertion from accident or forced training procedures; fatigue of muscles at the end of long races; muddy tracks; improper shoeing; and a horse that is too large for his foot structure. Signs of acute tendinitis appear swiftly. The horse shortly after injury, or even at the moment of injury, will pull up lame, holding the heel elevated to relieve pressure and not allowing the fetlock to drop because of pain. Heat, swelling, and pain are evident upon palpation.

In early acute stages, treatment by a veterinarian usually consists of an injection of corticoids and application of a light cast for about 2 weeks. If marked improvement is noted after this cast is removed, supportive bandages are applied for another 30 days. If there is little improvement, another cast may be tried. Finally, the horse is rested for at least one year, indicating the seriousness of the situation.

An alternative treatment used on less valuable horses that some lay people have been moderately successful with is *blistering*. Blisters are salve-like irritating substances that convert a chronic inflammation into an acute condition. This brings more blood to the part, hastening nature's reparative process. The hair is closely clipped from the affected area and the blistering agent applied by hand and rubbed into the pores of the skin. The horse should not be allowed to lick, rub, or bite the area. A rope halter is usually used to tie the head to prevent unwanted interference. Three days afterward, the affected area is bathed with warm soapy water, dried, and treated with Vaseline or sweet oil to prevent cracking of the skin. A rest of at least 1 year is still required.

Colic

This digestive problem is brought on by overeating, excessive drinking while hot, moldy feeds, and even infestation of roundworms. The intestine is blocked or impacted, creating pain where the horse is most sensitive. Signs are persistent movement, pain, sweating, rolling, and obvious discomfort. Rolling may lead to twisted gut, which is fatal. Horses should be haltered to prevent rolling. Other signs are curling up of the lip and refusal to eat. Treatment consists of walking the horse until the vet arrives. Mineral oils are often given by stomach tube to relieve compaction.

Founder (Laminitis)

The horny laminae of the hoof becomes congested with blood, preventing a normal "cushioned gait." An obviously painful lameness in the front feet, sometimes all four feet, suddenly appears followed by unusual growth of hooves that must be trimmed often. Founder (Figure 39-1) has been attributed to overeating, radical change in feed, lack of exercise, metritis (inflammation of the uterus) in recently foaled mares, and drinking very cold water when overheated.

Treatment may include standing in mud puddles or cold water to reduce swelling of the blood vessels, giving some relief until arrival of a veterinarian, if caught in the early states. Hypodermic medication may be effective, but in most cases the damage is irremediable and the only treatment is suitable shoeing.

Heaves

Heaves is characterized by coughing and wheezing while at work. It is thought to be due to allergies to dusty or moldy feed or to hereditary weakness. Feed-

Foundered   Normal   Foundered

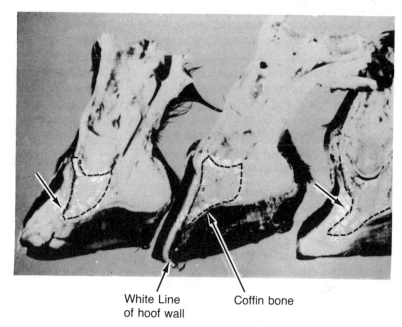

White Line
of hoof wall   Coffin bone

FIGURE 39-1.   Founder (laminitis), a complex disorder affecting the feet, causes irreparable damage if the coffin bone rotates away from the hoof wall. Note normal alignment of coffin bone with hoof wall in center cross section compared to foundered specimens. (Courtesy of the School of Veterinary Medicine, Texas A&M University)

ing clean forage with a little water sprinkled over it or all pelleted rations has shown good results in alleviating the condition. It is considered a serious fault in an otherwise healthy horse.

Influenza

This is a respiratory disease similar to flu in humans. It is caused by one of eight different viruses. The symptoms are coughing, nasal discharge (Figure 39-2), and a rapidly rising temperature that may reach 106 °F and last for 2 to 10 days. A complete rest for at least 30 days is recommended. Vaccines for two viruses are available to create immunity. Treatment by a veterinarian may include the use of antibiotics and/or sulfa drugs and rest.

Navel III

This infection of the navel cord of newborn foals leads to an inflamed navel, hot swollen joints, and lameness. It is caused by bacteria entering the bloodstream through the most obvious route. Swabbing the navel with iodine at

FIGURE 39-2. A nasal discharge may be the first sign to appear in upper respiratory infections, colds, or distemper. (Courtesy of the School of Veterinary Medicine, Texas A&M University)

birth is a common practice and a simple preventative. Treatment by a veterinarian may include a blood transfusion and use of antibiotics and/or sulfa drugs.

Poll Evil—Fistula Withers

Infection of a bursa in the poll or withers leads to ruptured, draining sores (Figure 39-3). It may be contagious, so infected animals should be isolated. The cause is not known, but bruises from poorly fitted saddles or headgear, possibly in combination with the same organisms that cause "lumpy jaw" and brucellosis in cattle, may be to blame.

Treatment should always be under direction of a veterinarian. Proper drainage and removal of dead tissue and/or use of caustic applications are common treatment methods.

Sleeping Sickness (Equine Encephalomyelitis)

There are two important types of virus, an Eastern strain and a Western strain. VEE (Venezuelan Equine Encephalomyelitis) has received great notoriety in Central America, Mexico, and the United States in recent years. The horse may stumble, fall, or stand with his head hanging down. Complete paralysis and death can occur. Mosquitoes and flies can transmit the virus. An

FIGURE 39-3.   A horse affected with poll evil. (Courtesy of the U.S. Department of Agriculture)

annual vaccination program, manditory in most states, appears to be the only safeguard, since animals either die or develop immunity once the disease is contracted. Inoculation is usually recommended before May, consisting of two injections 7 to 10 days apart. Immunity lasts about 6 months. Treatment is restricted to good care, feed, and water, although some veterinarians will also administer intravenous medicines. Figure 39-4 shows a horse with the disease.

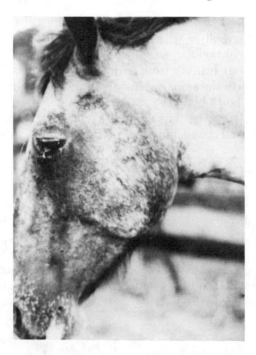

FIGURE 39-4.   Horses affected with sleeping sickness often take on a drowsy appearance and may fall asleep even while eating. The foam produced in the mouth of this victim was caused by saliva secreted as the result of a mouthful of oats, which were unchewed because of a sudden desire to sleep. (Courtesy of the School of Veterinary Medicine, Texas A&M University)

Strangles (Equine Distemper)

This contagious, bacterium-induced disease almost always occurs between 6 months and 5 years of age. Symptoms are high fever, cough, discharge of pus from the nose, and swelling of glands under the jaw, which usually break open and discharge pus. It is seldom fatal and can be treated with antibiotics, rest, and proper ventilation, with lifetime immunity the usual result.

Swamp Fever (Infectious Equine Anemia)

This virus is carried by flies and mosquitoes. The symptoms are high fever, labored breathing, pounding heartbeat, and exhaustion. There is no cure. Most affected horses (Figures 39-5 and 39-6) die within 30 days. Recovery results in immunity and apparent good health. It may be controlled by keeping horses out of low places and by properly draining land, but the disease is not restricted to low-lying or swampy areas as originally thought.

The Agar Gel Immuno Diffusion Test (AGID), more commonly known as the Coggins Test, has received wide acceptance as a method of determining carriers of the disease. A horse that has had swamp fever but has recovered becomes immune, but serves to spread the problem because of intermediate carriers such as mosquitoes. The Coggins Test involves drawing a blood sample and sending the serum portion (after clotting) to a federally approved diagnostic laboratory. Negative tests are required by many states for movement within the state or from other states. A negative Coggins Test certificate is

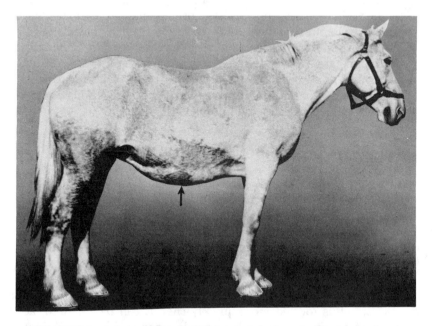

FIGURE 39-5. A horse affected with chronic infectious anemia (swamp fever). (Courtesy of the U.S. Department of Agriculture)

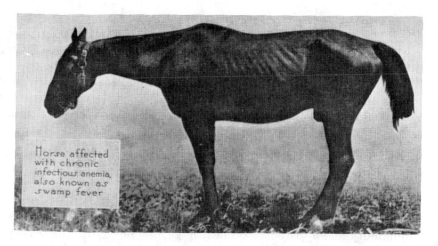

FIGURE 39-6. A case of the subacute form of infectious anemia showing extensive dropsical swellings on the abdomen. (Courtesy of the U.S. Department of Agriculture)

required at most of the larger horse shows and race tracks as a part of the health papers.

Tetanus (Lockjaw)

An organism that lives in soil, manure, and dust causes tetanus; therefore, the horse is most susceptible. The organism can only grow without air; thus, puncture wounds are most likely to harbor it. The tetanus organism produces a toxin (poison) 100 times as powerful as strychnine, causing muscle contraction. Signs are stiff legs, sensitivity to noise, and folding of inner eyelid over eye. Death may occur within 24 hours to 1 week. Treatments with muscle relaxants, antitoxins, and intravenous glucose injections are given but are seldom effective. A vaccine will prevent the disease.

Tying-Up Syndrome (Cording Up)

Tying-up is quite similar to azoturia. Some authorities make no distinction between the two problems. However, the main difference is that azoturia is characterized by muscle tremor, incoordination, and stiffness in the first few minutes of work. Tying-up has the same signs, but they do not occur until some time after muscular exertion, such as riding or racing. It would appear that tying-up is a mild form of azoturia.

The cause is generally related to moderate to heavy feeding of rations, an idle period, and continued feeding, followed by muscular exertion. Lactic acid, which is formed more quickly than it can be excreted, accumulates because of continued metabolism of glycogen and affects the muscles. It does not accumulate as rapidly, or in amounts that cause muscle destruction, as compared to azoturia. For this reason, it is less serious and recovery to normal is more likely in a shorter time. Signs are stiffness and improper flexing of the hind legs, a rigid back, and walking that is obviously painful. In numerous cases, the horse just collapses under the rider. If the horse can regain a standing position, it is usually with much incoordination and a dazed appearance.

Treatment is quite different from azoturia. Victims of tying-up are walked for 30 to 40 minutes, which usually relieves the signs (azoturia cases are not even moved). More severe reactions may be treated by intravenous corticoid injections and alkalinization of the bloodstream. Complete recovery is usually seen within 12 hours.

LEG AND FOOT PROBLEMS

There appears to be hardly any limit to the variety of problems that affect the soundness of limbs in horses. A brief discussion of the more common conditions follows. See Figure 39-7 for location of specific unsoundnesses.

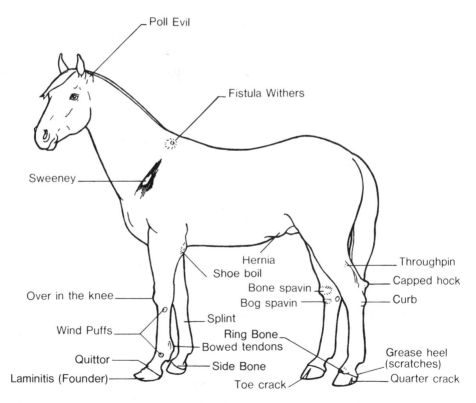

FIGURE 39-7. Anatomy and location of selected unsoundness in the horse.

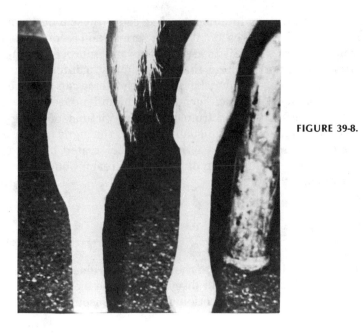

FIGURE 39-8. Bog spavin is a bulge or swelling on the front inner side of the hock (leg on left). Although unsightly, it normally consists of an accumulation of fluid at the hock joint capsule, which usually does not cause lameness. It is thought to be inherited, seen most often in straight-hocked horses. (Courtesy of the School of Veterinary Medicine, Texas A&M University)

Bog Spavin

Bog spavin (Figure 39-8) is a swelling of the joint capsule that occurs at the hock joint due to an accumulation of fluid. A large swelling is obvious at the anterior portion of the hock joint, although smaller protrusions may accompany it on either side. The condition does not usually result in permanent lameness. However, complete recovery is not to be expected in every case. Often the horse will appear lame until warmed up, especially during cold weather. The cause is usually faulty conformation (no cure), injury (as a result of quick stops, etc.), or rickets.

Bog spavin due to rickets can be treated by proper attention to dietary levels of vitamin A, vitamin D, calcium, and phosphorus, alone or in proper combination. Symptoms should disappear in four to six weeks in this case. Horses 6 months to 2 years of age are most affected by nutritionally caused bog spavin. Treatment of a trauma-induced condition varies. Two to three injections of a corticoid into the joint capsule given at weekly intervals is a common treatment by veterinarians. Horse owners have found rest and treatment with a stimulating linament and daily massage for 2 to 3 weeks to be effective in some cases. Firing or blistering is not recommended.

Bone Spavin

To the layman, this condition appears similar to bog spavin. However, this disorder is due to a bony enlargement and is located on the lower inside part

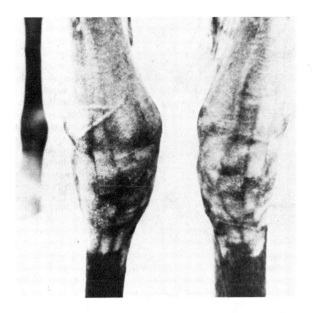

FIGURE 39-9. Bone spavin involves the inflammation of bone in the hock joint, an enlargement which can produce lameness. Rather than an accumulation of fluid as in bog spavin, bone spavin involves a serious disorder often producing lameness. Faulty conformation, pounding on hard pavement, or mineral deficiences may cause the disorder. (Courtesy of the School of Veterinary Medicine, Texas A&M University)

of the hock. Bone spavin (Figure 39-9) causes a reduction arc of the normal flight of the foot, reducing flexion to relieve pain and causing excessive wear of the toe. It is most common in horses used for quick stops, such as roping horses.

The cause is traced to hereditary weakness, bruises, sprains, or injuries. This is a serious condition that affects the usefulness of thousands of horses. Complete recovery should not be expected, but a usable horse is possible with proper care and warm-up.

The most consistently effective treatment is surgery or firing by a veterinarian. Pain to the horse is caused by tissues around the bone. Surgery to remove a portion of the cunean tendon and 60 days of rest usually relieves the pain and restore the majority of the soundness.

Firing (applying a hot iron or needle to a blemish), while of less value than tendon surgery, is sometimes prescribed. It is done to produce an acute reaction of a chronic inflammation in the hopes that it will undergo the natural process of healing. An effort is generally made to puncture the cunean tendon in firing to create the same effect as surgical removal.

Corns

This is similar to the condition that occurs in humans (Figure 39-10). It is caused by bruising the sole of the feet, generally due to rough terrain or poor shoeing. It is difficult to heal. Rest and qualified help may be necessary. Treatment usually consists of rest, special shoeing, poulticing, and strict sanitation.

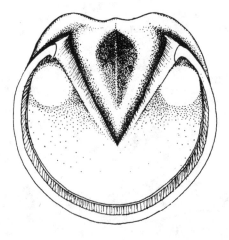

FIGURE 39-10. Corns, about the size of a quarter, on either side of the frog on the inner back side of the hoof, are the result of neglected feet, excessive trimming, improper shoeing, or other conditions that throw pressure on the feet. Corns are often discolored red or reddish yellow.

Gravel

Gravel, a lay term, supposedly is caused by the migration of a piece of gravel from the white line of the hoof to the heel area. Actually, a crack in the white line occurs first (see Figure 39-7) allowing infection of the sensitive structures. Inflammation is the result, and since the gravel (or other material) blocks the logical route, drainage breaks out at the heel area.

Excessively dry feet, creating a crack in the white line or sole, is the most common cause, although founder and injury are also suspect. Lameness appears before drainage is visible, although it may go undetected until after drainage. Horses often modify their gait to obtain relief, and it may not be noticeable except to one familiar with the mount. If caught before drainage, recovery is almost a certainty. Recovery chances after drainage are good but less optimistic.

Treatment involves cleaning and proper draining of the foot. Soaking the foot in epsom salts is recommended even if drainage has begun. Tincture of iodine and antiphlogistic pastes may be applied daily under a bandage for a week and once every 3 to 4 days afterward, provided that the foot stays dry. Treatment should continue until completely healed. Tetanus antitoxin should be administered.

Grease Heel (or Scratches)

Pasterns and fetlocks become inflamed with a mange-like skin condition (Figure 39-11). It is caused by poor sanitation, although some horses are more susceptible than others. Treatment consists of cleaning the fetlocks with mild soap and water. The hair is clipped closely around affected areas, and astringent, antiseptic substances are then applied at regular intervals. Clean, sanitary quarters are a necessary part of treatment routine.

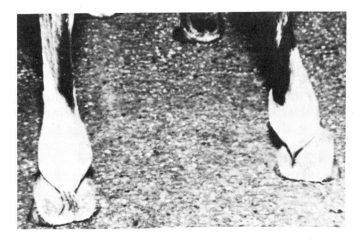

FIGURE 39-11. Grease heel (scratches) is a mange-like condition affecting pasterns and fetlocks. Note the rough scab-like infection caused, in this case, by unsanitary conditions in a muddy lot. (Courtesy of the School of Veterinary Medicine, Texas A&M University)

Hoof Cracks

Faulty conformation, brittle hooves, or foot injury can cause vertical cracks or splits from the coronet down (Figure 39-7). These may be mild or very painful (Figure 39-12). Treatment usually involves trimming the hooves and special shoeing.

FIGURE 39-12. Hoof cracks are commonly caused by droughty conditions. Hoof dressings and proper shoeing may be necessary to prevent lameness. (Courtesy of the School of Veterinary Medicine, Texas A&M University)

Navicular Disease

Inflammation of the small navicular bone and bursa of the front feet is one of the most important causes of lameness in horses. It does not occur in the hind feet. As the disease progresses, the inflamed bursa, bone, and surrounding tissue may undergo calcification. Rest, in the early stages, produces improvement, but the condition reappears when training or work resumes. Both front feet may be affected but usually only one foot is the case. Early signs are repeated lameness following a rest of a few days and pointing with the affected foot (shifting weight to the healthy foot and extending slightly the affected hoof). In riding, the toe of the foot strikes before the heel, shortening the stride and causing many owners to diagnose the injury as a shoulder problem.

The problem has been attributed to an inheritable condition resulting from upright conformation or a weak navicular bone. Although inheritance plays a part, it is not the actual cause and any horse may develop the problem. The chief aggravating factor is concussion of the front feet, especially where horses are worked on a hard surface or subjected to hard work such as racing, cutting, calf-roping, and so on.

Treatment consists of special shoeing and/or injection of corticoids. Both are usually only temporary solutions. The only permanent solution that has received acceptance is blockage of the nerves to desensitize that part of the foot. This can add several years of useful service.

Quittor

Quittor is a chronic inflammation of a collateral cartilage at the coronet (Figure 39-7). Swelling, heat, and pain over the coronary band are characteristic signs. It is manifested as a deep-seated running sore, usually confined to front feet, although occasionally seen in the hind feet. The infection may arise from a puncture wound or traumatic bruise that reduces circulation in the area.

Drainage and antiseptics have been successively used, although surgery is usually recommended to remove necrotic tissue. Incisions are not sutured, but a bandage and a poultice such as Denver Mud (Demco Company, Denver, Colorado) is applied to speed healing.

Ringbone

This is a new bone growth on the pastern bone of the front foot (Figure 39-13), only occasionally seen on the hind feet. It causes lameness and a stiff ankle. The condition is rare in Thoroughbreds but quite common in other breeds. Two types afflict horses, high ringbone and low ringbone, both similar except for location of the new bone growth.

Trauma is the usual cause of ringbone. Pulling, bruising, or a direct blow to the phalanges leads to heat, swelling, and pain in addition to the new bone

FIGURE 39-13. Ringbone is an enlargement of bony growth of the pastern bone resulting in constant lameness. It is usually caused by a direct blow or pounding on hard surfaces. (Courtesy of the School of Veterinary Medicine, Texas A&M University)

growth. Lameness is constant, and the horse will flinch when finger pressure is applied to the active area. X-rays are needed for positive diagnosis, however. Cold water bandages give some temporary relief. For more permanent relief, blistering (an irritant applied to the affected area), firing (the use of a hot iron or needle), or severing of the nerve leading to the area is used.

Sidebones

Sidebones (Figure 39-14) are quite common in horses, except Thoroughbreds. Lameness may or may not occur. It is more common in the forefeet and involves an ossification of the lateral cartilages just above and to the rear quarter of the coronet. The growth may occur on either or both sides of the foot.

The cause is probably concussion, causing trauma to cartilages. Poor shoeing or wire cuts have also been cited as causes. Unless pain and heat accompany sidebones, it does not often cause lameness, but mechanical inter-

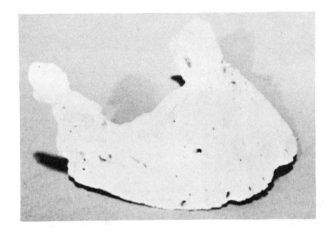

FIGURE 39-14. Buildup of bone occurring just above the coronet on either side of the front foot is characteristic of sidebone. The buildup may be pronounced as in this dramatic illustration but seldom results in lameness. (Courtesy of the School of Veterinary Medicine, Texas A&M University)

ference with foot movement does become a problem due to massive bone formation.

Treatment, if sidebones are definitely a cause of lameness, consists of cold water bandages for temporary relief. Corrective trimming and shoeing to roll the foot on the affected side are a more permanent solution. The hoof is also grooved to aid in expansion of the quarters of the foot, relieving pain.

Splints

Splints (Figure 39-7) are new bony growths on the cannon bone, usually on the inside and usually on the foreleg. When found on rear legs, they are usually on the outside of the cannon. The condition is usually associated with hard training, malnutrition, or faulty conformation in young horses under 2 years old. Horses 3 or 4 years old are occasionally affected. Trauma caused by slipping, running, falling, or jumping can induce the initial trauma leading to ossification. In mild cases, the lameness is not evident in the walk. Other gaits exhibit the problem although this condition does not always cause lameness.

Treatment of splints is more simple and effective than most leg disorders. Usually 30 days of rest will cause them to heal of their own accord, but some owners have hastened healing with hot and cold applications to affected limbs, and then an antiphlogistic pack is applied to reduce swelling. Firing and blistering have been tried with some success but are generally thought to be unnecessary since time and rest will relieve symptoms eventually.

Stringhalt

Stringhalt (Figure 39-15) is an involuntary flexion of the hind legs, most evident when backing a horse. The foot is markedly jerked toward the abdomen and may actually hit the abdominal wall. The cause is unknown but it is speculated that nervous diseases, degeneration of nerves, and/or spinal cord disorders could be among the inital phases of the problem. Surgery is the only

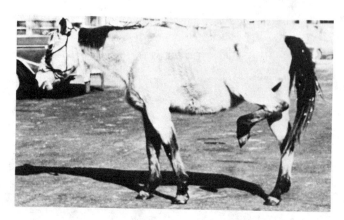

FIGURE 39-15. Stringhalt is an involuntary flexion upward of the hind legs. The foot may just as violently descend to strike the ground. (Courtesy of the School of Veterinary Medicine, Texas A&M University)

Chap. 39 DISEASES OF HORSES

FIGURE 39-16. Clinical signs of thrush are a degeneration of the frog of the foot, black discharge, and an offensive odor. Thrush is most often caused by poor sanitation and improper foot care. (Courtesy of the School of Veterinary Medicine, Texas A&M University)

treatment and consists of removal of a portion of a tendon that crosses the lateral surface of the hock joint.

Throughpin

Throughpin (Figure 39-7) is a puffy swelling that can be confused with bog spavin. It is found in the web of the hock and can be distinguished by its movement under finger pressure to the opposite side of the leg. The condition is not serious, and treatment with linaments and massage may reduce the swelling. However, drainage and an injection of corticoids is the most effective treatment.

Thrush

Thrush (Figure 39-16) is a degeneration of the frog of the foot caused by poor sanitation. It is most commonly found in the hind feet. The first sign of trouble is a strong offensive odor. Treatment consists of trimming away the infected part of the frog, sanitation, and the use of an antiseptic. Many thrush remedy preparations are used by owners; among them are iodine, formalin, creolin, and carbolic acid.

Wind-Puffs (Windgall)

A "puff" or windgall (Figure 39-7) is a harmless swelling of a bursa (fluid sac). The bursae just above the pastern joints in fore and rear legs are affected. There is no heat or pain associated with them. The condition is attributed to working on a hard surface, racing, and so on. It is quite common in any hard-working horse, especially on rough terrain. Treatment with cold packs and blistering agents is recommended. Linaments and massage may also be effective.

PARASITES

Flies and Mosquitoes

Not only are these insects irritating to animals, but also they may be *vectors* (carriers) of disease. Insecticide sprays are somewhat effective but are of short duration if the horse sweats and removes the protection. A clean stable and removal of manure heaps help prevent accumulation of flies. Spraying with an electric fogging machine or constructing a stable with good ventilation also helps.

Horse Bots

The botfly (Figure 39-17) lays eggs on the hairs around a horse's fetlocks, knees, throat, jaws, and lips. The fly, somewhat smaller than and similar in appearance to the honeybee, does not sting the horse as popularly thought. The eggs hatch in 7 to 10 days, and the young larvae enter the horse's mouth where they molt and grow for 2 to 4 weeks. Next the larvae migrate to the lining of the stomach and attach themselves, feeding on blood until maturity (Figure 39-18). They then release their hold, pass out with the droppings, enter the pupal or resting stage for 20 to 70 days, and eventually emerge as flies ready to lay eggs once again.

Signs of bot infestation are frequent digestive problems and even colic, emaciation, anemia, and lack of energy. Prevention is carried out mostly by good grooming. Clipping of the attached eggs may be necessary in a control program.

Treatment does not begin until 30 days after the first killing frost in the fall, which is nature's way of control until the spring. Inspection of the horse for botfly eggs is made, and the eggs are clipped and washed off at least 30 days before administering a recommended vermifuge. This treatment normally is administered via stomach tube. A veterinarian should be in charge because the preparations given (usually derivations of trichlorfon) are toxic to the horse if given in excess.

Mites

Microspic creatures that burrow under the skin to lay eggs causing mange are called mites. If kept off the skin, they die within an hour to two weeks so

A

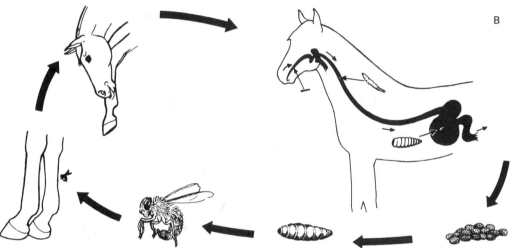

FIGURE 39-17. The common botfly (a) lays eggs in the spring on the leg hairs. (b) Eggs are hatched by the stimulation of the horse's lips, producing larvae that migrate to the stomach, where they attach and mature during the winter. Larvae are expelled with feces, pupate in the soil, and hatch out the following spring to produce another bot fly in the continuing cycle.

B

FIGURE 39-18. An autopsy revealed this severe infestation of bots, the larvae stage of the botfly, attached to the lining of the stomach wall. (Courtesy of Shell Chemical Co., San Ramon, California)

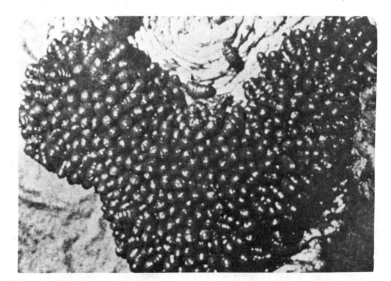

583

A

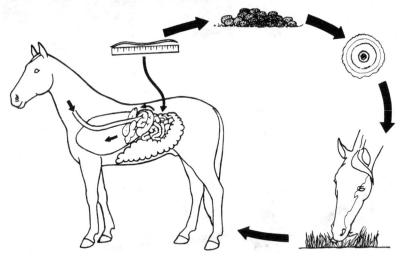

B

FIGURE 39-19. (a) Ascarids (roundworms) in the intestines of this young foal caused a rupture, peritonitis, and death. The life cyle (b) differs from strongyles in that damage is done in the liver and lungs. Larvae coughed up and swallowed mature in the stomach, produce eggs, and complete the cycle. (Courtesy of Shell Chemical Co., San Ramon, California)

control is easy with modern insecticides. Good grooming normally prevents them.

Worms

Ascarids (large roundworms) (Figure 39-19), stomach worms, and strongyles (bloodworms) are the most serious (Figure 39-20). Signs include poor growth, anemia, dull coat, listlessness, and digestive problems including colic. Pinworms (Figure 39-21) are less serious but a definite nuisance since the horse will rub its tail persistently.

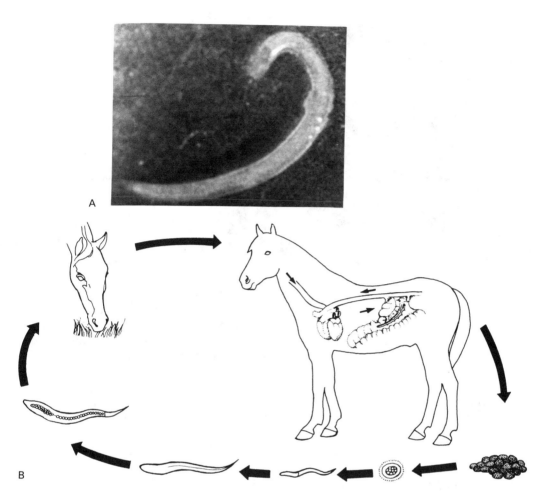

FIGURE 39-20. Large strongyles (a) are worms that inhabit the large intestine and lay eggs that mature to larvae in the excreted manure. (b) Larvae crawl up grass stems and are ingested by the horse to complete the cycle. Migration takes place through the stomach and circulatory system back to the large intestine, where eggs are laid.

The cause of infestation of all types of worms is generally related to a natural cycle that is difficult to break. Although the cycles vary, generally it can be summarized in the following way:

1. Eggs, larvae, or infected flies are swallowed by the horse in grazing, feeding, or drinking.
2. Young worms develop internally causing irritation and/or damage.
3. Mature worms lay eggs that are expelled in the feces.
4. Eggs mature and hatch into larvae or flies that reinfest the horse.

Pastures, once infested, may remain so for a long time because temperature changes and environmental conditions have little effect on the cycle.

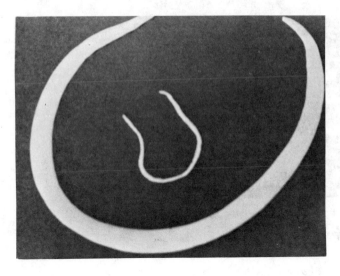

FIGURE 39-21. Pinworms (oxyuris) are whitish, may measure up to almost an inch in length, and can be seen in the feces of infested horses. The male (center) is much smaller than the female. (Courtesy of the U.S. Department of Agriculture)

Gathering manure from pastures or stables can be helpful if allowed to decompose in its own compost heap. Pasture rotations are helpful in reducing infections, but treatment with a worming agent via stomach tube is the only effective way to break the cycle. The problem is so persistent that horse owners have their horses wormed routinely once or twice a year.

INJURIES

Teeth Wear

Any horse may wear its rear teeth or molars down, producing sharp points that make it painful to eat. Veterinarians *float* (file off sharp points) teeth to correct the condition.

Wounds

Loose nails, projecting boards, sharp objects, barbed wire, and the heels and teeth of other horses cause great damage every year to horses. The horse will often panic in a critical moment, causing even further harm. Proper cleaning of wounds, suturing when needed, and tetanus shots should be considered.

STUDY QUESTIONS

1. Stable temperatures usually should be _____ as outside temperatures.

2. The normal rectal temperature for a healthy horse is _____.

3. Azoturia is brought on by irregular _____ without reducing _____ intake.

4. Colic could be caused by _____, _____, _____.

5. If colic is suspected a horse should be _____ to prevent twisted gut.

6. Stiffness in the legs and joints and fast-growing hooves are characteristic of _____.

7. A recommended treatment for founder is to stand the horse in _____ _____.

8. Heaves may be an _____ reaction.

9. Grease heel is caused by poor _____.

10. The navel of newborn foals should be treated with _____.

11. A running sore from the withers is called _____.

12. This infection of the _____ is _____ (contagious, noncontagious).

13. A notable type of sleeping sickness called _____ invaded the United States in the early 1970s.

14. The best safeguard against VEE is _____ because there is no _____.

15. High fever, coughing, and discharge of pus from the nose could be _____ or _____.

16. Low-lying areas, mosquitoes, and flies are prime conditions for _____ _____.

17. If a horse gets stiff in the legs, is sensitive to loud noise, has muscles twitch, and its inner eyelid folds over the eye, it probably has _____.

18. Lockjaw is _____ (seldom, usually) fatal even when treated.

19. _____ is a degeneration of the _____ of the foot characterized by an offensive odor.

20. The _____ fly lays eggs which hatch into _____ that feed on _____.

21. Internal treatment for bots should begin after a _____ because the larvae pupate in the warm _____.

22. A constant tail rubber probably has _____.

23. _____ are serious forms of worms.

24. To float a horse's teeth means to _____ them.

25. A wound on a horse should be _____ if needed and _____ and _____ shots given.

DISCUSSION QUESTIONS

1. What are the main internal parasites of horses, and what are the recommended preventive measures?
2. Which diseases or conditions are likely to result in digestive disturbances?
3. What are the major respiratory diseases?
4. Name the contagious diseases and give recommended treatment or preventive measures.
5. Discuss all the points of occurrence of foot and leg problems, their causes and treatment.
6. Define the following:
   a. Coggins Test
   b. blistering
   c. firing
   d. vectors
   e. bots

Chapter Forty

THE RABBIT INDUSTRY

The rabbit industry, termed rabbitry, is important in the world today. Most rabbits are commercially raised for meat, fur, wool, and laboratory usage. A few rabbit breeds, however, were developed strictly for their aesthetic value and are kept as pets. These fancy rabbit breeds have very little meat or fur value. Various shows are held for fancy rabbit breeds, being judged on aesthetic traits. Many of these breeds have traits such as variegated colored fur, extraordinarily long ears, and so on. These breeds are kept as pets in small numbers and are not covered in this book. Although the general rabbit management principles and practices discussed in these chapters will apply to such fancy breeds, our focus will be on rabbitry for meat, fur, and wool.

Rabbits have many advantages as meat producers. First they are excellent converters of feed to meat and will naturally produce more meat per pound of female live weight than most other animals. A sow, for example, who weans two litters of eight pigs each will produce 100% of her live weight in marketable pork. A 1000-lb cow weaning a 500-lb calf per year produces 50% of her live weight as beef. A rabbit doe can wean thirty rabbit fryers weighing 4 lb each per year or about 1000% of her live weight. Today about 50 to 60 million pounds of rabbit meat is produced in the United States each year. Production of marketable meat from rabbits is also achieved in a short amount of time. Generally only 90 days are required from mating a rabbit doe (female) until fryer rabbits are ready for market.

Second, rabbits can be kept easily, without large investments of money or equipment. Rabbitry will range from a small backyard operation with three or four hutches (rabbit houses), to a large commercial operation of several hundred hutches. As a backyard project, rabbits are easily raised, do not make annoying noises, and occupy only a small area for production. Hence they are very popular as FFA or 4H beginning animal projects.

DOMESTICATION AND BREEDS OF RABBITS

The domestic rabbit was domesticated in Africa, first being considered for food in Asia about 3000 years ago. In Europe, rabbits have been used for food for over 1000 years. Rabbits were brought to the United States from Europe in the early nineteenth century.

Contrary to the belief of many new rabbit producers, no domestic rabbit breeds in America developed from the native here (jackrabbit—genus *Lepus*) or the North American cottontail (genus *Sylvilagus*). Each of these are of a different genus than the domestic rabbits (genus *Oryctolagus*), are quite biologically different, and will not interbreed and produce young. Hence all breeds present today were either imported from other parts of the world or developed from imported breeds.

Rabbit breeds vary in size, use, color, and length of fur. Rabbit sizes vary from mature weights of 3 lb to 15 lb. Breeds with mature weights of 3 to 5 lb are termed small breeds and contain mostly fancy or pet rabbits. Breeds used for meat, fur, wool, or laboratory animals are contained in the medium breeds (9 to 12 lb mature weight) and large breeds (over 13 lb mature weight). Color of fur varies from solids to spotted to variegated. Length of fur varies from short hair breeds to the Angora rabbit with hair 8 to 10 in. long.

Although there are about 30 breeds of rabbits and over 70 breed varieties in the United States, the following are those utilized for meat, fur, and wool production.

Meat Breeds

American Chinchilla. The American Chinchilla rabbit is of three varieties—the standard (mature weight 6 to 7 lb), the heavyweight (mature weight 10 to 11 lb), and the giant (mature weight 13 to 15 lb). All are used for both fur and meat production. It is a meaty rabbit of medium length and compares with other breeds in meat production, quality, and flavor of meat. This breed was developed from the animals imported from France in 1919. Pelt (fur) color is an important selection trait for most breeders with the ideal an overall grayish color fur (chinchilla color) with a blue undercolor.

Californian.   The Californian rabbit was bred in the United States in 1923. It has broad shoulders, meaty back, and a good dressing percentage. Californian rabbits have a white body with dark gray or black ears, nose, tail, and feet. The eyes are a bright pink. The ideal mature weight for bucks is 9 lb and does is 9½ lb.

Champagne d'Argent.   The Champagne d'Argent breed is also known as the French Silver breed. It is an old breed, grown in France for over 100 years. Although a good meat rabbit, its fur is still one of the leaders for use in the garment industry. Champagne d'Argent rabbits are born black, but take on a silver or skimmed milk color with a dark slate blue undercoat at 3 to 4 months of age. Mature weights are 10 lb for bucks and 10½ lb for does.

Checkered Giant.   The Checkered Giant rabbit is a very large breed that is preferred by some breeders as a meat producer. The breed is white with black markings down the back, the sides, and in circles around the eyes.

Flemish Giant.   The Flemish Giant rabbit is the largest of the domestic rabbits with mature weights in excess of 13 lb. They are noted for size and quality of fur. One of their main uses is in crossbreeding programs with other breeds to increase meat production. Flemish Giants are in a variety of colors ranging from steel gray to black, sandy, white, or blue.

Satin.   The Satin rabbit is a good meat breed because of type and size. Mature weights are 9 to 10 lb. Nine colors are recognized with white being the most popular. They are noted for vivid colors and a sleek coat.

New Zealand.   The New Zealand breed is the most popular breed in the United States today. It is a very good meat and fur breed of medium size. Mature weights are 9 to 11 lb. New Zealand rabbits are either red or white. The red New Zealand rabbits have reddish bodies with cream bellies. Reds have brown eyes; white New Zealand rabbits have red eyes.

Fur Breeds

Angora.   The Angora rabbit is used for "mohair" production. Angora rabbits are white with long wool growth. Growth of the wool is usually about 1 in. per month and is plucked every 3 months. Mature weights range from 6 to 8 lb.

Rex.   The Rex rabbit's fur lacks the stiff guard hairs usually found on other rabbits. The fur has a plush-velvet, soft, silky feel and is desired by the garment industry. Colors vary including beaver, black, chinchilla, lilac, and opal.

SELECTION OF BREEDING RABBITS

Selection of breeding stock is done by the rabbitry beginner and by the producer who is selecting replacement does and bucks for the herd. The beginner first selects the breed or breed combination based upon the intended use of the rabbit (meat, fur, wool), the available markets for the end products, and personal preferences. Once the breed is determined, rabbits must be selected from the breed to serve as foundation stock. This will vary from 2 to 10 does and one buck to several hundred, depending on the size of the operation. This selection process is the same whether it is selection of replacement rabbits, or selection of foundation stock of a future herd.

Breeding rabbits are selected based on their ability to produce plenty of offspring with compact, meaty bodies, thick hindquarters, loins, and shoulders. The offspring's body carcass should be meaty and solid, not flabby and loose.

As with other enterprises, production records are very valuable in the selection process. In selecting rabbits, individual doe, buck, and sibling records are used. The minimum records that should be considered are the average litter number and weaning weights. Some selection criteria are listed below:

1. Good production by raising a uniform litter of over six fryers per litter with four litters per year.
2. Milking ability—10 or more teats, litter weights of over 6 lb per litter at 3 weeks of age.
3. Mothering ability—Easy breeding, high offspring survival rates.
4. Ability to grow—Bunnies that are at least 1.25 lb at 4 weeks of age, and 4 lb at 8 weeks of age.
5. A high-yielding carcass—over 50% dressing percentage.
6. Good fur quality.
7. Good health and vigor.

As with other livestock selection, the more criteria used in the selection process, the more certain the success of the selected individual. Selection should be from reliable breeders, and based on a lot more than the eye appeal of a furry bunny.

RABBITRY MANAGEMENT

Success in rabbitry, like any other livestock enterprise, will depend upon the management of the rabbit herd. Management skills are always based upon knowledge of the animal to be managed—rabbitry is no exception. The basic principles already learned in animal nutrition, reproduction, and health apply to rabbitry, along with those unique to the rabbit. In this chapter we deal with management of the rabbit herd, based upon such knowledge.

HOUSING AND EQUIPMENT FOR RABBITRY

Hutches

Although the size of the rabbitry enterprise varies from two rabbits to thousands, all require the same basic housing and equipment for successful management. Rabbits are kept in cages called hutches for ease of handling, protection, and proper management.

Selection of the type of hutch depends upon the size of the operation, climate, investment potential of the owner, and the desired ease of cleaning. Double-compartment hutches may be desirable for the small backyard operator (Figure 40-1), while thousands of wire hutches kept in a building would meet the needs of the larger operator. The climate will dictate the type of hutch that is used. Adult rabbits can withstand very cold temperatures (below freezing) with a minimum of protection, but they cannot withstand very hot temperatures or prolonged direct exposure to sunlight. Rabbits need some indirect sunlight. A rabbit also cannot withstand prolonged exposure to drafts or dampness, but needs some ventilation. Young rabbits are born without fur and are extremely susceptible to the environment—needing extra protection. Hutches should provide indirect light and 8 to 10 changes of air per hour but protect the rabbit from exposure to strong drafts, rain, and prolonged sun-

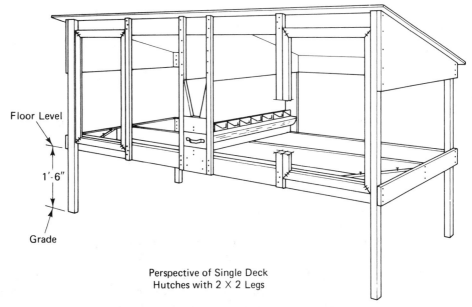

Floor Level

1'-6"

Grade

Perspective of Single Deck
Hutches with 2 X 2 Legs

FIGURE 40-1. Single-deck hutch.

light. Hutches should be constructed to allow convenience and ease of rabbit handling for the owner, ease of cleaning and proper sanitation, and maximum comfort for the rabbit. Such hutches can be very simple in construction, keeping investments to a minimum.

Sanitation of hutches is a must for rabbit health. Ease of cleaning must be considered in design and placement of rabbit hutches. Hutches are constructed from wire, wood, or both. Wire is usually more expensive and does not provide enough protection in adverse weather. Wood and wire hutches are easily constructed to meet the requirements based on the climate of the area.

Each compartment in a hutch should be large enough for one doe and her litter until weaning time, or for 10 to 15 weaned bunnies. A rule of thumb is 1 ft^2 of floor space for each pound of rabbit. Usually 2½ ft by 4 ft by 16 in. high is suitable for smaller breeds; and 2½ ft by 6 ft by 20 in. high suitable for larger breeds. For ease of handling of the rabbit and cleaning, the door of the hutch should be large enough to move nest boxes, feeders, and so on, in and out of the hutch easily. This will be a minimum of 13 in. by 16 in. Slotted wood (1- to 1½-in.-wide slots with ⅝ in. spacing) or ⅝-in. wire mesh floors allow feces to fall through and aid greatly in cleaning the hutches. Only one doe or buck is housed per compartment. Does can be housed in adjacent hutches, but a solid wall should always separate a doe from a buck rabbit. One compartment will house 10 to 15 fryers (weaned bunnies) in cool weather, or 10 to 12 fryers in hot weather. Two cages should be kept in a separate area for isolation of new or sick rabbits.

Nest Boxes

A nest box provides shelter and protection of newly born bunnies and seclusion of the doe when she gives birth (kindles). It should be large enough to prevent crowding but small enough to maintain body heat. The size of the nest box is from 12-by 12-by 16-in., to 10-by 16-by 8-in. with a 6-in. by 6-in. opening in the top or end (Figure 40-2). Ventilation holes should be placed in the sides of the nest box, with drainage holes placed in the floor. Nest boxes can be placed inside the hutch, or be a part of the floor of the hutch. Boxes inside the hutch also provide the doe with a space to escape the pestering young bunnies. If the box top is level with the floor, an escape board (1- by 6- by 12-in. board placed 9 in. above the floor) should be provided for the doe with growing bunnies.

Feeders and Waterers

Feeders and waterers vary from various types of containers, to self feeders and automatic watering systems. Containers used for feed and/or water should be 3 to 4 in. deep, 6 to 8 in. in diameter, easy to clean, and heavy enough to prevent rabbits from tipping them over. Hay mangers should be about 6 in. off the floor, and be filled from the outside of the hutch.

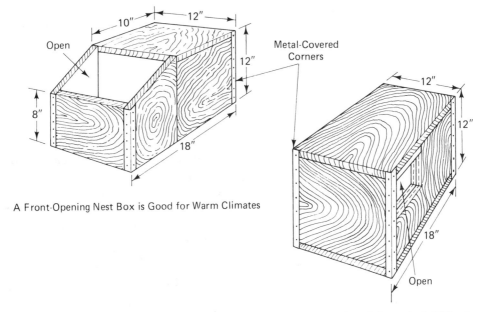

A Front-Opening Nest Box is Good for Warm Climates

A Covered Nest is Better in Cold Weather

FIGURE 40-2. Nest boxes.

Other Equipment

Other helpful equipment are measuring cups to measure out feed in ounces, scales to aid in performance records, brushes and combs for grooming, and a clean carrying basket or cart for rabbit handling.

FEEDING THE RABBIT HERD

Nutrient Requirements

The rabbit's digestive system could be compared to that of a horse (Chapter 37). It is a simple digestive system with an enlarged cecum and large intestines. This allows for the intake and utilization of high quality, leafy hays, grasses, and green vegetables. These are broken down by bacteria in the lower digestive tract as in the horse.

Unlike other animals, rabbits will eat their feces (called coprophagy). This is quite normal for rabbits and is based upon the construction of the digestive tract. Coprophagy is usually done late at night or in the early morning. Feces pellets that are light green and soft are eaten; those excreted in the day, which are brown and hard, are not eaten. This allows the rabbit to make full use of the bacteria digestion in the lower tract; i.e., conversion of forage protein to

TABLE 40-1. PERCENT COMPOSITION OF RABBIT RATIONS

Rabbit Type	Crude Protein	Fat	Fiber	Ash
Sexually active buck, pregnant does, does with litters, growing bunnies under 3 months of age	14–18	3–6	15–20	5–6
Dry does, bucks, maturing bunnies	12–14	2–4	20–28	5–6

high-quality bacteria protein, synthesis of B vitamins and breakdown of cellulose or fiber to energy. Thus a rabbit should be allowed to practice coprophagy.

Like all animals, rabbits require carbohydrates, fats, proteins, minerals, vitamins, and water. The amounts of each required depend upon the age of the rabbit, desired production, and rate of growth (Table 40-1). Clean water is a must. A doe can drink 1 gallon of water per day. A salt spool is used to supply needed minerals. Protein content of rations is usually simplified by using two rations, one 16%, the other 14%. These are fed to rabbits with high or low requirements (Table 40-1).

A suggested ration is 40% rolled oats, 25% wheat bran, 15% rolled oats or barley, 18% oil meal, 1% bonemeal, and 1% salt. This or a commercial ration should make up about 40% of the total feed, the other part from hay or green feeds.

Amount of Feed to Use

The amount of feed fed per day varies with the size of the rabbit and the production stage (Figure 40-3). Table 40-2 shows the amounts of feed to use as a guide. Green feeds or high-quality hays should be given along with the concentrate ration in Table 40-2.

Changing rabbits from one ration to another should be done gradually over a 7- to 10-day period. To do this, mix a small amount of the new feed with the old, increasing the proportion of the new feed over the 7- to 10-day changeover period.

Feeding Schedules

The regularity of feeding is more important than the number of times per day the rabbits are fed. Nursing does should be fed twice daily with either grain and hay at both feedings, or hay in the morning and grain and hay in the evenings. Rabbits fed once daily (dry does, etc.) should be fed in the late afternoon or evening since they eat mostly at night.

Types of Feeds

Concentrate or grain mixtures can be home mixed, or commercial supplements. At least two different grains should be used in the mix to enhance

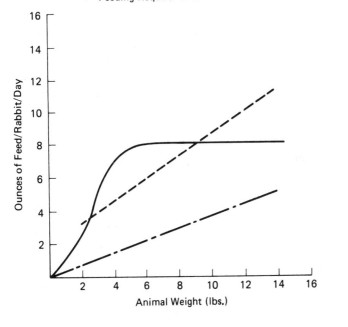

FIGURE 40-3. **Feeding requirements for rabbits.**

Legend:
—————— Developing Young and Lactating Does (Full Feed)
– – – – – – Pregnant Does
—·——·— Maintaining Bucks and Does

intake. Grains in order of preference by rabbits are rolled oats, wheat, grain sorghum, and corn. Cottonseed meal should not be used as a protein supplement due to gossypol content.

Hays that are used must be of high quality—high in nutrients, leafy, and clean. Legume hays, like alfalfa, are preferred. They are high in protein, clean, leafy, and relished by rabbits. They can be used as the only feed for dry does,

TABLE 40-2. FEEDING AMOUNTS FOR VARIOUS RABBITS

Rabbit Type	Feeding Regime
Young bunnies (birth to 3 weeks age)	Does' milk only
Bunnies over 3 weeks old	Whole grain (oats) free choice; some hay.
Weaned bunnies	Free-choice concentrate and hay ration.
Dry does, bucks	Full feed of hay plus 3 to 6 oz of concentrate per day if needed. Adjust to maintain proper weight. Do not let does become overfat.
Does at kindling	2 to 4 oz of concentrate first day. Gradually increase ½ to 1 oz/day to free choice in 5 to 7 days (4 to 7 oz/day). Feed 40% hay of the ration.
Lactating does	Free-choice concentrates, 40% hay.

bucks, and older rabbits, and up to 40% of the diet of lactating does. Grass hays are lower in quality and should not make up more than 10% of the diet.

Other green feeds can be used if available. This includes garden greens, roots, and tubers. If used, they should be fed only to bunnies over 3 months of age, and mature rabbits at a minimum level of 1.5% of their body weight. Greens include beets (roots and tops), cabbage, carrots (tops and roots), kale, potatoes, sweet potatoes (tubers and vines), turnips (tops and roots), and grass and legume leaves.

REPRODUCTIVE MANAGEMENT

Sexual Maturity

Puberty in rabbits is when a rabbit reaches sexual maturity. This is when the reproductive organs are fully developed and functional. Sexual maturity or puberty occurs at about 4 months of age in the small breeds, 6 to 7 months of age for medium breeds, and 9 to 12 months of age in large breeds. Rabbits can be mated for the first time at sexual maturity (5½ months for small breed does, 6½ months for bucks).

Mating

A doe's estrous cycle is different from other farm animals. Other farm animals have an estrus cycle in which the female has but a few days when she is in heat and will accept the male. The rabbit is almost the opposite. The doe has a 15- or 16-day estrous cycle. Of these 16 days, she is in heat 12 days and she will accept the buck when conception can occur. Only the first and last 2 days of the cycle are when a doe is not in heat and will not accept the buck.

Many litters of bunnies can come from one doe in a year. Proper management usually will allow for four to eight litters per year, depending on the production cycle. This will allow for proper weaning of a litter and proper conditioning of the doe prior to kindling (giving birth) of the next litter.

Young does are bred at 5½ months of age to an experienced buck. If she does not mate, breeding is tried again every 10 days until she is 6½ months old, then the young doe is force mated. Force mating is manually restraining the doe for the buck to mate.

Does are rebred when the litter is 3 to 7 weeks old, depending on the number of litters desired per year. As discussed earlier, chances are good that she will accept the buck. If she does not, try to breed the doe again in 3 to 4 days. Success usually follows. If the doe still fights the buck, she may need to be force mated.

Breed a doe in the morning or afternoon before feeding. This can be done by natural mating or artificial insemination. Natural mating is accomplished by taking the doe to the hutch of the buck, never vice versa. Introduction of a buck into the doe's hutch will result in the doe fighting off the buck. This happens not because the doe is not in heat, but because she is possessive of her hutch. Mating usually occurs fairly rapidly. Since the doe will ovulate about 8 to 10 hours after the first service, some producers will rebreed the doe at 8 to 10 hours after the first service. One buck is kept for 10 does, and a buck is not used more than once in 3 days. The reproductive life of a buck is 6 months to 2 years; the reproductive life of a doe is 2 to 3 years.

About 10 to 18 days after breeding, the doe should be tested for pregnancy. This can be done by test mating or palpation. In test mating the doe is returned to the buck. If she refuses the buck, she is thought to be pregnant.

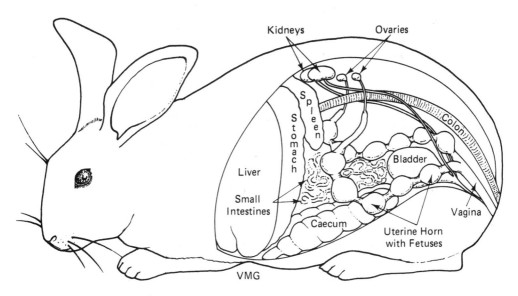

FIGURE 40-4. Pregnancy detection in the rabbit is relatively simple. With thumb on one side of the abdomen and fingers on the other, a gentle sliding motion up to the rib cage can detect marble size fetuses.

Palpation of does is done in 10 to 14 days after breeding. In palpation, the horns of the uterus are felt through the abdomen. Rabbits are easy to palpate and with a small amount of practice, this can be very accurate. To palpate a doe, restrain her by grasping a skin fold over her neck along with her ears with one hand. This leaves the other hand free for palpation. Pass the free hand, palm up, under the abdomen to the hind legs. With thumb on one side of the abdomen and fingers on the other, gently bring finger and thumb closer together, and slide them up to the rib cage. As you do this, the horns of the uterus should pass through your hand. The fetuses will be the size of marbles (Figure 40-4). Rebreed any open does.

Kindling

Kindling (giving birth to the young) normally occurs in 31 to 32 days after breeding. This can range from 28 to 35 days. Bunnies are born in litters of 2 to 15, with most litters 6 to 10. They are born in the nest box. They are born with their eyes shut and without fur. At one week, the eyes will open, and the fur begins to appear. At about 3 weeks, the bunnies will come out of the nest box and begin eating with the doe. Bunnies are weaned at 7 to 8 weeks (depending on the breeding cycle of the producer).

Management of the doe prior to, during, and after kindling involves:

1. Do not allow the doe to get too fat prior to kindling. Fat does have lower litter sizes and have complicated kindlings.
2. At 21 to 28 days of gestation, disinfect the nest box, fill it two thirds full with bedding, and place it in the doe's hutch. She will build a nest in it prior to kindling.
3. Prior to kindling, the doe will pull fur from her body and put it into the nest box. This will continue through kindling. Just prior to kindling, the doe will lose her appetite (1 to 2 days prior).
4. A doe usually kindles during the late evening or early morning. Parturition usually is complete within half an hour. Care should be taken to keep the doe calm during this time.
5. Examine the nest box after kindling to record births and to remove any dead bunnies.
6. If more than eight bunnies are born in a litter, foster extra bunnies to other litters. This is done by rubbing the bunny with the foster mother's fur.
7. Remove the nest box when all the young have left it, at about 3 weeks from kindling.
8. Cull does that lose or destroy their litters twice, or lack natural mothering ability.

Management of Weaned Bunnies

Bunnies can be weaned from doe milk as early as 4 weeks by placing them in another pen to be fed until slaughter. Usually, such early weaning results in

TABLE 40-3. EFFECT OF TIME OF REBREEDING AND WEANING ON NUMBER OF LITTERS PRODUCED PER YEAR BY DOES

Rebreeding[a] (days)	Weaning[a] (days)	Days between Litters	Litters/Doe/Year
14	28	45	8
21	35	52	7
28	42	59	6
42	56	73	5

[a]Days after kindling.

smaller and less meaty bunnies than those weaned at 7 to 8 weeks of age. Earlier weaning does, however, allow for more litters per year per doe (Table 40-3). Regardless of weaning strategy, bunnies are usually slaughtered for fryers at 8 weeks of age (56 days).

STUDY QUESTIONS

1. Rabbit production is termed _____ .
2. Rabbits are produced for _____ , _____ , _____ and _____ _____ .
3. A rabbit doe can produce _____ % of her weight in marketable meat per year.
4. Only _____ days are required from mating a doe until rabbit fryers are ready for market.
5. Some domestic rabbit breeds in the United States developed from the native hare and cottontail. (True, false) _____ .
6. Rabbit breeds vary in _____ , _____ , _____ and _____ .
7. Most rabbit breeds used for meat are of the _____ or _____ weight breeds.
8. The largest rabbit breed is the _____ _____ .
9. The most popular rabbit breed in the United States is the _____ _____ .
10. Rabbit "mohair" is produced by the _____ breed.
11. Rabbits are kept in _____ for ease of management.
12. Rabbits can withstand cold temperatures better than hot temperatures. (True, false) _____ .
13. In housing, a rabbit needs _____ ft^2 of floor space per pound of rabbit.
14. The rabbit's digestive tract is much like that of a _____ .
15. Puberty in rabbits occurs at about _____ months for smaller breeds to _____ months in larger breeds.
16. The estrous cycle of the doe is _____ days long of which she is in heat _____ of these days.
17. The doe ovulates _____ hours after service.
18. Giving birth in rabbits is termed _____ .
19. A doe will give birth about _____ days after service, and have _____ bunnies per litter.
20. Bunnies are usually slaughtered at _____ weeks of age and are called _____ at slaughter.

DISCUSSION QUESTIONS

1. Explain the differences between fancy and other rabbit breeds.
2. What are the advantages of rabbits for meat production?
3. Why are Rex rabbits desirable for fur production?
4. List five criteria you would use in selection of rabbits based on production records.
5. List the climate conditions a rabbit cannot withstand and must be protected from for good production.
6. What are the functions of a nest box and escape board?
7. Define and give the advantage of coprophagy.
8. List the green feeds that can be used for rabbits.
9. What makes the rabbit's estrous cycle different from that of other animals?
10. Describe how to palpate a doe.

Appendix A

CANADIAN BEEF AND PORK GRADING SYSTEMS

BEEF CARCASS GRADING REGULATIONS FOR CANADA (Revised 78 06 29)

Grade	Standards

1. **Canada A**
   (youthful; good to excellent quality)

   a. Maturity Class I (Divisions I and II)
   b. Lean: firm
      fine texture
      bright red color
      slight marbling
   c. Fat: firm
      white or slightly tinged
   d. Muscling: free from marked deficiency
   e. Described further by fat levels 1, 2, 3, and 4 for Canada A
      as follows:

Warm Carcass Weight (lb)	1	2	3	4
≤499	0.2–0.3	0.31–0.5	0.51–0.7	0.71 +
500–699	0.2–0.4	0.41–0.6	0.61–0.8	0.81 +
≥700	0.3–0.5	0.51–0.7	0.71–0.9	0.91 +

2. **Canada B**
   (youthful; medium
   quality)

   a. Maturity Class I (Divisions I and II)
   b. Lean: moderately firm
      somewhat coarse texture
      bright to medium dark red
      no marbling
   c. Fat: firm or slightly soft
      white to pale yellow
   d. Muscling: free from marked deficiency
   e. Described further by fat levels 1, 2, 3, and 4 for Canada B
      as follows:

Warm Carcass Weight (lb)	1	2	3	4
≤499	0.1–0.3	0.31–0.5	0.51–0.7	0.71 +
500–699	0.1–0.4	0.41–0.6	0.61–0.8	0.81 +
≥700	0.2–0.5	0.51–0.7	0.71–0.9	0.91 +

3. **Canada C, Class I**
   (youthful and intermediate age; medium to good
   quality)

   a. Maturity Classes I and II
   b. Lean: moderately firm
      bright to medium dark red
   c. Fat: firm or slightly soft
      white to pale yelllow
      light covering
      no excess proportion
   d. Muscling: low medium to excellent
   e. To include carcasses with *less than* fat level 1, Canada B
      with Canada B quality

Grade	Standards
4. Canada C, Class 2 (youthful and intermediate age; poor quality)	a. Maturity Classes I and II b. Lean: soft coarse and sinewy texture bright to extremely dark red c. Fat: firm to soft white to lemon yellow slight covering no excess proportion d. Muscling: excellent to deficient e. To include carcasses with less than fat level 1, Canada B, and less than Canada B quality
5. Canada D, Class 1 (mature select cows)	a. Maturity Class III b. Fat: firm white to pale yellow well over ribs and loins moderately over hips and chucks no excess proportion c. Muscling: excellent to good
6. Canada D, Class 2 (mature good to medium cows and steers)	a. Maturity Class III b. Fat: firm to slightly soft white to lemon yellow cover most of surface no excess proportion c. Muscling: medium
7. Canada D, Class 3 (mature fair to plain cows and steers)	a. Maturity Class III b. Fat: soft white to deep lemon yellow light to slight covering no excess proportion c. Muscling: fair
8. Canada D, Class 4 (excessively finished mature cows and steers)	a. Maturity Class III, but to include carcasses from Maturity Classes I and II with muscling and quality ranging from low medium to deficient b. Fat: an excess proportion c. Muscling and quality: any degree
9. Canada D, Class 5 (very thin mature cows and steers)	a. Maturity Class III, but to include Maturity Classes I and II, which are extremely thin and poor in quality b. Fat: little to no exterior finish c. Muscling and quality: very deficient
10. Canada E (mature stags and bulls)	a. Maturity Class III but may include more youthful animals if: b. Shows pronounced masculinity c. Lean: coarse and/or sticky texture dark red color

Source: Courtesy of W. R. Usborne, Associate Professor, Meat Science, University of Guelph.

Additional Points

1. Beef Carcass means the entire carcass of an animal of the bovine species, except the hide; that portion of the head and neck forward of the first cervical joint; that part of the foreshank below the knee joint; that part of the hindshank below the hock joint; the alimentary canal; liver; kidneys; spleen; genital tract and genitalia; mammary system; heart; lungs; membranous portion of the diaphragm; pillar of the diaphragm (hanging tender); spinal cord; internal fats, including channel fat, kidney fat, pelvic fat, and heart fat; external cod fat and udder fat; the tail at a point between the first and second coccygeal vertebrae or any portion of the beef carcass, the removal of which is required under the Meat Inspection Act or any regulations made thereunder.

2. The Canada A and Canada B quality grades are subdivided into four fat levels related to warm carcass weight, determined by taking one fat measurement; such measurement to be made between the eleventh and twelfth ribs after the carcass has been knife-ribbed at the minimum point of thickness in the fourth quadrant on the longitudinal axis of the longissimus muscle and perpendicular to the outside surface of the fat.

3. All beef carcasses which are graded must be branded as follows:

Canada A-1, A-2, A-3, and A-4	red ink
Canada B	blue ink
Canada C	brown ink
Canada D and E	black ink

4. Maturity classes
   a. Maturity Class I (Division I):
      (1) Bones that are soft, red, and porous when split
      (2) Pearl-like capping cartilage on all the vertebrae of the spinal column
      (3) Sacrum and sternum bones that show distinct divisions
   b. Maturity Class I (Division II):
      (1) Sacral vertebrae that are partially or completely fused
      (2) Lumbar vertebrae that have evidence of cartilage on the tips
      (3) Caps on the thoracic vertebrae that have slight ossification
   c. Maturity Class II:
      (1) Sacral vertebrae that are almost or completely fused
      (2) Lumbar vertebrae that have at least a red line present on the tips
      (3) Cartilaginous caps on the thoracic vertebrae that are partially ossified
      (4) Chine bones that show varying degrees of redness as the blood cells recede towards the periphery of the dorsal vertebrae
   d. Maturity Class III:
      (1) Sacral vertebrae that are completely fused
      (2) Chine bones that are generally hard, white, and flinty when split

(3) Cartilaginous caps on the ends of the lumbar vertebrae that are completely ossified and devoid of any red color

(4) Ribs that are wide, flat, and white

(5) Advanced ossification in the cartilaginous caps of the thoracic vertebrae and in the sternum bone

THE CANADIAN HOG CARCASS GRADING/SETTLEMENT SYSTEM*

Starting January 2, 1979, hog carcasses graded by federal government graders will be settled for on the grade indices shown in the accompanying table. This table is a revision of the one which has been in use since January 2, 1978. Indices above 90, have been increased in weight classes 6 and 7 and decreased in weight classes 8 and 9.

Duration

This table of indices will apply until December 31, 1979 unless there is mutual agreement by the Canadian Pork Council and the Meat Packers Council to recommend changes to it during the year.

A joint committee of the two councils has been set up to monitor and review the effect of the revised system. If after a period of adjustment, the revised table has not produced the result of arresting, or reversing, the trend to increased percentages of heavier carcasses, the committee will consider what action to recommend.

Use of Table of Indices

The following is an example of use of the grade/settlement system: Assume that a carcass weighed 170 lb and that total fat as measured at the loin and shoulder was 3.0 in.; and that the bid price was 65 cents per pound. The index grade would have been 105. Settlement would have been as follows:

$$170 \times \left(0.65 \times \frac{105}{100}\right) = \$116.02$$

*Courtesy of the Canadian Pork Council.

1979 TABLE OF DIFFERENTIALS

		Weight Class										
		1	2	3	4	5	6	7	8	9	10	11
		90	125	130	140	150	160	170	180	190	200	210+
		124	129	139	149	159	169	179	189	199	209	
							lb					
Fat class	Inches						Grade Index[a]					
1	-1.9	87	105	107	108	110	113	114	113	112	90	80
2	2.0–2.1	87	103	105	107	108	112	113	112	110	90	80
3	2.2–2.3	87	102	103	105	107	110	112	110	108	90	80
4	2.4–2.5	87	100	102	103	105	108	110	108	107	90	80
5	2.6–2.7	87	98	100	102	103	107	108	107	105	90	80
6	2.8–2.9	87	97	98	100	102	105	107	105	103	90	80
7	3.0–3.1	87	95	97	98	100	103	105	103	102	90	80
8	3.2–3.3	87	93	95	97	98	102	103	102	100	90	80
9	3.4–3.5	87	92	93	95	97	100	102	100	98	90	80
10	3.6–3.7	87	90	92	93	95	98	100	98	97	90	80
11	3.8–3.9	87	88	90	92	93	97	98	97	95	90	80
12	4.0–4.1	87	87	88	90	92	95	97	95	93	80	80
13	4.2–4.3	85	85	87	88	90	93	95	93	92	80	80
14	4.4–4.5	83	83	85	87	88	90	90	90	90	80	80
15	4.6–4.7	82	82	83	85	87	88	88	88	88	80	80
16	4.8–4.9	80	80	82	83	85	87	87	87	87	80	80
17	5.0–	80	80	80	80	80	80	80	80	80	80	80

DEMERITS

Type: Subnormal belly, and roughness—less 3 index points.
Quality: Abnormal fat, color, or texture—less 10 index points.
Trimmable: The actual weight reduction from the hot carcass weight if the demerit is of farm origin.
Ridglings: Index 67.

[a] Grade Index

Appendix B

COMPARISON
OF FARM ANIMALS

	Cattle	Sheep	Goats	Swine	Horses	Rabbits
Name						
Mature male	Bull	Ram	Buck (billy)	Boar	Stallion (stud)	Buck
Immature male	Bullock	Ram lamb	Buck kid	Shoat	Colt	Fryer
Castrated male	Steer	Wether	Wether	Barrow	Gelding	—
Mature female	Cow	Ewe	Doe (nanny)	Sow	Mare	Doe
Immature female	Heifer	Ewe lamb	Doe kid	Gilt	Filly	Fryer
Newborn	Calf	Lamb	Kid	Pig	Foal (colt)	Bunnies
Name of group	Herd	Flock	Band	Drove	Herd	Hutch or Herd
Type of digestive system	Ruminant	Ruminant	Ruminant	Nonruminant	Nonruminant (ruminant-like)	Nonruminant (ruminant-like)
Reproduction system						
Female						
Age at puberty (months)	8–10	5–7	7–10	4–7	15–24	4–12
Estrous cycle (days)	21	16	20	20	22	18–16
Estrus length	12–18 hr	24–36 hr	34–38 hr	48–72 hr	4–8 days	12 days
Time of ovulation	10–12 hr after estrus	Late estrus	Late estrus	Mid-estrus	1–2 day before estrus	From day 2 to day 14
Best time bred	Late estrus	Mid-estrus	Mid-estrus	Second day	Every other day starting second day	Any time doe will accept buck, usually day 2–14 of cycle
Gestation (days)	283	148	148	115	346	28–35
Act of parturition	Calving	Lambing	Kidding	Farrowing	Foaling	Kindling

Offspring/parturition	1	1	2	8–10	1	2–15 (av. 6)
Parturition to estrus	35 days	35 days	35 days	3–5 days	11 days	3–7 weeks
Mammary glands	4	2	2	4–9 pair	2	5 or more pair
Male						
Age at puberty (months)	10–12	4–6	4–6	4–8	13–18	5–6
Ejaculate Volume (cc)	5–7	0.75–1.5	1	200–300	75–150	—
Sperm/cc	1.2 billion	2.0 billion	2.0 billion	100–200 million	150 million	—
Total sperm (billion)	7	3	2	45	9	—
Females bred/insemination with AI	500–700	30–40		6–12	8–12	—
Female/male, normal breeding	25–40	35–60	30–50	25–50	20–30	10
Animal health						
Normal temperature (°F)	100.4–103.1	100.9–103.8	101.7–105.3	100.4–104	99.5–101.3	102.5
Respiration rate/min	18–28	12–24	12–20	15–24	8–16	—
Heart rate/min	60–70	70–80	70–80	60–80	32–44	—
Average expected productive life (years)	10	6–8	6–10	6–8	20–25	2–3

INDEX

Fowl (*See* Poultry)
Fowl pox, 455
Fowl typhoid, 454–55
Freeze branding, 186
Fructose, 107

G

Galactose, 107
Galloway, 49
Gasconne, 49
Geese:
  breeding of, 478
  breeds of, 477
  definition of terms, 474
  feeding of, 478
  history of, 474–77
Gelbvieh, 49, 51
Gelding, definition of, 501
Gilts, 347, 367, 381–82
Glucose, 107
Glycine, 105
Goat milk, 323
Goats:
  dairy breeds of, 317–19, 322–26
  diseases of, 326
  feeding of, 323–24
  history of, 316–17
  importance of, 315
  meat, 320, 329–30
  mohair breeds of, 319, 328–29
  reproduction in, 324–25
  selecting, 320–21
Goiter, 204
Grades and grading:
  of beef carcasses, 151–53, 156, 158–59
  of eggs, 463
  of pork, 386, 388, 607–8
Grass staggers, 205
Grass tetany, 204–5, 308
Gravel, 576
Grease heel, 576
Grease wool, 302
Grooming equipment, 188
Growth-promoting substances, 172–73
Guernsey, 221, 223
Guinea fowl:
  characteristics of, 487–88
  feeding of, 488
  history of, 486–87
Gumboro disease, 456

H

Hackney, 503–4
Ham–loin index, 386
Hampshire, 264
Hand mating, 347
Hardware disease, 118
Hatching, 433–34
Hays Converter, 51
Heaves, 567–68
Heifers (*See* Dairy cattle)
Helicopter, use of the, 10–11
Hemorrhagic septicemia, 213–14
Hereford:
  cattle, 29
  Polled, 40
  swine, 336

Heterosis, 165
Hogs (*See* Swine)
Holstein-Friesian, 223–24
Hoof cracks, 577
Horses:
  breeding of, 524–26
  directory of associations and sources of information, 520–23
  diseases of, 565–82
  draft, 510–11, 513–14
  feeding of, 533–41
  foaling, 526–30
  growth of, in America, 9–10
  evolution and history of, 491–99
  miniature, 518–19
  parasites in, 582, 584–86
  population data for, 499–500
  saddle and harness, 501–10
  selection of, 543–62
  terminology used for, 501
  type and classes of, 516–18
  zoological classification, 499
Hot iron branding, 186
Hybrid vigor, 165
Hydrolysis, 104, 105
Hyperkeratosis, 215–16
Hypoglycemia, 206, 408
Hysteria, 458

I

Inbred Livestock Registry Association, 340
Indian Runner, 471
Infectious bronchitis, 450
Infectious bursal disease, 456
Infectious conjunctivitis, 210–11
Infectious coryza, 451–52
Infectious equine anemia, 571–72
Infectious keratitis, 210–11
Infectious synovitis, 455–56
Influenza, 568
Insemination, 87–88
Institute of Biosciences and Technology center, 12
Interferon, 100
International Federation of Homing Pigeons, 486
Intradermal test, 205

J

Jacks, 515
Jennets, 515
Jersey, 223
Johne's disease, 205
Judging:
  beef cattle, 127–32
  dairy cattloe, 240, 242–43, 245
  horses, 543–62
  sheep, 288–92
  swine, 370–72, 374

K

Karakul, 276
Ketosis, 206, 308
Khaki Campbell, 471
Kidding, 325
Kindling, 600